# THERMOPHYSICAL PROPERTIES OF FREONS

# NATIONAL STANDARD REFERENCE DATA SERVICE
## OF THE USSR: A Series of Property Tables

# THERMOPHYSICAL PROPERTIES OF FREONS

## Methane Series, Part 1

**V. V. Altunin**
**V. Z. Geller**
**E. K. Petrov**
**D. C. Rasskazov**
**G. A. Spiridonov**

**Theodore B. Selover, Jr.**
*English-Language Edition Editor*

Springer-Verlag Berlin Heidelberg GmbH

**Thermophysical Properties of Freons, Part 1**

Originally published by Standards Publishers, Moscow, 1980 as Teplofizicheskiye Svoystva
Freonov, Vol. 1, in the Monograph Series of the National Standard Reference Data Service of
the USSR; State Committee on Standards of the Council of Ministers of the USSR.

English translation by J. I. Ghojel

1 2 3 4 5 6 7 8 9 0   B C B C   8 9 8 7

This book was set in English Times.
BookCrafters, Inc. was printer and binder.

**Library of Congress Cataloging in Publication Data**

Teplofizicheskie svoĭstva freonov. English.
  Thermophysical properties of freons.

  (National standard reference data service of the USSR)
  Translation of: Teplofizicheskie svoĭstva freonov.
  Includes bibliography.
  1. Chlorofluorocarbons—Thermal properties.
2. Chloroform—Thermal properties. 3. Fluoroform—
Thermal properties. I. Altunin, V. V. (Vlktor
Vladimirovich), date. II. Selover, Theodore B., date.
III. Ghojal, J. I. IV. Title. V. Series.
QD305.H6T4613  1987    547'.411    87-125
ISBN 978-3-662-30485-3      ISBN 978-3-662-30483-9 (eBook)
DOI 10.1007/978-3-662-30483-9

**DISTRIBUTION OUTSIDE NORTH AMERICA:**

# CONTENTS

# PREFACE TO THE SERIES

This treatise is part of a continuing series on thermodynamic properties of technologically important fluids. These very important contributions by scientists and engineers working through the aegis of the Soviet National Service for Standard Reference Data have been released to Hemisphere Publishing Corporation for translation to make them available to the English reading technical community. The authors are Soviet experts in the field.

While a team of translators was involved in producing the English versions, the overall series is being published under the technical editorship of T. B. Selover, Jr.

Each volume presents a comprehensive survey of the world's literature up to its publication date. A special effort has been made to give a thorough presentation of Russian as well as other work. Many studies not previously known to Western counterparts are included. The results have been to broaden the range of applicability of data and to improve upon equations of state to provide accurate computation methods for generating smoothed tables of properties.

For some volumes there are no equivalent comprehensive surveys available in English. Thus, a valuable service has been fulfilled to workers in the fields of process design, equipment development, custody transfer, and safety.

Each volume is set up in the same way with Part I dealing with a study of all necessary aspects of experimental data interpretation and analysis. Then in Part II the fundamental constants, symbols with units, and data tables can be found. The use of SI units is consistent throughout.

The section on experimental data is particularly important because it covers, in detail, the key studies. Possible errors in measurement or data analysis that could have led to inaccurate tabular results in publications are pointed out.

Methods of constructing the equation of state and procedures for computing the data tables are covered thoroughly. There is a detailed error analysis of data generated from the equation of state relative to literature values.

Properties of each fluid in phase equilibria are treated at the freezing curve and the saturation curve with tables given for both temperature and pressure dependence. Properties cover the range of triple point to critical point in the condensed phase. Single phase properties in each volume cover a range of temperature and pressure wide enough to include most practical applications. Ideal gas property equations and calculation methods are included. The scope of properties generated and covered in the tables is more comprehensive than is typically found in any one English-language treatise. This gives the advantage of internal consistency.

An extensive bibliographic listing is included with each volume. All Russian citations have been translated. In some cases the availability in English of translated Russian sources is given.

Frequent reference to key English-language references has been made to help in providing correct translations. In some cases the original symbols have been changed to avoid confusion within the text or to avoid misuse of terms commonly found in English. Where mistakes in the original Russian text have been found, they have been corrected and so noted as editor's changes. Added descriptive clarification has been used in some places for tables, figures, and text to clarify meaning.

Although we recognize that key review papers have been published in 1986 for some of the fluids in this set, the series stands on its own merit. It represents a vast accumulation of knowledge never before available to Western countries. Through careful study of these volumes, workers in the field will develop an appreciation for both the scope of studies and where differences exist. Hopefully, this will lead to more dialogue between the authors and their Western counterparts.

*Theodore B. Selover, Jr.*

# PREFACE

In this book, published experimental data, tables, and equations for four freons are systematized and critically assessed. The freons of the methane series are: freon-20, 21, 22, 23. Based on the analysis of the most reliable data, the authors compiled equations from which comprehensive tables of thermophysical properties of the said substances are computed in a temperature interval from normal boiling point to 473 K and in a pressure interval from 0.1 to 20 MPa. The overwhelming majority of these tables are published for the first time.

The book is intended for the staff of science-research and design-construction organizations; for researchers in thermophysics, cryogenics and chemical technology; and could be used by educators, graduate students, and students of physicochemical engineering disciplines.

This book contains 53 tables and 38 illustrations.

*V. V. Altunin*
*V. Z. Geller*
*E. K. Petrov*
*D. C. Rasskazov*
*G. A. Spiridonov*

# FOREWORD

The acceleration of scientific and technological progress is one of the major tasks of the tenth Five-Year Plan, promulgated by the Twenty-fifth Congress of the Communist Party of the Soviet Union. In light of this task, it is important to develop and expand the sphere covered by standardization, since such expansion would allow accumulation of the latest scientific and technological achievements, organically combine pure and applied sciences, and promote rapid and practical implementation of scientific achievements.

The State Service for Standard Reference Data oversees one of the new trends in standardization, that of standardization of the most reliable data available on the physical constants and properties of materials and substances.

The development of standard reference data is a multifaceted scientific and technological task that must be based on the Soviet and worldwide practice of selection and estimation of the reliability of data. An important stage in this work consists of preparing standard reference publications that contain not only data with an estimated reliability but also modern methods of obtaining such estimates. It is precisely these such publications that should serve as a methodological basis for the development of official tables of standard reference data on the properties of materials and substances.

The experience accumulated in working out tables of data on the thermodynamic properties of gases and liquids is of great interest from this point of view. This work has been particularly developed in the USSR. The work of Soviet investigators in thermophysics, which has established the basis for Soviet and international tables of thermophysical properties, has been acclaimed worldwide. Currently, this work is carried out by over 40 research organizations in the USSR under the auspices of the scientific program of the

State Service of Standard Reference Data and the Working Group on Thermodynamic Tables of the Soviet National Committee on Numerical Scientific and Technological Data of the Presidium of the USSR Academy of Sciences.

The publication of this series of books will supply reliable reference information to a wide circle of engineers and scientists and will serve as a basis for the development and practical utilization of pertinent official tables of standard reference data. The use of such data, obtained on the basis of the most up-to-date and accurate methods of study, is one of the conditions necessary for improving the level of scientific and developmental work. It ensures effective monitoring of industrial processes and quality of industrial output and promotes efficient utilization and accounting of the consumption of raw and finished materials, fuel, and energy. It is thus obvious that supplying reliable data on the properties of substances and materials is an important economic activity. The State Service of Standard Reference Data has been established primarily to supply this need.

*V. V. Boytsov*

# NOTATION AND UNITS

$T$  temperature, K

$p$  pressure, MPa

$\rho$  density, kg/m$^3$

$z$  coefficient of compressibility

$c_p$  specific heat at constant pressure, kJ/(kg·K)

$c_v$  specific heat at constant volume, kJ/(kg·K)

$r$  heat of vaporization, kJ/kg

$h$  enthalpy, kJ/kg

$s$  entropy, kJ/(kg·K)

$w$  speed of sound, m/s

$\mu$  adiabatic Joule-Thomson effect K/MPa

$\alpha$  coefficient of thermal expansion, K$^{-1}$; $\alpha = 1/v[(\partial v/\partial T)_p]$

$\gamma$  thermal coefficient of pressure, K$^{-1}$; $\alpha = 1/v[(\partial v/\partial T)_p]$

$\eta$  dynamic viscosity, Pa·s

$\nu$  kinematic viscosity, m$^2$/s; $\nu = \eta/\rho$

$F$  Helmoltz function, kJ/(kg·K)

$\lambda$  thermal conductivity, kW/(m·K)

$a$  thermal diffusivity, m$^2$/s; $a = \lambda(\rho C_p)$

Pr  Prandtl number

$\sigma$  surface tension, Pa·m

$k_A$  adiabatic index

$\tau$  reduced temperature $= T/T_{cr}$

$\pi$  reduced pressure $= P/P_{cr}$

$\omega$  reduced density $= \rho/\rho_{cr}$

## Log Notation

lg    $\log_{10}$
ln    $\log_e$

## Subscripts

cr    critical
0    triple point
$f$    freezing point
$s$    saturation conditions

## Superscripts

(')    liquid at saturation
(")    gas at saturation

# INTRODUCTION

There are several systems for designating freons, which are given in detail in Ref. [0.17]. The chemical compounds under consideration in this handbook have the following names (per Geneva Nomenclature) and symbols (per ISO-68 nomenclature):

| | |
|---|---|
| $CHCl_3$ | Trichloromethane (F-20 or R-20) |
| $CHFCl_2$ | Fluorodichloromethane (F-21 or R-21) |
| $CHF_2Cl$ | Difluorochloromethane (F-22 or R-22) |
| $CHF_3$ | Trifluoromethane (F-23 or R-23) |

Computational and theoretical investigation of thermodynamic properties of freons until recently have been conducted with state parameters covering only the traditional range of their applications. Therefore, modeled and generalized equations of state having relatively few empirical constants had been used.

Thus, in 1940, Benning and McHarness, on the basis of their experimental $pvT$ data for Freons-11, -21, and -22 ($T \approx 300-450$ K, $p \approx 0.03-2.1$ M.Pa), derived constants for the Beattie–Bridgeman equation of state. For a long time, tables calculated from these equations were presented in reference books [0.7, 0.45, and others]. Table 1 shows that Eq. (0.1) contains only five controllable constants, but the assumption $\epsilon = 0$ for freons decreases the power of the equation relative to $v$ to the third. The application range of the Beattie–Bridgeman equation of state is narrow and even at $\epsilon \neq 0$ the density stays in the limit $\omega = 0-0.4$ [0.5].

In 1955–1959, Martin and coworkers developed a relatively simple method for determination of coefficients for the equation of state (0.2).

Twelve coefficients for the Martin–Hou equation of state can be computed using analytical relations if the critical parameters, Boyle point temperature,

**TABLE 1.** Main Types of Equations of State for Gaseous Freons

| Form of equation | | Commentary | Title of equation |
|---|---|---|---|
| $$P = \frac{RT(1-\varepsilon)}{v^2}(v+B) - \frac{A}{v^2}$$ | (0.1) | $B = B_0\left(1 - \dfrac{b}{v}\right)$ $A = A_0\left(1 + \dfrac{a}{v}\right)$ $\varepsilon = C/vT^3$ | Beattie–Bridgeman |
| $$p = \frac{RT}{v-b} + \sum_{i=2}^{5}\frac{F_i(T)}{(v-b)^i}$$ | (0.2) | $F_i = A_i + B_i T + C_i \times$ $\times \exp(-kT)$ | Martin–Hou |
| $$p = RT\varrho + (B_0 RT - A_0 - C_0/T^2)\varrho^2 + (bRT-a)\varrho^3 + a\alpha\varrho^6 + c\varrho^3(1+\gamma\varrho^2)\times$$ $$\times(1/T^2)\exp(-\gamma\varrho^2)$$ | (0.3) | | Benedict–Webb–Rubin (BWR) |
| $$\pi = Wa \cdot \omega\tau + (A_2+B_2\tau+C_2/\tau^2+D_2/\tau^4)\omega^2 + (A_3+B_3\tau+C_3/\tau^2)\omega^3 +$$ $$+ (A_4/\tau^2+A_5/\tau^4)\omega^3(1+\beta\omega^2)\exp(-\beta\omega^2)+A_6\omega^6$$ | (0.4) | $Wa = (RT/pv)_{cr}$ $\omega = \varrho/\varrho_{кр}$ $\pi = p/p_{кр}$ $\tau = T/T_{кр}$ | Modified BWR (BWRM) |
| $$p = RT\varrho + (B_0 RT - A_0 - C_0/T^2 + D_0/T^3 - E_0/T^4)\varrho^2 + (bRT-a-d/T)\varrho^3 +$$ $$+ a(a+d/T)\varrho^6 + c\varrho^3(1+\gamma\varrho^2)(1/T^2)\exp(-\gamma\varrho^2)$$ | (0.5) | | Modified BWR (BWRC) |
| $$\pi = \alpha_к(\omega) + (\tau-1)\sum_{i=1}^{6}b_i\omega^i + \frac{(\tau-1)^2}{\tau}\sum_{i=2}^{4}c_i\omega^i$$ | (0.6) | $\alpha_к = 1-(1-\omega)^5\{1+[a_0\omega + [a_1(a_2-\omega)\omega^2]/[1+ +4(1-\omega)^2]]\}$ | Rombusch |
| $$z = 1 + \sum_{i=1}^{r}B_i(T)\varrho^i$$ | (0.7) | $B_i = \sum_{j=0}^{s_i}b_{ij}\frac{1}{\tau^j}$ | Virial |
| $$z = 1 + \varrho\sum_{i=0}^{m}(a_0\varrho + \beta_0)^i\left[\sum_{j=0}^{n}c_{ij}\left(\frac{\alpha_1}{T}+\beta_1\right)^j\right]$$ | (0.8) | | — |

temperature dependence of the saturated vapor pressure, and one isochore in the liquid phase at $\omega \geqslant 1.5$ are known [3.50]. Different versions of this method were employed when studying the thermodynamic properties of Freon-22 [3.50] and Freon-23 [0.39, 4.25, 4.26]. In Ref. [1.58] nine coefficients of Eq. (0.2) for gaseous Freon-20 are given, and thermodynamic tables are calculated in the interval $\tau = 0.52-1.4$ for $\pi \leqslant 3.6$.

Some investigators [0.5, 0.42] think that the Martin–Hou equation is best used for $\omega \leqslant 1.0-1.2$. It is possible to widen the range of applicability of Eq. (0.2) and increase its accuracy somewhat by decreasing the number of zero values of coefficients $A_i$, $B_i$, and $C_i$ [0.47, 1.1] or by adding an extra term of the form $F_6(T)/\exp (av)$ [3.50].

Equation (0.3), also called the Benedict–Webb–Rubin equation, contains eight constants and is recommended for the evaluations of thermodynamic properties of little known gaseous freons and their mixtures [0.23, 0.39]. Normally, for the calculations of thermodynamic tables and diagrams, complicated variations of the BWR equation are employed. In Table 1, two modifications of an equation of this type with 11 constants are given. Equation (0.4) was proposed by Morsy [0.42] and Eq. (0.5) by Starling [0.48].

Coefficients of BWRM were determined in several works: [0.41] for Freon-22 and [0.39, 0.42, 0.43] for Freon-23. For Freon-22 in Ref. [0.41] this equation is recommended for use in the gaseous phase when $\tau = 0.46-1.28$ and $\omega \leqslant 1.6$. The mean error in calculating the pressure is evaluated as $\pm 0.54\%o$. For Freon-23 in Refs. [0.39, 0.42, 0.43] an almost identical method of determination of equation coefficients is used, but the composition of the initial data is different and, hence, the area of application of the equations appreciably differs. From the methodical point of view, Morsy's method [0.43] is the most interesting, whereby coefficients of Eq. (0.4) are derived for eight gases including Freons-12, -13, -14, -23, and C-318.

The BWRC equation was first tested on nonpolar gases and their mixtures. It is most commonly used in a generalized form [0.49]. The recommended formulas in Ref. [0.49] for the determination of coefficients of Eq. (0.5) are simple functions of critical parameters and of Pitzer's acentric factor ($T_{cr}$, $\rho_{cr}$, $p_i$). In a more general case, it is imperative also to know, for each pair of mixture components, what is termed the constant of binary interaction $K_{ij}$.

The BWRC generalized equation for freons and their mixtures is verified in [0.22, 0.25, 1.12]. It was discovered that unlike the correlation of Starling–Han, where $K_{ii} = 0$ for pure components, better results are achieved for many freons when $K_{ii} \neq 0$. In Ref. [0.22] the adjusting parameters $K_{ii}$ and $K_{ij}$ are determined for a large group of pure freons of the methane series and their mixtures.

Other variations of the BWR equation, differing from (0.3)–(0.5) by the form of temperature dependence for $\rho^2$, $\rho^3$, and $\rho^6$, are analyzed in Refs. [0.23, 0.44].

The Rombusch equation (0.6) also involves 11 controllable constants, but, judging from [0.46], these constants are determined "manually." Coefficients of Eq. (0.6) are also given there for Freons-11, -12, -13, and -21. The equation is recommended for use in the region $\tau = 0.5-2.0$ and $\omega = 0-1.8$.

Theoretically, the Rombusch equation can be used to calculate the thermodynamic properties in the liquid phase. But this equation is used in Ref. [0.46] and then in [0.11] only to calculate the tables for the values $v$, $h$, $s$ of superheated and saturated vapors of Freon-21 at temperatures from 213 to 473 K and pressures up to 5 MPa. To determine the thermodynamic properties of the saturated liquids in Ref. [0.46], as is the case in the majority of other works, an autonomous system of equations including the Clausius–Clapeyron equation

$$r = T(dp_s/dT)(v'' - v') \qquad (0.9)$$

and empirical formulas for temperature dependencies of saturated vapor pressure and orthobaric liquid density are employed.

In the majority of works, experimental data on saturated vapor pressure of freons are approximated by equations of the following forms:

$$\ln \pi_s = A + B/\tau + C \ln \tau + D\tau^k \qquad (0.10)$$

$$\ln \pi_s = \frac{1}{\tau} \sum_{j=0}^{n} a_j \tau^j + a_{n+1} \ln \tau \qquad (0.11)$$

Normally $k = 3-6$ [0.28, 0.46] and $n = 3-4$ [0.26, 4.25, 4.28, and others]. But for Freon-22, for example, in Refs. [0.18, 0.51] $n = 5$, $a_{n+1} = 0$ and, apart from this, in [0.51] the righthand side of Eq. (0.11) includes a term with a fractional power. In Refs. [0.39, 0.41, 4.34] for Freon-23 a generalized Riedel equation is used

$$\ln \pi_s = \alpha_c \ln \tau - 0.0838(\alpha_c - 3.75)(36/\tau - 35 - \tau^6 + 42 \ln \tau), \qquad (0.12)$$

in which the parameter $\alpha_c$ is taken as a function of temperature. Clearly, Eq. (0.12) cannot be applied to polar substances without correction. But the value $\alpha_c = (d \ln \pi_s/d\tau)_{\tau=1} = R_i$ is considered a fundamental characteristic of the substance which is used as a determining criterion in the theory of thermodynamic similarity [1.17]. Therefore, it is preferable to correct the last term of Eq. (0.12) as was done in Ref. [1.15].

For the approximation of experimental data on the density of the saturated liquid, an equation of the following form is used:

$$\omega' = 1 + \sum_{j=1}^{m} d_j (1 - \tau)^{j/3} \qquad (0.13)$$

in which $m = 3-4$ [0.26, 0.46, 0.51, 4.25, 4.28, and others].

The system of equations (0.9)–(0.13) has been used to calculate the thermodynamic properties of liquid freons on the saturation line for 20 years. Selection of the coefficients of these equations for Freons-21, -22, and -23 is given in a number of experimental and theoretical works.

As regards the thermodynamic functions in the ideal gas state, in the overwhelming majority of contemporary work for freons of the methane series, emphasis is on the data of Barho [0.37] and these data are approximated by an equation of the form

$$c_p^0 = \sum_{j=0}^{n} c_j v^j \qquad (0.14)$$

where $n = 3-5$ [0.18, 0.21, 0.28, and others]. In [0.21] the coefficients of this equation for Freons-20, -21, and -23, which are computed from the data in [0.37] in the interval $T = 100-500$ K, are given.

In view of the accumulation of experimental data, increase in their accuracy, and expansion of the studied region of state, more complicated equations of state (see Table 1) are being used for the calculation of thermodynamic equations of state. The techniques for determining the constants of equations of state have also been developed constantly. Since the 1960s, computers have been used extensively to approximate the experimental data. Up to the present time, effective algorithms and programs have been developed enabling even computers of medium power (types BaESM-4, M-220) to successfully solve the problems of statistical analysis of massive homogeneous and heterogeneous thermodynamic data with the purpose of building multiparametrical equations of state. A detailed review of the progress made in this field of applied thermophysics is given in [0.5].

In later years, for all freons of the methane series, equations of state in the form of (0.7) or (0.8) have been developed. These equations are in the form of polynomial expansion in the power of density and temperature. From Table 1 it is clear that these equations are open ended and are capable, depending on the range of the independent variables and degree of accuracy of approximation, of containing different numbers of empirical constants. It is important to emphasize that the construction of equations of state of the form (0.7) and (0.8) dictates the use of computer methods. For the initial data, results of accurate $pvT$ measurements [0.1, 0.26, 0.28, 2.1, 3.1, 3.66, 4.3, and 4.14] or the total sum of experimental $pvT$ and $c_p(pT)$ data [3.1, 3.56] should be used.

Details of applicable methods of statistical analysis of experimental data about thermophysical properties of gases and liquids are given in Refs. [0.4, 0.5, 0.29, 0.32, 1.3, 1.4, 3.2, 3.3, and 4.3]. The derivations of experimentally substantiated equations of state of the form (0.7) and (0.8) for Freons-21, -22, and -23 are given and discussed in the following chapters of the present handbook. Short commentary is given on references in which equations of state of nonconventional structure are derived for freons of the methane series under consideration. The specific searching techniques for the coefficients are also given there.

Table 1 indicates that the recommended equations of state (0.7) and (0.8) in the present work are reduced to

$$z = 1 + \sum_{i=1} \sum_{j=0} b_{ij}\, \rho^i/\tau^j \qquad (0.15)$$

For the transition from the equation of state with a displaced center of expansion ($\beta_0 \neq \beta_1 \neq 0$) to the polynomial of virial form with $\alpha_0 = \alpha_1 = 1$, the following recommended formula can be used:

$$b_{rk} = \sum_{i=p}^{m} (\beta_0)^{i-p} \frac{l!}{(l-p)!\,p!} \left[ \sum_{j=k}^{n_i} \tilde{b}_{ij} (\beta_1)^{j-k}\ \frac{j!}{(j-k)!\,k!} \right] \qquad (0.16)$$

where $p = r - 1$ and $r$ = series order for $\rho(0 \leqslant r \leqslant m + 1)$.

For the polynomial equation (0.15), the program for calculating thermodynamic properties can be substantially reduced since it is possible, in this case, to limit the set of arithmetic operators

$$P_k = 1 + \{K_{ij}^{(k)}\} = 1 + \sum_{i=1} \sum_{j=0} K_{ij}^{(k)}\, b_{ij}\, \rho^i/\tau^j \qquad (0.17)$$

where $K_{ij}^{(k)}$ = a complex formed from the indices of summations of $i$ and $j$. Obviously, to find the numerical values of the arbitrary operator, it is sufficient to program expression $P_k$ in a general form with tuning for the corresponding $K_{ij}^{(k)}$. The algorithm of such a program and examples of entering the calculated correlation for thermodynamic values were discussed earlier in a paper by V. V. Altunin and G. A. Spiridonov [0.2]. An analogous method is described in a later work of I. I. Perelshtein and E. B. Parushin [0.29], where different symbols for operators are employed.

In Table 2 a selection of calculated correlations using 10 particular values of operator $P_k$ are given:

$$P_1 = 1 + \{i\}; \quad P_6 = 1 + \{i + j\}$$

$$P_2 = 1 + \{1/i\}; \quad P_7 = 1 + \left\{ \frac{1+i}{i} \right\}$$

$$P_3 = 1 + \{j/i\}; \quad P_8 = 1 + \left\{ \frac{j-1}{i} \right\}$$

$$P_4 = 1 + \{1 + i\}; \quad P_9 = 1 + \left\{ \frac{i+j}{i} \right\}$$

$$P_5 = 1 + \{1 - j\}; \quad P_{10} = 1 + \left\{ \frac{j(j-1)}{i} \right\}$$

Should the need arise, the enumeration of thermodynamic values calculated from the equation of state can be easily expanded using the recommendations in Refs. [0.2, 0.5].

For the approximation of thermodynamic functions in the ideal gas state, they are written here in the form of a power series in reciprocal temperature:

$$(H_T^0 - H_0^0)/RT = \sum_{j=0}^{m_1} a_j/\tau^j \qquad (0.18)$$

$$c_p^0/R = \sum_{j=0}^{m_2} \beta_j/\tau^j \qquad (0.19)$$

$$s_T^0/R = \sum_{j=0}^{m_3} \gamma_j/\tau^j \qquad (0.20)$$

Saturated vapor pressure dependence on temperature is described by a power series polynomial of the form

$$p_s = \sum_{j=0}^{m_4} A_{sj}\,\tau^j \qquad (0.21)$$

Notice that, theoretically, formulas (0.21) and (0.19) have no advantage whatsoever over formulas (0.11) and (0.14). They are, however, more convenient to use for operative calculations.

As a starting point for enthalpy calculations, it is necessary to adopt the state of stable crystal at 0 K. Then, in the lefthand side of formula (0.18) we find the value

$$(H_T^0 - H_0^0)_{GV} = (H_T^0 - H_0^0)_G + \Delta H_{s_0}^0 \qquad (0.22)$$

TABLE 2. Calculation Formulas for the Fundamental Thermodynamic Values

| Value | Initial relation | Calculation formula |
|-------|------------------|---------------------|
| Helmholtz function | $F = -\int pdv$ | $\left[\begin{array}{l} F - F_0 + \\ + RT\left[\ln\left(\dfrac{\rho}{\rho_{st}}\right) + P_2 - 1\right]\end{array}\right.$ |
| Entropy | $s = -\left(\dfrac{\partial F}{\partial T}\right)_\rho$ | $s = s_0 + R\left(P_8 - 1 - \ln\dfrac{\rho}{\rho_{st}}\right)$ |
| Internal energy | $u = F + Ts$ | $u = u_0 + RT(P_3 - 1)$ |
| Enthalpy | $h = u + \dfrac{p}{\varrho}$ | $h = h_0 + RT\ (P_9 - 1)$ |
| Gibbs function | $\Phi = h - Ts$ | $\Phi = \Phi_0 + \\ + RT\left(\ln\dfrac{\rho}{\rho_{st}} + P_7 - 1\right)$ |

TABLE 2. Calculation Formulas for the Fundamental Thermodynamic Values (*Continued*)

| Value | Initial relation | Calculation formula |
|-------|-----------------|--------------------|
| Volatility (fugacity) | $f = \exp\left(\dfrac{1}{RT}\displaystyle\int vdp\right)$ | $= \varrho RT \exp\ (\mathbf{P}_7 - 1)$ |
| Heat capacity at constant volume | $c_v = T\left(\dfrac{\partial s}{\partial T}\right)_\varrho$ | $c_v = c_v^0 - R(\mathbf{P}_{10} - 1)$ |
| Heat capacity at constant pressure | $c_p = c_v - T\dfrac{\left(\dfrac{\partial p}{\partial T}\right)_\varrho^2}{\left(\dfrac{\partial p}{\partial v}\right)_T}$ | $c_p = c_p^0 + R\left(\dfrac{\mathbf{P}_5^2}{\mathbf{P}_4} - \mathbf{P}_{10}\right)$ |
| Speed of sound | $a = \sqrt{-v^2\left(\dfrac{\partial p}{\partial v}\right)_s}$ | $a = \sqrt{RT\left(\mathbf{P}_4 + \dfrac{\mathbf{P}_5^2}{\dfrac{c_p^0}{R} - \mathbf{P}_{10}}\right)}$ |
| Isothermal Joule–Thomson effect | $\delta = -\left(\dfrac{\partial h}{\partial p}\right)_T$ | $\delta = \dfrac{1 - \mathbf{P}_6}{\varrho \cdot \mathbf{P}_4}$ |
| Adiabatic Joule–Thomson effect | $\mu = \left(\dfrac{\partial T}{\partial p}\right)_h$ | $\mu = \dfrac{1}{\varrho}\left[\dfrac{1 - \mathbf{P}_6}{c_p^0\mathbf{P}_4 + R(\mathbf{P}_5^2 - \mathbf{P}_4\mathbf{P}_{10})}\right]$ |
| Thermal pressure | $p_t = \left(\dfrac{\partial p}{\partial T}\right)_\varrho$ | $p_t = \varrho \bar{R}TP_5$ |
| Coefficient of thermal expansion | $\alpha = \dfrac{1}{v}\left(\dfrac{\partial v}{\partial T}\right)_p$ | $\alpha = \dfrac{1}{T}\dfrac{\mathbf{P}_5}{\mathbf{P}_4}$ |
| Coefficient of isothermal compression | $\beta = -\dfrac{1}{v}\left(\dfrac{\partial v}{\partial p}\right)_T$ | $\beta = \dfrac{1}{\varrho RT\mathbf{P}_4}$ |
| Thermal coefficient of pressure | $\gamma = \dfrac{1}{p}\left(\dfrac{\partial p}{\partial T}\right)_v$ | $\gamma = \dfrac{1}{T}\dfrac{\mathbf{P}_5}{\mathbf{P}_1}$ |
| Ratio of heat capacities | $\chi = \dfrac{c_p}{c_v}$ | $\chi = 1 + \dfrac{\mathbf{P}_5^2}{\left(\dfrac{c_p^0}{R} - \mathbf{P}_{10}\right)\mathbf{P}_4}$ |
| Adiabatic index (isentropic) | $k_A = -\dfrac{v}{p}\left(\dfrac{\partial p}{\partial v}\right)_s$ | $k_A = \dfrac{\mathbf{P}_4}{\mathbf{P}_1} - \dfrac{\mathbf{P}_5^2}{\left(\dfrac{c_p^0}{R} - \mathbf{P}_{10}\right)\mathbf{P}_1}$ |

where $\Delta H_{s_0}^0$ = heat of sublimation of the crystal at 0 K.

The numerical values of $\Delta H_{s_0}^0$ for Freon-22 and Freon-23 are known and are given in Table 3. For Freon-21, instead of the unknown value of $\Delta H_{s_0}^0$ we used the arbitrary value $h_0$ = 500 kJ/kg and, for Freon-20, this constant is assumed so as to give $h_{273.15}$ = 400 kJ/kg.

Some values of the characteristic parameters stated in Table 3 are not directly linked with the recommended system of equations. They are rather the product of the critical analysis of literature data* to be discussed in the following sections of this book. This point is related primarily to the parameters of critical points. It is established that for the accurate determination of thermodynamic properties of substances in the vicinity of critical points, nonanalytical types of equations must be employed [0.5, 0.6]. In addition, the applicability of the scaling law to adequately describe critical anomalies is proven. The fundamental conditions of the scaling law (static and dynamic) are sufficiently elaborated in review [0.6].

Digressing from the discussion of details, we note that according to Scofield, for example, the polar coordinates $\bar{v}$ and $\theta$, which are linked with $\Delta\rho^* = |\omega - 1|$ and $\Delta T^* = \tau - 1$ and which characterize the distance from the critical point, are the nonanalytical parameters of equation of state $p_{sc} = f(v, \theta)$.

Transfer to the analytical equation of state $p_A$ can be achieved with the help of a special switching function as, for example, is done for carbon dioxide [0.5]:

$$p = f(\bar{r})p_A + [1 - f(\bar{r})]p_{sc}, \qquad (0.23)$$

where the switching function

$$f(\bar{r}) = 1 - \{1 - \exp[-(k_1/\bar{r})^{n_1}]\}\{1 - \exp[-(k_2/\bar{r})^{n_2}]\}. \qquad (0.24)$$

According to the equilibrium scaling law, the equations of compressibility and two-phase curves in the vicinity of the critical point have the form

$$1 - \pi_s = c_1(1 - \tau) + c_2(1 - \tau)^{2-\alpha} + \rho_{cr}c_3(1 - \tau)^{2\beta\delta-1}, \qquad (0.25)$$

$$\rho^\pm = \rho_{cr} + b(1 - \tau) + B_1^\mp(1 - \tau)^\delta + B_2^\mp(1 - \tau)^{\beta+\Delta}. \qquad (0.26)$$

Averaged values of the critical indices $\alpha$, $\beta$, and $\delta$ are $\alpha$ = 0.100, $\beta$ = 0.355, and $\delta$ = 4.352. From the experimental data of U. R. Chashkin and others [4.17] for Freon-23, we find $\beta$ = 0.355 and $\Delta$ = 0.315.

Several polynomial-type analytical equations of state for Freon-22 [0.28, 3.1, 3.66] are known. In order to deduce these equations, efforts were made to fulfill the classical critical conditions:

$$p = p_{cr}; \ (\partial p/\partial v)_{T,cr} = (\partial^2 p/\partial v^2)_{T,cr} = 0;$$
$$(\partial p/\partial T)_{v,cr} = (dp_s/dT)_{cr}.$$

* The values of dipole moments shown in Table 3 are derived from data in Ref. [0.27].

TABLE 3. Characteristic Parameters for Freons of the Methane Series

| Parameter | Numerical value for | | | | |
|---|---|---|---|---|---|
| | Freon-20 | Freon-21 | Freon-22 | Freon-23 | |
| Temperature at the critical point $T_{cr}$, K | $536.6 \pm 0.2$ | $451.60 \pm 0.05$ | $369.30 \pm 0.02$ | $299.00 \pm 0.05$ | |
| Pressure at the critical point $p_{cr}$, MPa | $5.47 \pm 0.05$ | $5.190 \pm 0.005$ | $4.988 \pm 0.002$ | $4.82 \pm 0.02$ | |
| Density at the critical point $\rho_{cr}$, kg/m$^3$ | $500 \pm 10$ | $528 \pm 3$ | $514 \pm 2$ | $525 \pm 3$ | |
| Temperature at the normal boiling point $T_{NBP}$, K | $334.3 \pm 0.1$ | $282.05 \pm 0.1$ | $232.35 \pm 0.05$ | $191.00 \pm 0.05$ | |
| Temperature at the triple point $T_0$, K | $209.5 \pm 0.5$ | $138.1 \pm 0.5$ | $115.74 \pm 0.01$ | $117.97 \pm 0.05$ | |
| Heat of sublimation of the crystal at 0 K $\Delta H_{s0}^0$, kJ/kg | — | — | $355.5 \pm 0.2$ | $359.3 \pm 0.2$ | |
| Dipole moment $\mu_D$, $D$ | $1.04 \pm 0.02$ | $1.29$ | $1.409 \pm 0.03$ | $1.645 \pm 0.09$ | |
| Molecular mass | | | | | |
| On the chemical scale $M_{ch}$, kg/kmol | $119.390$ | $102.933$ | $86.476$ | $70.019$ | |
| On the carbon scale $M_C$, kg/kmol | $119.3849$ | $102.9285$ | $86.4723$ | $70.0160$ | |
| Gas constant $R_{ch}$, kJ/(kg · K) | $0.0696402$ | $0.0807743$ | $0.096146$ | $0.118744$ | |
| $R_C$, kJ/(kg · K) | $0.069643$ | $0.080777$ | $0.096150$ | $0.118749$ | |

The coefficients of equations of the type (0.7) or (0.8) in previously cited works were determined by the least-squares method in the initial basis of expansion. To satisfy the critical condition, the method of Lagrange's undetermined multipliers was used. In Ref. [3.1], $T_{cr}$ and $\rho_{cr}$ were fixed and $p_{cr}$ was determined. An alternative calculation algorithm is also possible: fix $T_{cr}$ and $p_{cr}$ and determine the critical density $\rho_{cr}$ for the point at which critical conditions are fulfilled [0.28]. As evident from Table 30, the calculated values of $\rho_{cr}$ [0.28] differ substantially from experimental data. Our personal experience shows that both methods of seeking parameters of the "calculated critical point" are of little use.

From the scaling law mentioned above, it follows that within the boundaries of polynomial expansions of the type (0.7) or (0.8), coordination of $pvT$ data in a wide single-phase region and on the saturation line for the simultaneous implementation of all critical conditions at the "experimental critical point" is not feasible. All the same, when deducing the analytical equation of state, it is helpful to make use of at least the second and third critical conditions in order to confidently "cut off" the upper boundary of the two-phase region [0.5, 3.1].

In the works of V. V. Altunin and O. G. Gadetsky [3.1] and Oguchi et al. [3.56], when approximating experimental data for liquid and gaseous Freon-22, equality of chemical potentials in the two-phase region ($\varphi' = \varphi''$) was achieved. It is known that this condition is a characteristic sign of a single equation of state for gaseous and liquid phases.

To solve the said problem not only $pvT$ data in a single-phase region were used, but also data about $p_s(T)$, $\rho_s(T)$, and, in Ref. [3.1], also data about heat capacity of the liquid.

In less favorable situations, the reasonable alternative to the monostructural single equation of state could be a compounded single equation which A. V. Kletsky [0.19] calls a system of coordinated equations of state. The main equation of the system

$$p = p_0 + T \sum_{i=1}^{m} \sum_{j=0}^{s_i} a_{ij}(1 - T_s/T)^i(1 - \rho/\rho_{cr})^j \qquad (0.27)$$

must describe, by partly changing the structure, both zones in which the thermal surface is divided by the critical constant-volume line. When $\rho > \rho_{cr}$ (this region of state parameters is arbitrarily called the liquid zone)

$$p_0 = p_s = \exp\left[\sum_{l=1}^{4} B_l T_s^l\right] \qquad (0.28)$$

$$T_s = T_{cr} + \sum_{i=3}^{n} A_i(1 - \rho/\rho_{cr})^i$$

when $\rho < \rho_{cr}$ (in the vapor zone)

$$p_0 = \sum_{i=0}^{s_i} a_{ij}(1 - \rho/\rho_{cr})^j$$

$$T_s = T_{cr} \qquad (0.29)$$

Mutual transitions of zone equation and limit transitions are guaranteed by definite restrictions which are superpositioned on a part of the coefficients $a_{ij}$ [0.19]. Using this method, A. V. Kletsky (1979) derived specific equations. Analytical methods of compiling tables of viscosity and thermal conductivity of liquid and gaseous freons made progress only in recent years. Until the mid-1960s compilers of tables had at their disposal only experimental information about the temperature dependence of viscosity and thermal conductivity of some gaseous freons at atmospheric pressure ($\eta_T$, $\lambda_T$) and liquid freons at saturation pressure ($\eta_l$, $\lambda_l$). For the past several years experimental work on a grand scale has been carried out in some laboratories, mainly in the Soviet Union. This work concerns the dependence of viscosity and thermal conductivity on pressure for several freons in the gaseous and liquid phases in a wide interval of temperature at pressures up to 50–60 MPa.

Most of the experimental data about viscosity and thermal conductivity of compressed Freons-21, -22, and -23 in the single-phase region were obtained by V. Z. Geller, C. I. Ivanchenko, and V. G. Peredri [0.15, 2.4, 2.5, 3.9, 3.10, 3.26, 4.2] and by D. C. Raskazov, U. M. Babikov, and H. Y. Filatov [2.13, 4.7, 4.8, 4.9]. These data are a basis for the equations and tables of transport properties recommended in the present book.

To describe the temperature dependence of viscosity and thermal conductivity of individual freons at atmospheric pressure the following equations are often used:

$$\eta_T = \sum_{j=0}^{1} a_{\eta, j} \tau^{j/2} \qquad (0.30)$$

$$\lambda_T = \sum_{j=0}^{1} a_{\lambda, j} \tau^{j} \qquad (0.31)$$

Generalized equations (for groups of freons) are proposed in Refs. [0.50, 1.18, 4.11, and others]. Some of them are given in Chap. 1.

In Refs. [1.49, 2.22, 2.30, and others] viscosity of liquid freons of the methane series on the saturation line is approximated by an equation of the type

$$\ln \eta_L = \sum_{j=-1}^{n} A_j \tau^{j}, \qquad (0.32)$$

where $\eta = 0-2$. A more complex equation is used in Ref. [0.40].

Several types of equations are proposed to describe the experimental data about viscosity and thermal conductivity of compressed freons of the methane series.

It is well known that in some regions of state of compressed and liquefied gases, the dependence of excess viscosity $\Delta\eta = (\eta_{p,T} - \eta_T)$ and excess thermal conductivity $\Delta\lambda = (\lambda_{p,T} - \lambda_T)$ can be described with simple functions of density. Therefore, when analyzing experimental data for the calculation of tables, the following formulas are often used:

$$\eta = \eta_T + \sum_{i=0}^{m_1} b_{\eta, i} \, \rho^i \tag{0.33}$$

$$\lambda = \lambda_T + \sum_{i=1}^{m_2} b_{\lambda, i} \, \rho^i \tag{0.34}$$

where $m_1 = 4-8$ and $m_2 = 4$. The coefficients of Eqs. (0.33) and (0.34) for Freons-21, -22, and -23 were repeatedly determined in different density intervals and covered by different experimental data [0.13, 0.15, 1.3, 1.4, 3.20, 3.29, 3.60].

In Ref. [0.15], on the basis of analysis of experimental data for Freons-21, -22, and -23 in the region of reduced density $\omega \leqslant 1.9-2.0$, the coefficients of the following generalized equation are determined

$$[(\eta - \eta_T)/\eta_{0.7}] \cdot 10^{-4} = \sum_{i=1}^{9} b_{\eta, i} \, \omega^i \tag{0.35}$$

where $\eta_{0.7}$ is viscosity at $T = 0.7 \, T_{cr}$ and $p = 0.7 \, p_{cr}$. Numerical values of the coefficients $\overline{b}_{\eta, i}$ are given in Table 4 and the assumed values $\eta_{0.7}$ are given in the corresponding sections of this reference book.

**TABLE 4.** Coefficients of the Generalized Equations (0.35) and (0.41)

| $i$ | $\overline{b}_{\eta, i}$ | $\overline{a}_{\eta, i}$ | $\overline{\beta}_{\eta, i}$ |
|---|---|---|---|
| 0 | — | 0.397133 | 0.001449 |
| 1 | 0.02500 | 1.414087 | 0.020085 |
| 2 | 0.01390 | −2.156796 | −0.069557 |
| 3 | −0.13069 | 2.405557 | 0.168403 |
| 4 | 0.12846 | −1.381979 | −0.200158 |
| 5 | −0.08264 | 0.275277 | 0.145676 |
| 6 | −0.07428 | 0.070019 | −0.077080 |
| 7 | 0.16461 | −0.039094 | 0.030819 |
| 8 | −0.09023 | 0.004662 | −0.007762 |
| 9 | 0.01605 | — | 0.000846 |

In Ref. [0.16], for the determination of the viscosity of dense fluids (polar and nonpolar) in the region $\omega = 0-2.6$, it is recommended that the generalized equation

$$\eta = \eta_T \left[ 1 + \sum_{i=1}^{6} \sum_{j=0}^{3} a_{\eta,\ ij}\ \omega^i / \tau^j \right] \tag{0.36}$$

be used with the help of tables of $\eta/\eta_T$ for $\pi = 0.2-30$, $\tau = 0.6-5.0$, and $z_{cr} = 0.23-0.26$ that have been compiled. These tables could be useful for the evaluation of viscosity of little known freons (e.g., Freon-20).

As far as equations of type (0.36) with individual constants are concerned, they are not widely used, probably because of the large number of controllable constants (up to 20–25 or more).

Comparative analysis in [0.3, 0.32, 3.2] of different viscosity equations in $\omega\tau$ variables showed that from the computational and physical viewpoints, preference should be given to equation

$$\eta = \eta_T \ \exp\left[ \sum_{i=1}^{m} \sum_{j \neq 0}^{n} a_{\eta,\ ij}\ \omega^i / \tau^j \right] \tag{0.37}$$

These equations are compact, applicable in the whole range of the single-phase state, and reproduce separation of isotherms of excess viscosity at moderate density values, where $(\partial \Delta\eta/\partial T)_\rho > 0$. At high density values $(\partial \Delta\eta/\partial T)_\rho < 0$ and appropriate conditions, extrapolation outside the limits $\omega > 2.5-3.0$ is allowed with confidence.

The separation character of the isotherms of excess thermal conductivity is different. At high density, values in the liquid phase $(\partial \Delta\lambda/\partial T)_\rho > 0$ and in the wide vicinity of the critical point $\Delta\lambda$ greatly increase and $(\partial \Delta\lambda/\partial T)_\rho < 0$.

The region of critical anomaly of thermal conductivity, according to Senger's evaluation, is considerably wider and includes the interval $|\Delta T^*| = 0.2$ at $\rho = \rho_{cr}$ and the interval $|\Delta\rho^*| = \pm 0.7$ at $T = T_{cr}$ [0.38].

Dynamic scaling law predicts, quite accurately, the asymptotic behavior of thermal conductivity [0.5, 0.6], but it has been unable, thus far, to compete with the semiempirical equations, one of which is proposed in Ref. [0.38]

$$\lambda = \lambda_0(T) + \Delta\lambda_{reg}\ (\rho,\ T) + \Delta\lambda_{sc}\ (\rho,\ T), \tag{0.38}$$

where $\lambda_0 = \lambda_T$, $\Delta\lambda_{reg}$ is the regular part of excess thermal conductivity which is independent of the proximity to the critical point, and $\Delta\lambda_{sc}$ is the separate scaling part of excess thermal conductivity which is substantial only in the vicinity of the critical point, and

$$\Delta\lambda_{sc} = c_1\ \Delta\lambda_c\ F\ (\Delta T^*,\ \Delta\rho^*) = c_1\ \Delta\lambda_c\ \omega^{1/2}$$
$$\times \exp\ [-A\ (\Delta T^*)^2 - B(\Delta\rho^*)^4], \tag{0.39}$$

where $\Delta\lambda_c \sim T\rho(c_p - c_v)/\eta$ and the constants $c_1$ and $B$ are near unity.

Equations of this type have been tested on well-studied substances such as $H_2O$, $CO_2$, and Ar. The detailed study of thermal conductivity in the critical region of freons is left to the future. But it is already clear that the simplest approximations which were popular in the past have limited value. We think that the following empirical equation could be a suitable model for an analytical single equation of thermal conductivity for the region in which $\Delta T^*$ and $\Delta\rho^*$ exceed $0.05-0.1$:

$$\lambda = \lambda_T \exp\left[\sum_{i=1}^{m} \sum_{j=0}^{n} a_{\lambda,\,ij}\ \omega^i/\tau^j\right] \tag{0.40}$$

which, as was shown for $CO_2$, can reproduce the function $\lambda(\rho, T)$ fairly well in the gaseous and liquid phases up to the solidification line and also part of $\Delta\lambda_{sc}$ [0.5, 0.32].

Equations of the types (0.37) and (0.40) were used [3.2, 3.3] for the approximation of experimental data and calculation of table of viscosity and thermal conductivity for Freon-22.

The lack of reliable data about density is an obstacle to the application of equations of the type $\Delta\eta(\omega, \tau)$ or $\eta/\eta_T(\omega, \tau)$. In particular, for Freon-22, there were no measurements of $\rho(p, T)$ at temperatures lower than 280 K and pressures higher than $p_s$ until recently. Densities of liquid Freon-21 at high pressures were studied only at $T > T_{NBP}$. This situation served as an incentive for the elaboration of equations of the form $F(\bar\eta, \bar p, T)$ indirectly linking experimentally measured values.

The works of C. L. Rivkin [0.30, 0.31] established that within a wide region of $pT$ variables, lines of constant viscosity of liquids are almost linear. Therefore, for the description of viscosity of liquid freons at densities $\omega \geq 1.9-2.0$, a generalized equation of the following form could be used:

$$\tau = 0.7 \sum_{i=0}^{m} \bar\alpha_{\eta,\,i}\,(\eta^*)^{-i} + \pi \sum_{i=0}^{m} \bar\beta_{\eta,\,i}\,(\eta^*)^{-i} \tag{0.41}$$

where $\eta^* = \eta/\eta_{0.7}$. The coefficients of Eq. (0.41) for the group of freons under consideration are found in Refs. [0.12, 0.15, 0.24]. The values of $\bar\alpha_{\eta,\,i}$ and $\bar\beta_{\eta,\,i}$ in Table 4 were taken from Ref. [0.15].

It is pertinent to mention that Eq. (0.41) describes practically the whole region of the liquid state (single-phase and along the saturation line) from $\omega = 2.0$ to the line of crystallization. In the zone of $\omega = 1.9-2.0$ the equation merges with Eq. (0.35).

As we get nearer to the critical point, the constant-viscosity lines in the $pT$ diagram begin to get noticeably distorted, and in the region of superheated vapor they even have a maximum. In view of this, in the more general case, equation $T = f(\eta, \bar p)$ must be more complex, namely:

$$\tau = \sum_{i=0}^{m} \sum_{j=0}^{n} a_{\eta,\,ij} \, \bar{p}/\eta^{j} \tag{0.42}$$

where parameter $\bar{p}$ is the same as $p$ or $(p - p_s)$ [0.36].

In Ref. [0.20], to describe the viscosity of liquid and gaseous Freon-22, an equation of the following type is used:

$$p = \sum_{i=0}^{m} \sum_{j=0}^{n} a_{ij}(T - T_{cr})^{i}(\eta - \eta_i)^{j} \tag{0.43}$$

in which $m = 2$ and $n = 11$. This equation describes a wide region of the single-phase state, but it is fairly complex and contains 24 empirical constants.

For the calculation of tables of thermal conductivity of freon at $\omega \leqslant$ 1.9–2.0, equations of the type (0.34) are used in the present work.

At higher densities, separation of isotherms of the excess thermal conductivity in the coordinates $(\Delta\lambda, \rho)$ is noticed. Therefore, the theoretical equations must become more complex.

In Ref. [3.3] an equation of the type (0.40) with four coefficients is recommended for Freon-22. In [0.35], for the calculation of tables of thermal conductivity of Freons-12, -13, -14, and -22, the following equation is used:

$$\lambda = \lambda_T + A \exp(nt) \, [\exp(c_1\omega) - \exp(-c_2\omega^2)] \tag{0.44}$$

In the present work, for the calculation of tables of thermal conductivity of liquid freons, the following system of equations is used

$$\lambda = \lambda_s \sum_{i=0}^{9} \sum_{j=0}^{6} \bar{a}_{\lambda,\,ij} \, \tau^{i} \, (\pi/10)^{j} \tag{0.45}$$

$$\lambda_s = \lambda_{cr} + \sum_{j=0}^{n} c_j \, (T_{cr} - T)^{j/3} \tag{0.46}$$

This system was developed by V. Z. Geller and coworkers [0.13, 1.4] on the basis of experimental data for Freons-21, -22, and -23.

The temperature dependence of thermal conductivity on the saturation line is given by equations of type (0.46) with individual constants $c_j$ for each freon. The pressure dependence of thermal conductivity is determined by the generalized equation (0.45).

The coefficients of the generalized equation of thermal conductivity of liquid freons are presented in Table 5, and the coefficients of Eqs. (0.46) for

**TABLE 5.** Coefficients of Generalized Eq. (45)

| $j$ | Values $\bar{\alpha}_{\lambda 1ij}$ for $i$ and $j$ | | | |
|---|---|---|---|---|
| | 0 | 1 | 2 | 3 |
| 0 | —1.10704 | —0.68156 | 1.40965 | —0.944264 |
| 1 | —0.33919 | 2.25599 | —5.05850 | 4.452905 |
| 2 | —3.67110 | 21.18587 | —32.63517 | 16.91989 |
| 3 | 9.77897 | —56.31672 | 83.18183 | —49.13515 |
| 4 | —10.37926 | 59.36453 | —79.97439 | 51.61741 |
| 5 | 5.12614 | —29.14068 | 35.19539 | —24.96465 |
| 6 | —0.97589 | 5.52094 | —5.29561 | 4.663276 |

Freons-20, -21, -22, and -23 are given in the corresponding sections of this book.

The system of equations (0.45)–(0.46) is applicable for the calculation of thermal conductivity of liquid freons right up to the solidification curve. For Freon-20, $pT$ parameters of the solidification curve at high pressure are known, and for other liquids they can be evaluated using the recommendations in [0.14].

One of the important physical constants is surface tensions at the liquid–vapor border. Temperature dependence of $\sigma$ for freons of the methane series has not been sufficiently studied, and for Freons-20 and -23, for example, only few experimental points are known [0.34]. It is, therefore, useful to adopt the generalized equations [0.33, 1.2, 1.8]. The most universal equation is derived in Ref. [0.33]:

$$\sigma^* = \sigma B = \sum_{m=1}^{5} \beta_m \, (1-\tau)^m \sum_{n=0}^{3} \alpha_n A^n \cdot 10^{-8} \qquad (0.47)$$

where $B = (RT)^{-1/3} \, p_{cr}^{-2/3}$ [unit of $B$ is mol$^{1/3}$/(Pa $\cdot$ m)]
$A = 100(\pi_s)_{\tau=0.625}$; $\beta_1 = 1.1239$; $\beta_2 = 9.1160$; $\beta_3 = -29.0038$;
$\beta_4 = 51.1109$; $\beta_5 = 35.1050$; $\alpha_0 = 7.56938$; $\alpha_1 = -2.57629$
$\alpha_2 = 0.711868$; $\alpha_3 = -0.07567$

The parameter $A$ was introduced by L. P. Filipov [1.17] as a criterion for thermodynamic similarity. For the group of freons under consideration its value is about 1.9.

In conclusion, we note that in the last 20 years quite a few reference books with tabular data about thermophysical properties of freons were published [0.7, 0.8, 0.9, 0.10, 0.18, 0.28, 0.34, 0.45, 0.51, and others]. The most actively updated reference data are those for Freon-22 [0.28, 0.51, 1.2, 1.4, 3.1, 3.2, 3.3, 3.20, 3.31, and others]. But new tabular data about thermophysical properties of Freons-20 and -23 were not published in the 1970s, and they are not included at all in the widely used references [0.9, 0.10]. Also, in the overwhelming majority of reference books, no equations are given. Nevertheless, equations represent a compact and universal way of storing and using information about thermophysical properties of substances, and in the present conditions they are not of less practical significance than tables. Therefore, in the following chapters of the present book, the numerical values of the coefficients for most of the equations considered here, which were derived as a result of the consistent analysis of experimental data, are given.

# THERMOPHYSICAL PROPERTIES OF FREON-20

Freon-20 (trichloromethane, chloroform) is a high-boiling cooling agent, and its thermophysical properties at elevated pressures have been considerably less scrutinized than other freons of the methane series. At low pressures, there are relatively extensive experimental data about the thermodynamic and transport properties of Freon-20. But this information is not generalized, and comprehensive reference data about the thermophysical properties of Freon-20 are lacking.

## 1.1. A REVIEW OF PUBLISHED DATA
## ABOUT THERMODYNAMIC PROPERTIES

From Tables 6 and 7 it follows that thermodynamic characteristics of Freon-20 at low pressures have been determined for gaseous and liquid phases and on the liquid–vapor saturation line. But the discrepancies among diverse groups of experimental data often exceed the error of measurement mentioned by the authors of the experimental work. This observation is addressed, in particular, to the experimental data about the second virial coefficient of equations of state (Fig. 1) and to the data about saturated vapor pressure. The agreement of experimental data for the heat capacity of saturated liquids is good, but the direct measurements were made at temperatures below the normal boiling point ($T_{NBP}$ = 334.3 K). The comparative analysis of experimental data about the orthobaric density shows that the temperature dependence of density of the saturated liquid can be accepted with certainty in the interval from the triple point to the

1

**TABLE 6.** Experimental Study of the Thermodynamic Properties of Freon-20 at Low Pressure

| Year | Authors | Measured value | Temperature, K | Phase[a] | Number of experimental points | Reference |
|------|---------|----------------|----------------|----------|-------------------------------|-----------|
| 1913 | Dolezalek and Schulze | $c_p$ | 223—293 | L | — | [1.27] |
| 1922 | Magnus and Schmidt | $B_1$ | 371 | G | 1 | [1.32] |
| 1925 | Williams and Daniels | $c_p$ | 295—319 | L | — | [1.70] |
| 1949 | Aihara | $c_v$ | 279 | G | 2 | [1.19] |
| 1949 | Lambert et al. | $B_1$ | 323—383 | G | 4 | [1.42] |
| 1955 | Francis and McGlashan | $B_1$ | 316—398 | G | 8 | [1.32] |
| 1955 | Staveley et al. | $c_p$ | 284—329 | L | — | [1.60] |
| 1957 | Harrison et al. | $c_p$ | 246—303 | L | 8 | [1.34] |
| 1965 | Zaalishvili et al. | $B_1$ | 353—383 | G | 4 | [1.11] |
| 1965 | Fort and Moore | $\varrho$; $w$ | 293 | L | 1; 1 | [1.31] |

[a] G, Gas; L, liquid.

critical point, and the temperature dependence of density of dry saturated vapor from $T_{NBP}$ to $T_{cr}$ (Fig. 2).

Experimental data on saturated vapor pressure are generalized in Refs. [1.58, 1.61]. In Ref. [1.58] the following equation is recommended:

$$\lg p_s = A + B/T + C \lg T + DT, \qquad (1.1)$$

**TABLE 7.** Experimental Study of the Thermodynamic Properties of Freon-20 on the Liquid–Vapor Equilibrium Line (Saturation)

| Year | Author | Measured value | Temperature, K | Phase[a] | Number of experimental points | Reference |
|------|--------|----------------|----------------|----------|-------------------------------|-----------|
| 1902 | Kuenan and Robson | $p_s$ | 518—536 | V | — | [1.41] |
| 1912 | Herz and Rathman | $p_s$ | 300—334 | V | — | [1.36] |
| 1913 | Fletcher and Turi | $r$ | 273—314 | L–V | — | [1.61] |
| 1914 | Beckman and Liesche | $p_s$ | 291—343 | V | — | [1.22] |
| 1915 | Drucker and Jumeno | $p_s$ | 209—263 | V | — | [1.28] |
| 1916 | Rex | $p_s$ | 273—303 | V | — | [1.52] |
| 1921 | Schulze | $\varrho$ | 278—299 | L | 3 | [1.57] |
| 1923 | Herz and Neukirch | $\varrho$ | 483—536 | L, V | 15 | [1.37] |
| 1926 | Mathews | $r$ | 334 | L–V | 1 | [1.46] |
| 1926 | Schmidt | $p_s$ | 273—373 | V | 8 | [1.56] |
| 1928 | Smyth and Morgan | $\varrho$ | 213—333 | L | 10 | [1.59] |
| 1932 | Wright | $\varrho$ | 331—373 | V | 7 | [1.71] |
| 1937 | Coop | $\varrho$ | 193—293 | L | 4 | [1.26] |
| 1938 | Scatchard and Raymond | $p_s$ | 308—333 | V | 6 | [1.55] |
| 1952 | Patton | $\varrho$ | 298 | L | 1 | |
| | | $r$ | 298 | L–V | 1 | [1.54] |

[a] L, Liquid; V, vapor.

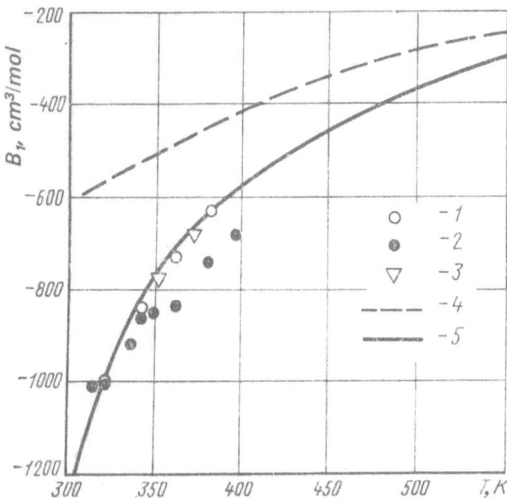

**FIG. 1.** Second virial coefficient of Freon-20. Experimental data: 1, Lambert et al. [1.42]; 2, Francis, and McGlashan [1.32]; 3, Zaalishvili et al. [1.11]. Calculated data: 4, Eq. (1.15); 5, present work.

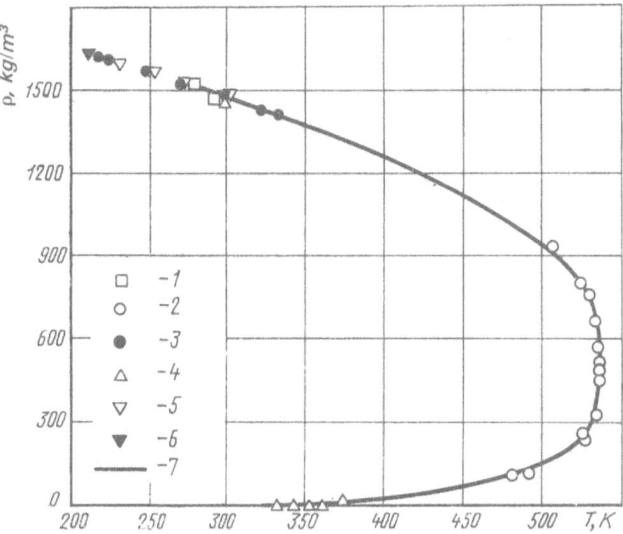

**FIG. 2.** Orthobaric density of Freon-20. Experimental data: 1, Schulze [1.57]; 2, Herz and Neukirch [1.37]; 3, Smyth and Morgan [1.59]; 4, Wright [1.71]; 5, Coop [1.26]; 6, Scatchard and Raymond [1.55]. Calculated data: 7, present work.

the coefficients of which are given for two temperature intervals (I—$T$ = 275–340 K and II—$T$ = 340–536.6 K) and, for the case, when $p_s$ is in atmospheres:

$$A_I = -2.9593966 \cdot 10^1 \qquad A_{II} = 1.4088080 \cdot 10^1$$
$$B_I = -1.0258492 \cdot 10^3 \qquad B_{II} = -1.9585807 \cdot 10^3$$
$$C_I = 1.4881037 \cdot 10^1 \qquad C_{II} = -3.3974131 \cdot 10^0$$
$$D_I = -1.4702661 \cdot 10^{-2} \qquad D_{II} = 1.0602881 \cdot 10^{-3}$$

The mean deviation of calculated and experimental values of $p_s$ in interval I is equal to 0.75% and in interval II is equal to −0.58%. The maximum deviation does not exceed 1.52%.

At pressures higher than 0.04 MPa, Eq. (1.1) is based on the only measurements by Kuenan and Robson [1.41]. Therefore, the larger-than-usual error margin should be ascribed to the recommended values of $p_s$.

A number of important parameters which are key fundamental points on the saturation line and experimental data of certain references are reviewed and shown in Table 8.

Large tolerances for the values of critical parameters adopted in this book are only natural since $p_{cr}$ and $\rho_{cr}$ were experimentally determined only in Refs. [1.37, 1.41]. In the absence of accurate empirical formulas for the analytical determination of the values of $p_s$ and $\rho^l$ for freons, we recommend the use of the generalized equations obtained in Ref. [1.15]

$$\ln \pi_s = \text{Ri} \ln \tau + (\text{Ri} - 4 + p_\alpha) \psi (\tau), \qquad (1.2)$$

$$\ln \omega' = [1.4 + 0.03 \ln (p_{cr} R T_{cr}^{3/2}) + 0.03 \text{ Ri} - 0.2 p_\alpha + L_\alpha + L_\beta] (1 - \tau)^{1/3} + [0.68 - 0.07 \text{ Ri} - 0.5 p_\alpha + L_\alpha] s (\tau), \qquad (1.3)$$

where Ri = $(d\ln \pi_s / d\tau)_{\tau=1}$ is the Riedel criterion, $p_\alpha$, $L_\alpha$, and $L_\beta$ are adjusting constants, and the temperature functions are

**TABLE 8.** Parameters of Fundamental Points on the Saturation Liquid–Vapor Equilibrium Line for Freon-20

| Year | Authors | Critical parameters[a] | | | $T_{\text{NBP}}$, K | $T_f$, K | Source |
|------|---------|------|------|------|------|------|------|
| | | $T_{cr}$, K | $p_{cr}$, MPa | $\rho_{cr}$, kg/m³ | | | |
| 1878 | Zaeichevsky | 533.15 | 5.56 | — | — | — | |
| 1895 | Picktet and Altschul | 531.95 | — | — | — | — | |
| 1902 | Kuenan and Robson | 536.0 | 5.46 | — | — | — | [1.40] |
| 1923 | Herz and Neukirch | 535.6± ±0.2 | — | 496 | 334.30± ±0.05 | — | |
| 1935 | Harand | 536.65 | — | — | — | — | |
| 1943 | Fisher and Reichell | 536.75 | — | — | — | — | |

TABLE 8. Parameters of Fundamental Points on the Saturation Liquid–Vapor Equilibrium
Line for Freon-20 (*Continued*)

| Year | Authors | Critical parameters[a] | | | $T_{NBP}$, K | $T_f$, K | Source |
|------|---------|-----------|------|------|------|------|--------|
| | | $T_{cr}$, K | $p_{cr}$, MPa | $\rho_{cr}$, kg/m³ | | | |
| 1947 | Stull | 533,15 | 5.56 | — | 334.45 | 209.65 | [1.61] |
| 1954 | Svitoslavsky and Kregleivsky | 536,3 | — | — | — | — | [1.40] |
| 1965 | Zaalishvili et al. | — | — | — | 334,25 | — | [1.11] |
| 1968 | Seshadri et al. | 536.71 | 5.47 | 500 | — | — | [1.58] |
| 1968 | Seshadri et al. | 536,5 | 5,47 | 498 | 334,2 | 210 | [1.13] |
| 1970 | Phillips and Murphy | — | — | — | 334,35 | — | [1.50] |

[a] The values of the parameters adopted in this book are given in Table 3.

$$\psi(\tau) = 4(\tau - 1) \tau - 2.68 \ln \tau + s(\tau), \qquad (1.4)$$

$$s(\tau) = (\tau - 1)\{0.2(\tau + 1)^2 + 0.5\}. \qquad (1.5)$$

For freons of the methane series, Ri = 6.5–7.2 and the adjusting constants vary
in the range $p_\alpha$ from $-0.2$ to $-0.06$; $L_\beta$ ranges from 0 to 0.1, and $L_\alpha$ from
$-0.05$ to 0. For Freon-22, for example, Ri = 6.7964, $p_\alpha$ = 0.1644, and $L_\alpha$ =
$L_\beta$ = 0. Consequently, the adjusting constants $L_\alpha$ and $L_\beta$ are, generally
speaking, considered to be of little significance. The constants $p_\alpha$ can be evaluated with the help of empirical correlation in [1.15] as follows:

$$\ln z_{cr} = -0.088(\text{Ri} - p_\alpha) - 0.728, \qquad (1.6)$$

if the following relations obtained by L. P. Filipov [1.17] are taken into account:

$$\text{Ri} = 4.919 \text{ Pi} + 5.811 = 7.78 - 3.27 \lg A_\Phi, \qquad (1.7)$$

where $\text{Pi} = -(\lg \pi_s)_{\tau=0.7} - 1$, $A_\Phi = 100 \ (\pi_s)_{\tau=0.625}$.

At the indicated values of $\tau$ the pressure of saturated vapor is measured fairly
accurately and the numerical values of $A_\Phi$, Pi, and Ri are reliable.

For the calculation of orthobaric vapor densities in Ref. [1.17] the following
formula is suggested:

$$\lg (1 - z_G) = (0.094 \lg A_\Phi + 0.254) + (0.705 - 0.0668 \lg A_\Phi) \lg(z_L) \quad (1.8)$$

Liquid surface tension $\sigma$ at the saturation line can be calculated using the generalized equations (0.47) or (1.9):

$$\sigma^* = \sigma_0^*(1 - \tau)^n \tag{1.9}$$

$$\sigma^* = \sigma B^{-1} = \sigma(\sqrt[3]{KT_{cr}\,P_{cr}^{\,2}})^{-1}$$

where $K$ = constant.

The accuracy of Eq. (1.9) is somewhat less but is simpler and gives, as shown in Refs. [1.2, 1.8], sufficiently satisfactory results for freons. According to experimental data for freons of the methane series $n = 1.249$, and $\sigma_0^* = 12.35$ [1.2]. Consequently, for Freon-20 the constant $B = 5.911 \cdot 10^{-3}$ N/m. In Bridgeman's work [1.24, 1.25] at pressures up to 2500 MPa, the fusion temperature and volume expansivity $\Delta v_{fusion}$ of solid Freon-20 were measured. From these experimental data the researchers in [1.20] determined the constants of Simon's equation:

$$p - p_0 = a[(T/T_0)^c - 1] \tag{1.10}$$

where $T_0 = 209.7$ K, $p_0$ = pressure of $T_0(\sim T_{tr})$, $a = 8330$ bar, and $c = 1.52$. Another set of constants is obtained in Ref. [1.9]. In a much later work by M. K. Zhukhovsky [0.14] a more universal equation for the fusion curve is proposed, and its parameters for many substances, including Freon-20, are obtained.

For the approximation of Bridgeman's experimental data about $\Delta v_{fusion}$ the following equation was used in Ref. [1.10]

$$\Delta v_{fusion} = \Delta v_0 \exp\left[-b\,(T/T_0 - 1)\right] \tag{1.11}$$

For Freon-20 the constant $b$ was found to be equal to $0.58 \pm 0.02$ in the interval $p_{fusion} = 1000\text{--}2500$ MPa, but at pressures 400–1200 MPa, $b = 0.91 \pm 0.05$.

The thermodynamic functions of Freon-20 in the ideal gas state were calculated in several works. As a rule, calculations are carried out by approximating them to the model of a rigid rotary-harmonic oscillator (RRHO). The deviation from harmonic oscillations was taken into account only in the work of Barho [0.37], the correction being found by semiempirical method.

Table 9 shows the values of heat capacity $C_p^0$, enthalpy $(H_T^0 - H_0^0)$, and entropy $S_T^0$ from the results in Refs. [0.37, 1.7, 1.33, 1.39, 1.54]. From Table 9 it appears that data in [1.39, 1.54] are practically identical and the values of $C_p^0$ in Barho's work are on the average 0.5–1% higher than those calculated from the RRHO model. In Chap. it is shown that the calculated data for Freon-22 [0.37] correspond well with the results of direct calorimetric measurements. This confirms the accuracy of Barho's calculation method, and preference should therefore be given to his data.

**TABLE 9.** Values of Thermodynamic Functions for Freon-20 in the Ideal Gas State by Different Researchers

| $T$, K | $C_p^0$, cal/(mol·K) | | | | $H_T^0 - H_0^0$, cal/mol | | | | $S_T^0$, cal/(mol·K) | | | |
|---|---|---|---|---|---|---|---|---|---|---|---|---|
| | [1.33] | [0.37] | [1.39] | [1.54] | [1.33] | [1.7] | [1.39] | [1.54] | [1.33] | [1.7] | [1.39] | [1.51] |
| 100 | 9.62 | — | 9.637 | 9.637 | 836 | — | 837 | 836 | 57.19 | — | 57.180 | 57.180 |
| 150 | 11.13 | — | — | 11.414 | 1363 | — | — | 1364 | 61.43 | — | — | 61.432 |
| 200 | 13.02 | 13.04 | 12.978 | 12.978 | 1975 | — | 1975 | 1974 | 64.94 | — | 61.933 | 64.934 |
| 250 | 14.47 | 14.49 | — | — | 2663 | 3396 | — | — | 68.00 | 70.723 | — | — |
| 298.16 | 15.71 | — | — | 15.627 | 3390 | — | — | 3383 | 70.66 | — | — | 70.628 |
| 300 | 15.76 | 15.79 | 15.671 | 15.671 | 3419 | — | 3412 | 3412 | 70.76 | — | 70.724 | 70.725 |
| 400 | 17.83 | 17.90 | 17.747 | 17.747 | 5104 | 5106 | 5088 | 5088 | 75.58 | 75.643 | 75.532 | 75.532 |
| 500 | 19.34 | 19.45 | 19.266 | 19.266 | 6967 | 6967 | 6943 | 6942 | 79.74 | 79.787 | 79.664 | 79.661 |
| 600 | 20.44 | 20.60 | 20.378 | 20.378 | 8959 | 8954 | 8928 | 8927 | 83.37 | 83.407 | 83.280 | 83.280 |
| 700 | 21.27 | 21.47 | 21.217 | 21.217 | 11040 | 11040 | 11009 | 11009 | 86.59 | 86.621 | 86.487 | 86.487 |
| 800 | 21.91 | 22.16 | 21.872 | 21.872 | 13210 | 13197 | 13165 | 13165 | 89.47 | 89.500 | 89.365 | 89.365 |
| 1000 | 22.88 | 23.19 | 22.831 | 22.831 | 17690 | 17676 | 17641 | 17641 | 94.47 | 94.494 | 91.355 | 94.356 |
| 1300 | 23.77 | — | 23.751 | 23.751 | 24690 | 24677 | 24640 | 24640 | 100.59 | 100.615 | 100.471 | 100.471 |

The following equations were obtained in Ref. [0.21] to represent the ideal gas functions with temperature dependence:

$$C_p^0 = a_0 + 2a_1 T + 3a_2 T^2 + 4a_3 T^3 \tag{1.12}$$

$$H_T^0 - H_0^0 = a_0 T + a_1 T^2 + a_2 T^3 + a_3 T^4 \tag{1.13}$$

$$s_T^0 = a_0 \ln T + 2 a_1 T + 1.5 a_2 T^2 + 1{,}333 a_3 T^3 + a_4 \tag{1.14}$$

When $C_p^0$ is in units of kJ/(kmol $\cdot$ K), the coefficients $a_j$ for Freon-20 are equal to

$$a_0 \cdot 10^{-2} = 0.2291506$$
$$a_1 = 0{,}09485$$
$$a_2 \cdot 10^3 = -0.05482929$$
$$a_3 \cdot 10^6 = 0.009470524$$
$$a_4 \cdot 10^{-2} = 1.157499$$

For the calculation of thermodynamic properties of Freon-20 in the single-phase region at elevated pressures, Seshadri et al. [1.58] used an equation of the following type:

$$p = \frac{RT}{v - b_0} + \sum_{i=2}^{5} \frac{A_i + B_i T + C_i \exp\,(-5{,}475\ \tau)}{(v - b_0)^i} \tag{1.15}$$

The coefficients of this equation are obtained by a method proposed by Martin and Hou (1955), and when $v$ is in liters per mole, $p$ is in atmospheres, and $T$ is in Kelvins, they are equal to

$b_0 = 6{,}054802 \cdot 10^{-2}$;      $A_4 = -1.93684745 \cdot 10^{-1}$;
$A_2 = -21.5847159$;      $B_4 = 0$;
$B_2 = 1.31856963 \cdot 10^{-2}$;      $C_4 = 0$;
$C_2 = 30{,}306821$;      $A_5 = 0$;
$A_3 = 3.17370469$;      $B_5 = 1.1943037 \cdot 10^{-5}$;
$B_3 = -1{,}4828215 \cdot 10^{-3}$;      $C_5 = 0$.
$C_3 = -5{,}4014024$;

The tables of the values of $v$, $h$, and $s$ for gaseous Freon-20 were calculated in Ref. [1.58] using Eq. (1.15) at $T = 280-750$ K and $p = 0.1-200$ atm. It is essential to note that the numerical values of the constants $A_i$, $B_i$, and $C_i$ in the Martin–Hou equation are strongly dependent on the values of the characteristic parameters. Additionally, the analytical relations used to determine these constants can be diverse; see, for example, Ref. [1.1]. It is possible to evaluate the reliability of the calculated tables in Ref. [1.58] using, for example, the generalized BWRC equation, which was discussed in the introduction.

We note that the analytical expressions $f(T_{cr}, p_{cr}, \text{Pi})$ for the determination of individual coefficients of Eq. (0.5) are given in Ref. [0.49], and the adjusting constants $K_{ij}$ for freons of the methane series are given in Refs. [0.22, 1.12].

The numerical values of the Pi criterion are also shown there for several freons.

At relatively low (less than the critical) pressures the simple generalized equation which was derived by I. I. Perelshtein [1.14] on the basis of the modified equation (0.1) gives entirely satisfactory results for freons of the methane series.

On the basis of analysis of experimental data for normal liquids, L. P. Filipov [1.17] found a simple formula to calculate the isothermal compressibility

$$\beta_T \, RT_\rho/M \;=\; \rho(\partial_\rho/\partial p)_T \, RT_\rho/M \;=\; 72/\omega^{7.5} \qquad (1.16)$$

Integration gives the following equation for the isotherm of the liquid

$$\rho \;=\; \rho_0[1 \;+\; 612(p \;-\; p_0)(\rho_{cr}/\rho_0)^{7.5}1/(\rho_0 RT)]^{1/8.5}, \qquad (1.17)$$

which is applicable in the interval of reduced density values $\omega = 2.2-3.2$. For the initial parameters $p_0$ and $\rho_0$ the pressure and density of the liquid on the saturation line can be adopted.

Thus, the system of generalized equations described above allows the calculation of thermodynamic properties of insufficiently studied substances in a wide region of state using a limited number of initial experimental data. However, the thermodynamic tables for Freon-20 in the single-phase region at elevated pressures are not presented in this handbook. This is so because of the absence of experimental data essential to verify the equations and the tables.

## 1.2. A REVIEW OF PUBLISHED DATA
## ABOUT VISCOSITY
## AND THERMAL CONDUCTIVITY

The tables of viscosity of gaseous Freon-20 at atmospheric pressure are given in monographs [1.5, 1.6]. These tables cover the temperature interval $T = 273-700$ K, and the difference between the data amounts to ~6%. A comparison of the tabular data in [1.5, 1.6] with the results of direct measurements (Fig. 3) shows that tables in [1.6] are inaccurate and that the earlier ones are based on experimental results obtained before 1933 (Table 10).

To determine the viscosity of Freon-20 at atmospheric pressure, the following equation can be used to calculate gas viscosity:

$$\lg (\eta_T/\eta_{T_{cr}}) \;=\; \left( \sum_{j=0}^{3} a_j \, \tau - j \right) \lg \tau, \qquad (1.18)$$

where $\eta_{T_{cr}} = 17.95$ μPa · s
$\quad a_0 = 0.74739;$ $\qquad\qquad a_2 = -0.21477;$
$\quad a_1 = 0.41890;$ $\qquad\qquad a_3 = 0.03636.$

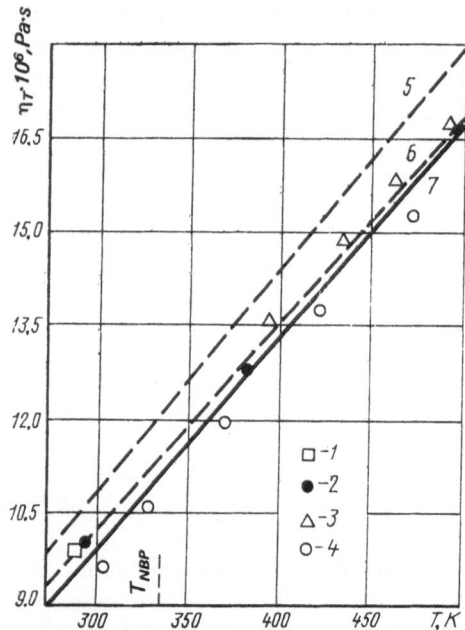

**FIG. 3.** Viscosity of gaseous Freon-20 at low pressures. Experimental data: 1, Suhrmann [1.62]; 2, Braune and Linke [1.23]; 3, Titani [1.65]; 4, Pal and Barua [1.48]. Calculated data: 5, Golubev and Gnezdilov [1.6]; 6, Golubev [1.5]; 7, present work.

**TABLE 10.** Experimental Studies of Viscosity of Freon-20

| Year | Authors | Temperature, K | Pressure, MPa | Number of experimental points | Phase[a] | Method[b] | Reference |
|------|---------|---------------|--------------|------------------------------|----------|-----------|-----------|
| 1897 | Thorpe and Rodger | 273—330 | $p_s$ | 4 | L | OD | [1.64] |
| 1908 | Tsakolotes | 293; 313 | $p_s$ | 2 | L | Ca | [1.66] |
| 1910 | Rappanecker | 290—485 | 0,1 | 3 | G | Ca | [1.51] |
| 1914 | Vogel | 273 | 0.1 | 1 | G | Ca | [1.68] |
| 1923 | Suhrmann | 287 | 0,1 | 1 | G | OD | [1.62] |
| 1924 | Miller | 281; 293 | $p_s$ | 2 | L | Ca | [1.47] |
| 1927 | Zimmermans and Martin | 288—303 | $p_s$ | 2 | L | RB | [1.44] |
| 1928 | Kolosovsky | 283—296 | $p_s$ | 4 | L | Ca | [1.44] |
| 1930 | Braune and Linke | 293—618 | 0.1 | 7 | G | OD | [1.23] |
| 1933 | Titani | 394—581 | 0.1 | 7 | G | Ca | [1.65] |
| 1935 | Bridgeman | 303/348 | 0,1—800 | 4 | L | RB | [1.5] |
| 1960 | Rabinovich and Ponomarenko | 283—323 | $p_s$ | 5 | L | CA | [1.44] |
| 1968 | Pal and Barua | 304—573 | 0,1 | 5 | G | OD | [1.48] |
| 1970 | Phillips and Murphy | 210—356 | $p_s$ | 11 | L | Ca | [1.50] |

[a] G, Gas; L, liquid.
[b] Ca, Capillary; OD, oscillating disk; RB, rolling ball.

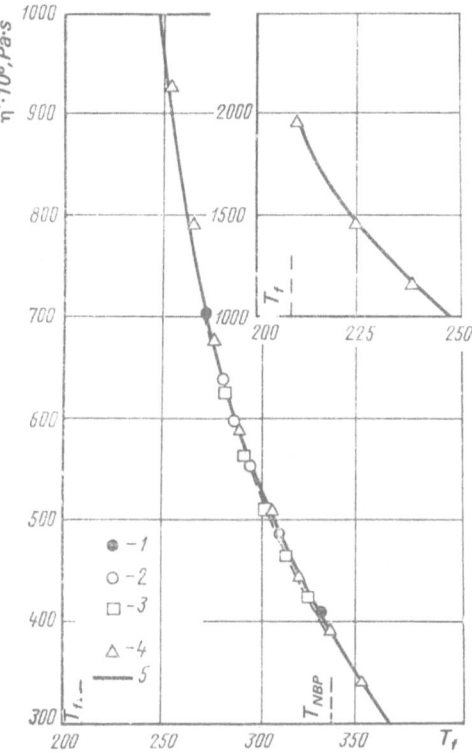

**FIG. 4.** Viscosity of liquid Freon-20 in the vicinity of the saturation line. Experimental data: 1, Thorpe and Rodger [1.64]; 2, Kolosovsky [1.44]; 3, Rabinovich and Ponomarenko [1.44]; 4, Phillips and Murphy [1.49, 1.50]. Calculated data: 5, present work.

**TABLE 11.** Experimental Studies of Thermal Conductivity of Freon-20

| Year | Authors | Tempera-ture, K | Pressure, MPa | Number of experimen-tal points | Phase[a] | Method[b] | Refer-ence |
|------|---------|-----------------|---------------|--------------------------------|----------|-----------|------------|
| 1895 | Weber | 282—288 | $p_s$ | 2 | L | F | [1.69] |
| 1913 | Eucken | 373 | 0,1 | 1 | G | F | [1.30] |
| 1940 | Riedel | 293 | $p_s$ | 1 | L | F | [1.53] |
| 1941 | Bates et al. | 293—323 | $p_s$ | 3 | L | F | [1.21] |
| 1945 | Hutchinson | 291 | $p_s$ | 1 | L | CC | [1.38] |
| 1948 | Held and Drunen | 289 | $p_s$ | 1 | L | NH | [1.35] |
| 1950 | Lambert et al. | 339—358 | 0,1 | 2 | G | H | [1.43] |
| 1954 | Vines and Bennett | 333—422 | 0,1 | 4 | G | H | [1.67] |
| 1954 | Mason | 303—323 | $p_s$ | 2 | L | CC | [1.45] |
| 1954 | Filipov | 288—363 | $p_s$ | Graphic | L | CC | [1.16] |
| 1964 | Djalalian | 216—293 | $p_s$ | 7 | L | H | [1.29] |
| 1967 | Tauscher | 214—348 | $p_s$ | 10 | L | NH | [1.63] |

[a] G, Gas; L, liquid.
[b] CC, coaxial cylinder; F, flat plate; H, heated filament; NH, nonstationary method of heated filament.

On the basis of analysis of a relatively large number (about 200 points) of experimental data covering the viscosity of gaseous freons in the interval $\tau = 0.6-1.9$, N. G. Sagaedakova [4.11] formulated the generalized equation for computing gas viscosity:

$$\tau = a + b(\eta_T/\eta_{T_{cr}}) + c(\eta_T/\eta_{T_{cr}})^2 \tag{1.19}$$

The probable error in the calculated values of $\eta_T$ from this equation apparently does not exceed $\pm 1.5\%$.

For liquid Freon-20, tables in [1.5] include the temperature interval from 273 K to $T_{NBP}$ and do not contradict experimental results. But it is apparent from Table 10 and Fig. 4 that the viscosity of liquid Freon-20 is scrutinized in a far wider temperature interval at the present time.

For the calculation of viscosity of liquid Freon-20 in the region $\omega = 0-1.9$, use of the generalized equation (0.35) is recommended.

At $\omega \geq 1.9$ the generalized equation (0.41) is recommended, the coefficients for which are given in Table 4. The viscosity of Freon-20 at the reference point for Eq. (0.35) is found from the data of Phillips and Murphy [1.49, 1.50] and equals $\eta_{0.7} = 295 \ \mu Pa \cdot s$.

Practically all the experimental studies of thermal conductivity of gaseous and liquid Freon-20 are enumerated in Table 11.

Figures 5 and 6 show that the "temperature dependence" of thermal conductivity can be represented by straight-line functions in a wide temperature interval for gaseous and liquid Freon-20. In the first case, the agreement of experimental data are admittedly good, whereas in the second case there is a need to select more reliable data.

For calculation of thermal conductivity of gaseous Freon-20 at low pressures, the following equation is recommended:

$$\lambda_T \cdot 10^4 = \alpha_0 + \alpha_1 T \tag{1.20}$$

where $\lambda$ is in W/(m $\cdot$ K); $\alpha_0 = -5.050 \cdot 10^{-3}$; $\alpha_1 = 4.076 \cdot 10^{-5}$. This equation describes the experimental data with an error less than 2%.

For freons of the methane series, O. B. Tsvetkov et al. [1.18] suggested the generalized equation

$$\lambda_T/\lambda_{T_{cr}} = -0.14091 + 1.36468\tau - 0.2232\tau^2(c_v/c_{v, \ T_{cr}}) \tag{1.21}$$

which approximates a relatively large number (about 100 points) of experimental data covering thermal conductivity at atmospheric pressure in the interval $\tau = 0.5-1.3$ with a mean error around 2%.

The thermal conductivity of liquid Freon-20 was first determined by Weber [1.69] using the method of the flat plate. The experiments were carried out

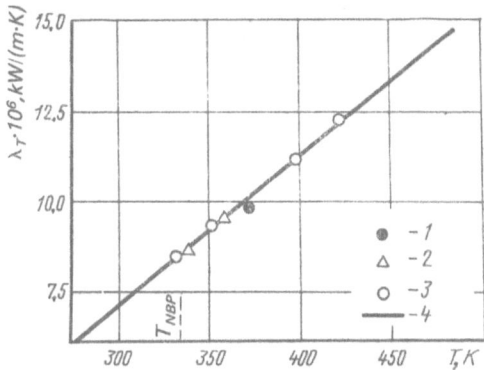

**FIG. 5.** Thermal conductivity of gaseous Freon-20 at low pressures. Experimental data: 1, Eucken [1.30]; 2, Lambert et al. [1.43]; 3, Vines and Bennett [1.67]; 4, present work.

extremely carefully at various measuring tolerances, and the divergence with the new experimental data did not exceed 4–6%.

Riedel [1.53] employed the absolute flat plate method to determine λ for Freon-20 at $T = 293$ K. The quality of these data is high since the results for a large number of substances agree well with the most reliable data by other researchers. In Refs. [1.21, 1.38] the same method was used, but data there compared with other results are 10–15% higher. The results of the remaining measurements generally agree with each other within the limits of experimental error.

The temperature dependence of thermal conductivity of liquid Freon-20 on the saturation line is established, based on the analysis of conforming measurements. These were carried out in Refs. [1.16, 1.29, 1.35, 1.45, 1.53, 1.63].

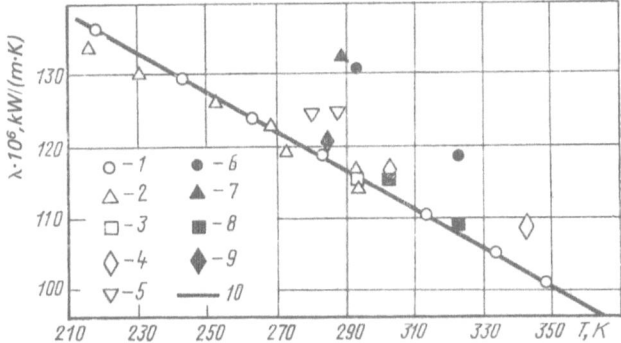

**FIG. 6.** Thermal conductivity of liquid Freon-20 near the saturation line. Experimental data: 1, Tauscher [1.63]; 2, Djalalian [1.29]; 3, Riedel [1.53]; 4, Filipov [1.16]; 5, Weber [1.69]; 6, Bates et al. [1.21]; 7, Hutchinson [1.38]; 8, Mason [1.45]; 9, Held and Drunen [1.35]; 10, present work.

As an approximating function, Eq. (0.46) is used. The coefficients of this equation for Freon-20 at $\lambda_{cr} = 400 \cdot 10^{-4}$ W/(m $\cdot$ K) are given below.

$$c_1 = -8{,}56330 \cdot 10^{-3}; \qquad c_4 = 6{,}41577;$$
$$c_2 = 2{,}58130; \qquad c_5 = -3{,}02624;$$
$$c_3 = -6{,}05146; \qquad c_6 = 0{,}54782.$$

To determine the thermal conductivity of liquid Freon-20 at elevated pressures, the use of the generalized equation (0.46) is recommended. The coefficients of this equation are determined in Refs. [0.13, 1.4] from the experimental data for Freon-21, -22, and -23 and are given in Table 5. The generalized equations obtained in Refs. [1.3, 1.4] are suitable for the calculation of viscosity and thermal conductivity of liquid freons of the methane series at pressures up to 50–60 MPa.

## 1.3. TABLES OF THERMOPHYSICAL PROPERTIES

The recommended listing of data covering the thermophysical properties of Freon-20 on the liquid–vapor saturation line are given in Table 12. The values of $\rho$, $h$, $s$, and $r$ were calculated using the thermal equations of state (0.9), (1.1), (1.3), and (1.12). Tabular data for $\eta$, $\lambda$, and $\sigma$ were calculated using the generalized equations (0.35), (0.41), (0.45), and (1.9).

TABLE 12. Thermophysical Properties of Freon-20 on the Saturation Line

| $T$ | $p$ | $\rho'$ | $\rho''$ | $h'$ | $h''$ | $s'$ | $s''$ | $r$ | $\eta' \cdot 10^6$ | $\eta'' \cdot 10^6$ | $\lambda' \cdot 10^6$ | $\sigma \cdot 10^3$ |
|---|---|---|---|---|---|---|---|---|---|---|---|---|
| 275 | 0.0090 | 1524.5 | 0.473 | 401.9 | 672.4 | 4.004 | 4.988 | 270.5 | 797.0 | 10.18 | 121.0 | 29.75 |
| 280 | 0.0115 | 1515.1 | 0.592 | 405.4 | 675.1 | 4.017 | 4.980 | 269.7 | 748.0 | 10.35 | 119.6 | 29.04 |
| 290 | 0.0186 | 1494.8 | 0.920 | 413.6 | 680.4 | 4.046 | 4.966 | 266.8 | 662.0 | 10.69 | 116.9 | 27.63 |
| 300 | 0.0293 | 1475.6 | 1.40 | 423.0 | 686.2 | 4.077 | 4.955 | 263.1 | 587.0 | 11.03 | 114.2 | 26.24 |
| 310 | 0.0435 | 1456.5 | 2.03 | 432.3 | 691.3 | 4.109 | 4.945 | 259.0 | 523.0 | 11.37 | 111.4 | 24.86 |
| 320 | 0.0620 | 1438.6 | 2.79 | 442.2 | 696.7 | 4.141 | 4.936 | 254.6 | 468.0 | 11.70 | 108.7 | 23.50 |
| 330 | 0.0879 | 1419.6 | 3.88 | 451.7 | 702.2 | 4.170 | 4.929 | 250.5 | 421.0 | 12.03 | 105.9 | 22.15 |
| 340 | 0.1224 | 1398.0 | 5.26 | 460.8 | 707.7 | 4.197 | 4.923 | 246.9 | 381.0 | 12.35 | 103.2 | 20.82 |
| 350 | 0.1682 | 1375.3 | 7.10 | 469.8 | 713.2 | 4.222 | 4.917 | 243.3 | 348.0 | 12.68 | 100.4 | 19.50 |
| 360 | 0.2255 | 1353.9 | 9.33 | 478.9 | 718.6 | 4.247 | 4.913 | 239.0 | 319.0 | 13.00 | 97.7 | 18.21 |
| 370 | 0.2979 | 1331.1 | 12.10 | 487.9 | 724.0 | 4.272 | 4.910 | 236.1 | 294.0 | 13.33 | 94.9 | 16.93 |
| 380 | 0.3830 | 1307.3 | 15.31 | 497.3 | 729.3 | 4.296 | 4.907 | 232.1 | 273.0 | 13.66 | 92.1 | 15.67 |
| 390 | 0.4823 | 1282.2 | 19.01 | 507.0 | 734.6 | 4.322 | 4.905 | 227.6 | 254.0 | 13.99 | 89.4 | 14.43 |
| 400 | 0.6039 | 1258.3 | 23.53 | 517.0 | 739.8 | 4.347 | 4.904 | 222.7 | 237.0 | 14.33 | 86.3 | 13.21 |
| 410 | 0.7417 | 1233.2 | 28.62 | 527.5 | 744.9 | 4.373 | 4.903 | 217.4 | 221.0 | 14.68 | 83.6 | 12.01 |
| 420 | 0.9058 | 1216.9 | 34.73 | 538.4 | 749.8 | 4.399 | 4.903 | 211.5 | 206.0 | 15.05 | 80.8 | 10.84 |
| 430 | 1.097 | 1178.5 | 42.17 | 549.5 | 754.6 | 4.425 | 4.902 | 205.4 | 191.4 | 15.44 | 77.9 | 9.69 |
| 440 | 1.339 | 1147.9 | 51.33 | 561.7 | 759.2 | 4.452 | 4.901 | 197.5 | 176.6 | 15.88 | 75.0 | 8.56 |
| 450 | 1.601 | 1117.9 | 61.73 | 574.4 | 763.2 | 4.480 | 4.900 | 188.8 | 165.7 | 16.58 | 72.4 | 7.47 |
| 460 | 1.880 | 1086.3 | 73.24 | 587.2 | 766.8 | 4.507 | 4.898 | 179.6 | 154.5 | 17.18 | 69.4 | 6.41 |
| 470 | 2.219 | 1051.9 | 87.59 | 600.0 | 770.0 | 4.534 | 4.895 | 170.0 | 141.0 | 17.86 | 66.8 | 5.38 |
| 480 | 2.600 | 1020.4 | 104.00 | 613.0 | 772.1 | 4.562 | 4.893 | 159.1 | 129.6 | 18.70 | 64.1 | 4.39 |
| 490 | 3.017 | 985.0 | 125.05 | 628.0 | 774.6 | 4.590 | 4.891 | 147.5 | 117.7 | 19.40 | 61.2 | 3.44 |
| 500 | 3.466 | 944.5 | 148.51 | 641.8 | 776.4 | 4.618 | 4.888 | 134.6 | 105.3 | 20.30 | 58.4 | 2.54 |
| 505 | 3.706 | 924.0 | 161.92 | 649.5 | 776.9 | 4.633 | 4.885 | 127.4 | 99.9 | 20.90 | 57.0 | 2.12 |
| 510 | 3.954 | 896.3 | 177.13 | 657.5 | 777.4 | 4.648 | 4.883 | 119.9 | 93.1 | 21.60 | 55.5 | 1.70 |
| 515 | 4.207 | 871.4 | 194.00 | 665.7 | 777.2 | 4.664 | 4.880 | 111.5 | 87.7 | 22.50 | 53.8 | 1.31 |
| 520 | 4.468 | 838.4 | 213.95 | 675.2 | 776.9 | 4.681 | 4.876 | 101.7 | 81.2 | 23.60 | 51.8 | 0.94 |
| 525 | 4.743 | 804.5 | 239.49 | 685.6 | 775.6 | 4.699 | 4.870 | 90.0 | 75.4 | 25.30 | 49.5 | 0.60 |
| 530 | 5.044 | 753.2 | 278.61 | 698.1 | 771.3 | 4.722 | 4.860 | 73.0 | 67.7 | 30.80 | 46.1 | 0.29 |
| 535 | 5.352 | 656.7 | 381.07 | 719.5 | 756.8 | 4.756 | 4.826 | 37.3 | 55.7 | 39.90 | 39.0 | 0.05 |
| 536.6 | 5.472 | 499.9 | 499.95 | 739.7 | 739.7 | 4.784 | 4.784 | 0.0 | — | — | — | 0.00 |

# THERMOPHYSICAL PROPERTIES OF FREON-21

## 2.1. A REVIEW OF PUBLISHED DATA
## ABOUT THERMODYNAMIC PROPERTIES

The fundamental experimental and computational investigations of thermody-namic properties of Freon-21 (fluorodichloromethane) were carried out in the late 1960s. However, the results obtained are not fully reflected in the reference handbooks [0.8–0.10 and others]. Thus, the tables for the values of $\rho$, $h$, and $s$ on the saturation line, which were published in paper [2.1], are presented in the reference survey [0.34] and the tables in the single-phase region in handbook [0.10]. The tables of thermodynamic properties on the saturation line (recom-mended in books [0.8–0.10]) repeat the data calculated by Rombusch and Giesen [0.46, 2.37], which are based on a small number of measurements [2.23–2.37].

In the present handbook significantly more complete information about the thermodynamic properties of liquid and gaseous Freon-21 is presented.

### Experimental Data in the Single-Phase Region

From Table 13 and Fig. 7, it is apparent that gas density (compressibility) at elevated pressures was measured in three laboratories, and liquid density was measured in one laboratory (at $p \leq 20$ MPa and $\omega \leq 2$–5).

To determine $pvT$ behavior of superheated and saturated vapor, Benning and McHarness [2.25] used the method of the nonballast piezometer with a mercury lock and pycnometer. The authors approximated the measurement re-

**TABLE 13.** Experimental Studies of Thermodynamic Properties of Freon-21 in the Single-Phase Region

| Year | Authors | Measured properties | Temperature, K | Pressure, MPa | Phase[a] | Number of experimental points | Reference |
|------|---------|---------------------|----------------|---------------|----------|-------------------------------|-----------|
| 1940 | Benning and McHarness | $\varrho$ | 326—446 | 0.3—2.1 | G | 22 | [2.25] |
| | | $\varrho$ | 247—296 | 0.02—0.15 | G | 10 | [2.25] |
| 1940 | Benning et al. | $c_p$ | 311—407 | 0.1 | G | 6 | [2.27] |
| 1967, 1969 | Sheludyakov et al. | $w$ | 244—463 | 0.1—10 | G | 269 | [2.15, 2.16] |
| 1968 | Kalafati et al. | $\varrho$ | 308—473 | 0.3—18.7 | G, L | 284 | [2.10] |
| 1969, 1971 | Shilyakov et al. | $\varrho$ | 309—473 | 0,2—6,8 | G | 292 | [2.19, 2.21] |
| 1969 | Shumskaya and Gruzdev | $c_p$ | 300—469 | 0,1 | G | 9 | [2.20] |
| 1969 | Meyer | $w$ | 196—275 | 0,1 | L | 18 | [3.52] |
| 1970 | Hajjar and MacWood | $B_1$ | 313—403 | 0,1 | G | 4 | [3.45] |
| 1975 | Gruzdev and Shumskaya | $c_p$ | 303—453 | 0,1—2 | G | 63 | [2.9] |

[a] G, Gas; L, liquid.

sults using the derived interpolation equations which describe the initial data in the single-phase region with a discrepancy of 1%. In Ref. [2.1] a joint analysis of experimental data from [2.10, 2.25] was performed. It was established that in the gaseous phase, 12 out of 32 points have deviations of up to 0.2%, 10 have deviations less than 0.4%, 7 have deviations less than 0.6%, and only 3 points have deviations up to 0.8%.

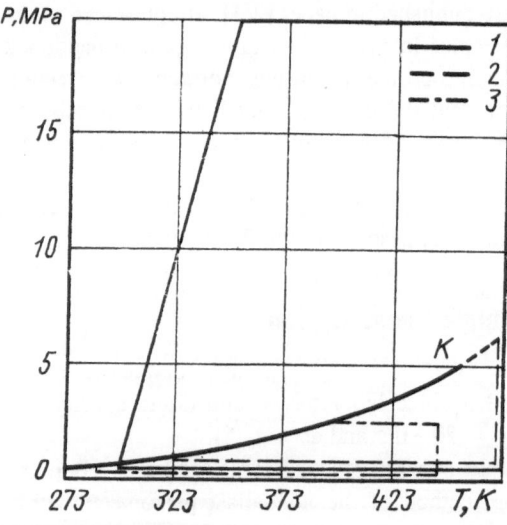

**FIG. 7.** Experimental studies of thermal properties of Freon-21 by: 1, Kalafati et al. [2.10]; 2, Shilyakov et al. [2.19]; 3, Benning and McHarness [2.25].

The most detailed studies of compressibility of gaseous and liquid freons were performed at MEI under the supervision of D. D. Kalafati and D. C. Raskasov [2.10, 2.11]. The authors of these works employed the method of the constant-volume unballasted piezometer. The amount of the investigated substance in the piezometer was determined by weighing the removable glass on analytical scales after freezing the freon out of the piezometer. The temperature was measured by a 10-ohm platinum resistance thermometer coupled with a low-resistance potentiometer of class 1 accuracy. The error of thermostat temperature measurement did not exceed 0.05 K. The pressure was measured by class 0.05 dead-weight piston manometers. To separate the investigated freon from the oil while measuring the pressure, a mercury-membrane zero indicator was used, the accuracy of which reached 20 mm water column. The experiments were carried out along the constant-volume lines so that the measurements ended in the two-phase region. Deformation of the piezometer due to temperature and pressure was corrected for. The maximum relative error in determining the specific volume was 0.2–0.3%, depending on the regions of state parameters.

In the work at MEI, the Freon-21 used had a purity of 99.954% (moisture, 0.032%; nonvolatile residue, 0.001%; hydrocarbons, 0.013%). During the experiments, chromatographic analyses were performed several times, the samples of freon being extracted before and after the experiments. No appreciable difference of results from the initial data of the analysis was noticed.

During this period, the results of detailed piezometric and acoustical measurements were published. These studies were performed at the Thermophysical Institute of the Siberian Branch of the Academy of Sciences of the USSR under the supervision of A. N. Soloviov and E. P. Sheludyakov [2.15–2.19]. In this investigation Freon-21 having a purity of 99.8% (moisture, 0.19%; nonvolatile residue, 0.004%) was used.

For the experimental investigation of compressibility of superheated and saturated vapors, the method of the constant-volume nonballasted piezometer was employed. The fundamental variables (pressure, temperature, and mass of the substance) were measured with instruments of the same class as those used in the work of MEI. The distinguishing feature of the piezometric rig was the use of a membrane zero indicator of pressure of the electrocontact type. The sensitivity of this instrument reached 1 mm water column, but the zero point drifted by 10 mm water column. The authors in Refs. [2.19, 2.21] evaluated the scatter of experimental points on the saturation curve as 0.4% for pressure, and the repeatability of presssure during the forward and reverse tracing along the constant-volume lines was 0.15%.

Direct measurements of the second virial coefficient were performed in only one work (see Table 13). The calculated values of $B_1$ from $pvT$ data [2.10, 2.25] were presented in paper [2.1], and their satisfactory agreement in the interval $T = 313–403$ K is noted. But, according to the data of MEI, the value of the derivative $dB_1/dT$ is greater, and discrepancies at low temperatures are considerable.

Constant-pressure heat capacity of gaseous Freon-21 was measured in three works. The most accurate and significant measurements were made by V. A. Gruzdev and A. I. Shumskaya [2.9, 2.20] in a flowing capillary calorimeter. The error of measurement, according to the authors, was not more than 0.4–0.7%. The purity of the Freon-21 used was 99.6–99.7% (moisture, 0.03%; Freon-12 and -13 combined, 0.013%; other gases, 0.05–0.3%). By graphically extrapolating the experimental isotherms, the authors obtained the temperature dependence of heat capacity in the ideal gas state. They also established that the experimental values of $C_p^0$ agree, within the measurement error margin, with the calculated results for whole ranges without taking into account the correction for anharmonic oscillations. The data of Barho [0.37] are somewhat higher.

In the present work, the initial values of thermodynamic functions of Freon-21 in the ideal gas state are taken from the data of Ref. [1.39].

E. P. Sheludyakov and coworkers [2.15, 2.16] measured the speed of sound in superheated and saturated vapors using the method of stationary waves in a resonator (at low frequency) with an estimated error of 0.5%. It is noted in Ref. [2.1], however, that the values of $\omega$ calculated by the thermal equation of state differ considerably from the experimental results (by 2–4% on the average), and that discrepancy is systematic in character. This discrepancy is, in particular, tied with the chosen values pf $C_p^0$ in Ref. [2.1].

Using the method of Pierce's acoustic interferometer, Meyer [3.52] measured the speed of sound in liquid Freon-21 and approximated the experimental data with an equation of type (3.1) having the coefficients

$$a_0 = 891.0; \qquad\qquad a_2 = 15.45 \cdot 10^{-4};$$
$$a_1 = -3.953 \cdot 10^0; \qquad a_3 = -0.764 \cdot 10^{-5}.$$

## Experimental Data on the Saturation Line

From Table 14, it follows that the saturated vapor pressure and the orthobaric density of vapor and liquid were studied more fully in the interval $T = 294–448$ K. However, the agreement between experimental data about $p_s$ and $\rho''$ is not very good (Figs. 8 and 9). Not shown in Fig. 8 are the experimental values of $p_s$ obtained by A. A. Shilyakov and coworkers [2.19, 2.21]. These data are systematically higher than the MEI data by 0.3% on the average. The maximum deviation is 0.9%, but at $T > 393$ K the discrepancy decreases to 0.07%. The higher values of $p_s$ and $\rho''$ in the experiments [2.19, 2.21] can be explained by the effect of dissolved impurities in the investigated substance [2.21]. The values of $\rho'$ obtained in Ref. [2.26] using the method of hydrostatic weighing are higher for Freon-21 (Fig. 9) and Freon-22 [3.1, 3.66].

Presented in Table 15 are the parameters of fundamental points which were experimentally determined on the saturation line and which are recommended in some reference books.

It is pertinent to mention that in the majority of previous works the values of $T_{cr}$ and $T_{NBP}$ were taken from reference [2.26]. The last two values are con-

**TABLE 14.** Experimental Studies of Thermodynamic Properties of Freon-21 on the Saturation Line

| Year | Authors | Measured properties | Temperature, K | Phase[a] | Number of experimental points | Reference |
|---|---|---|---|---|---|---|
| 1940 | Benning and McHarness | $p_s$ | 243—448 | V | 7 | [2.24] |
| | | $\varrho$ | 405—440 | V | 3 | [2.25] |
| | | $\varrho$ | 244—441 | L | 9 | [2.26] |
| 1940 | Benning et al. | $c_p$ | 261—338 | L | 4 | [2.27] |
| 1967, 1969 | Sheludyakov et al. | $w$ | 238—451 | V | 28 | [2.15, 2.16] |
| 1968 | Kalafati et al. | $p_s$ | 308—448 | V | 16 | [2.10] |
| 1968 | Vaskov and Panauty | $c_p$ | 209—332 | L | — | [0.11] |
| 1969 | Shilyakov et al. | $p_s$ | 294—449 | V | 26 | [2.19] |
| | | $\varrho$ | 305—445 | V | 22 | |
| 1969 | Lavrov et al. | $c_p$ | 307—363 | L | 29 | [2.12] |
| 1969 | Dorokhov et al. | $\sigma$ | 244—358 | L–V | 22 | [1.8; 3.13] |
| 1971 | Sinitsin et al. | $\sigma$ | 293—443 | L–V | 16 | [0.33; 1.2] |

[a] L, Liquid; V, vapor.

firmed by later measurements (Table 15), but the critical parameters are apparently nearer to the values obtained by A. A. Shilyakov [2.21] in special experiments using a moving-meniscus method.

Heat capacity of liquid Freon-21 was investigated in a wide interval of temperatures (Table 14), but experimental data are presented only in Refs. [2.12, 2.27]. In Ref. [0.11] the following interpolated formula is presented

$$c_s' = 1.0300 + 7.9550 \cdot 10^{-4} \cdot t + 3.3910 \cdot 10^{-6} \cdot t^2, \qquad (2.1)$$

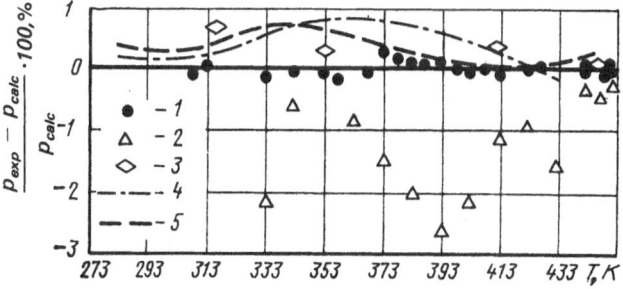

**FIG. 8.** Deviation of values of saturated vapor pressure of Freon-21 from the adopted values in the present work. Experimental data: 1, Raskazov et al. [2.10, 2.11]; 2, Sheludyakov and others [2.15]; 3, Benning and McHarness [2.24]. Tabular data: 4, Benning and McHarness [2.24]; 5, Rombusch and Giesen [2.37].

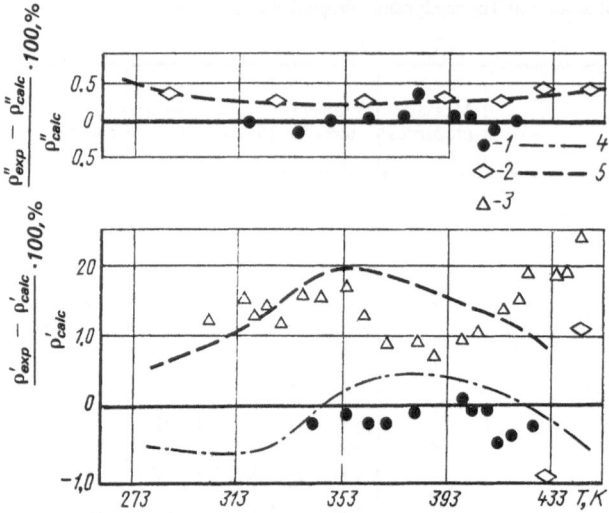

**FIG. 9.** Deviation of the values of density on the saturation line for Freon-21 from the adopted values in the present work. Experimental data: 1, Petrov et al. [2,10]; 2, Benning and McHarness [2.25, 2.26]; 3, Shilyakov and others [2.19]. Tabular data: 4, Benning and McHarness [2.25, 2.26]; 5, Rombusch and Giesen [2.37].

where $c_s'$ is in kJ/(kg $\cdot$ K).

Heat capacity at constant pressure in the liquid phase in Ref. [2.12] was determined by a comparative method, i.e., heating curves of the investigated and reference liquids were compared on a time basis. The error in the results did not exceed 0.5%, with the largest contribution being when the heating curves are differentiated. In the region of overlapping temperatures, the discrepancy in the indicated results does not exceed 0.7% [2.12].

The experimental data for surface tension of liquid Freon-21 cover the range

**TABLE 15.** Parameters of Fundamental Points on the Saturation Line for Freon-21[a]

| Year | Authors | $T_{cr}$, K | $p_{cr}$, MPa | $\rho_{cr}$, kg/m³ | $T_{NBP}$, K | $T_f$, K | Reference |
|------|---------|-------------|---------------|--------------------|--------------|----------|-----------|
| 1940 | Benning and McHarness | 451.65 | 5.168 | 522 | 282.05 | 138.2 | [2.26] |
| 1966, 1968 | Rombusch and Giesen | 451.50 | 5.181 | 525 | 282.04 | — | [0.46, 2.37] |
| 1967 | Sheludyakov et al. | 451.65 | 5.178 | — | — | — | [2.15] |
| 1968 | Kalafti et al. | 451.65 | 5.194 | — | — | — | [2.10] |
| 1968 | ASHRAE | 451.65 | 5.17 | 522 | 282.07 | 138.15 | [2.22] |
| 1965 | Gallant | 451.65 | 5.17 | 522 | 282.05 | 138.15 | [2.29] |
| 1969 | Woodline | — | — | — | 282.05 | 138.15 | [2.39] |
| 1971 | Shilyakov | 451.55 | 5.198 | 529.7 | — | — | [2.21] |
| 1970 | Phillips and Murphy | — | — | — | 282.05 | — | [1.50] |

[a] The values of the parameters adopted in this book are shown in Table 3.

$T = 244-443$ K (Table 14). These experimental data are approximated well by Eq. (1.9) with the coefficient $n = 1.249$ and $B = 5.5203 \cdot 10^{-3}$ N/m.

## Equations of State and Tables

In reference books [0.7, 0.45, and others] the tables of thermodynamic properties of Freon-21 which were compiled by Benning and McHarness (1939) are recommended.

Rombusch and Giesen [0.46] compiled a new system of equations including an equation of state (0.6) and the interpolated formulas (0.10), (0.13), and (0.19). They calculated the tables for the values of $\rho$, $h$, and $s$ on the saturation line for $T = 213-451$ K, and the $h$, $\lg p$ diagram in the region $T = 203-473$ K at pressures up to 10 MPa [2.37]. Tabular data [2.37] are presented in some reference books [0.8-0.10 and others]. In Ref. [0.11], with the help of the same system of equations, detailed tabular values are calculated for $v$, $h$, and $s$ for superheated vapor in the interval $T = 243-473$ K and $p = 0.01-5$ MPa. It should be pointed out that a large part of the tables under consideration cannot be looked at as experimentally substantiated.

V. V. Altunin et al. [2.1] suggested a thermal equation of state of type (0.7), which was obtained on the basis of computer analysis of experimental $pvT$ data for liquid and gaseous Freon-21 in the interval $T = 298-473$ K and $p = 0.3-19$ MPa. The calculated tabular values of $\rho$, $h$, and $s$ from this equation are partially reproduced in paper [2.1] and in Refs. [0.10, 0.34].

E. P. Sheludyakov and A. A. Shilyakov [2.18, 2.21], on the basis of graphical and analytical analyses of experimental data [2.19, 2.25], obtained the following equation of state for the superheated vapor of Freon-21

$$p = (RT/v) - a[v^q(T - T_0)]^{-1} \exp(-\alpha/v) \qquad (2.2)$$

where $p$ is in bar and $v$ is in m³/kg; $T_0 = 138.15$ K; $a = 0.21576$; $\alpha = 1.3367 \cdot 10^{-3}$; $q = 2.0630$. The authors assert that this equation approximates the initial $pvT$ data in the interval $\tau = 0.56-1.05$ and $\omega = 0.001-0.95$ with standard and maximum deviations of 0.12 and 0.35% respectively. In Ref. [2.17] this equation is used to calculate the tables of the values $v$, $h$, $s$, $c_p$, and $c_v$ of superheated and saturated vapor in the range $T = 243-473$ K and $p = 0.01-5$ MPa.

It should be pointed out that in Refs. [2.17, 2.18, 2.21] the thermodynamic properties of saturated vapor of Freon-21 were calculated using the empirical vapor pressure curve equation

$$\ln p_s = A + B \ln(T - T_0) + C(T_{cr} - T)^n \qquad (2.3)$$

the coefficients of which were found from the experimental data [2.19, 2.24] in the interval $T = 243-449$ K. But in view of the discrepancy between the experimental data [2.19] and the results [2.10] which were obtained with purer samples of Freon-21, the generalized equation (1.2) is recommended for the calculation of $p_s$ in the entire liquidus range $T_0-T_{cr}$. It is expedient to calculate the orthobaric density of a liquid from the generalized equation (1.3). Parameters Ri and $p_\alpha$ of Eqs. (1.2)–(1.3) for Freon-21 can be evaluated using correlations (1.6)–(1.7) or determined from experimental data in the interval $\tau = 0.625-0.7$.

In the works of A. P. Klimenko and coworkers [0.22, 1.12] a correcting constant ($K_{ii} = -0.025$) for the generalized equation of state of type (0.5) is obtained for Freon-21. They report that the BWRC equation approximates 280 experimental points [2.19] in the region $\tau = 0.68-1.04$, $\pi = 0.05-1.51$, and $\omega = 0.02-0.92$ with a standard deviation of 0.33%.

In paper [0.26], information about an algorithm for a computer program is given. This program was written at the Thermophysics Institute of the Siberian Branch of the Academy of Sciences of the USSR to calculate thermophysical properties of freons. As a mathematical model for the equation of state the following expression is adopted

$$z = 1 + \sum_{i=1}^{r}[K_{1i}+K_{2i}\,\tau+K_{3i}\,\exp\,(-K_4\,\tau)]\omega^{-i} \qquad (2.4)$$

The calculations of $p_s$, $\rho'$, $\sigma$, $c_p^0$, $H^0$, and $s^0$ are executed by equations (0.11), (0.13), (1.9), (1.12), (1.13), and (1.14) respectively. The tables of $p_s$, $\rho$, $h$, $s$, and $\sigma$ on the saturation line of Freon-21 (at $T = 273-451$ K) and $\rho$, $h$, $s$, $\gamma$, and $\omega$ in the gaseous phase (at $p = 0.1$ MPa, $T = 283-433$ K) presented in [0.26] are in good agreement with the recommended tables in the present work.

Thus, there are several equations of state which are suitable for the calculation of thermodynamic properties of gaseous Freon-21. For the compressed liquid, the equation of state and thermodynamic tables were compiled in Ref. [2.1] only.

## 2.2. SYSTEM OF EQUATIONS FOR THE CALCULATION OF THERMODYNAMIC PROPERTIES

To obtain the equation of state of Freon-21, a computer method for joint analysis of experimental data covering the thermal properties for liquid and gas was used [0.5, 2.1]. For the initial condition the experimental $pvT$ data of MEI [2.10] are adopted. These data cover the region $T = 308-473$ K and $p = 0.3-19$ MPa for gaseous and liquid phases and about 300 measured values of $z(\rho, T)$ and $p_s(T)$. The resultant equation of state is similar to (0.15) and has the coefficients presented in Table 16.

Given in Table 17 are the coefficients of the interpolated formulas (0.18)–(0.21) for the determination of temperature dependence of saturated vapor pressure and three function in the ideal gas state (enthalpy, heat capacity, and entropy).

**TABLE 16.** Coefficients of the Equation of State (0.15) for Freon-21 ($\rho$, g/cm³; $\tau = T/451.65$)

| $i$ | Value $b_{ij}$ at $j$ | | | | |
|---|---|---|---|---|---|
| | 0 | 1 | 2 | 3 | 4 |
| 1 | $0.233889633 \cdot 10^1$ | $-0.305403400 \cdot 10^1$ | $-0.217922813 \cdot 10^1$ | $0.110472338 \cdot 10^1$ | $-0.575342224 \cdot 10^0$ |
| 2 | $-0.268911716 \cdot 10^1$ | $0.145434544 \cdot 10^1$ | $0.792086048 \cdot 10^0$ | $0.150672581 \cdot 10^1$ | $0.962453130 \cdot 10^{-1}$ |
| 3 | $0.412827329 \cdot 10^1$ | $0.655474718 \cdot 10^0$ | $-0.487760676 \cdot 10^0$ | $-0.115257330 \cdot 10^0$ | — |
| 4 | $-0.315857655 \cdot 10^1$ | $-0.243289633 \cdot 10^1$ | $-0.188795402 \cdot 10^1$ | — | — |
| 5 | $0.406934804 \cdot 10^1$ | $-0.600400793 \cdot 10^0$ | $0.180448803 \cdot 10^0$ | $0.221726465 \cdot 10^0$ | — |
| 6 | $-0.167145539 \cdot 10^1$ | $0.217886870 \cdot 10^1$ | $-0.457427093 \cdot 10^0$ | — | — |

**TABLE 17.** Coefficients of Auxiliary Interpolated Equations (0.18)–(0.21) for Freon-21

| $j$ | $A_{sj}^*$ | $\alpha_j$ | $\beta_j$ | $\gamma_j$ |
|---|---|---|---|---|
| 0 | $-0.924375070 \cdot 10^2$ | $0.122267350 \cdot 10^2$ | $0.13635445 \cdot 10^2$ | $0.615635180 \cdot 10^2$ |
| 1 | $0.611097455 \cdot 10^2$ | $-0.959102930 \cdot 10^0$ | $-0.509841820 \cdot 10^1$ | $-0.465871910 \cdot 10^2$ |
| 2 | $0.371902056 \cdot 10$ | $0.730351950 \cdot 10^1$ | $-0.439020789 \cdot 10^0$ | $0.354966340 \cdot 10^2$ |
| 3 | $-0.759756227 \cdot 10^1$ | $-0.247756820 \cdot 10^1$ | $0.103994680 \cdot 10^1$ | $-0.139826580 \cdot 10^2$ |
| 4 | $-0.438698453 \cdot 10^{-1}$ | $0.339640600 \cdot 10^0$ | $-0.223798380 \cdot 10^0$ | $0.216379730 \cdot 10^1$ |
| 5 | $0.201709440 \cdot 10^0$ | — | — | — |
| 6 | $0.194881380 \cdot 10^0$ | — | — | — |
| 7 | $-0.641840765 \cdot 10^{-1}$ | — | — | — |
| 8 | $0.541764224 \cdot 10^{-2}$ | — | | |

* At $\bar{\tau} = T/100$ for $A_{sj}$; at $\tau = T/451.65$ for $\alpha_j$, $\beta_j$, and $\gamma_j$.

The standard deviations in the approximation of initial experimental data in [2.10] by the equation of state amount to 0.27%. The histogram of the deviations is presented in Fig. 10.

The calculation of thermodynamic properties on the saturation line (boiling and condensation line) is carried out using the independent equation-of-compressibility curve. The standard deviation for reproducibility of initial experimental data for the set $p_s$, $T_s$ by the polynomial (0.21) amounts to 0.09%. Agreement is also good between the values of $\rho'$ and $\rho''$, which are calculated by Eqs. (0.15) and (0.21), and the data of MEI (see Fig. 9).

Comparison with the experimental results of other researchers showed that

• data [2.19, 2.21] about the compressibility of vapor are generally lower and the deviation, on the average, amounts to 0.25%;
• the experimental [2.15, 2.16] and calculated values of speed of sound in the superheated vapor agree well at relatively low pressures (Table 18), but the discrepancies near the saturation line and in the vicinity of the critical point are large and reach 5–10%;
• deviations of the measured results [2.9] from the calculated values of heat capacity $c_p$ of superheated vapor do not exceed the experimental and calculation errors;
• the calculated values of heat capacity of saturated liquid agree fairly well with the experimental data [0.11, 2.12];
• the surface tension of liquid Freon-21 is calculated from the generalized equation (1.9) with an error of 1–1.5%.

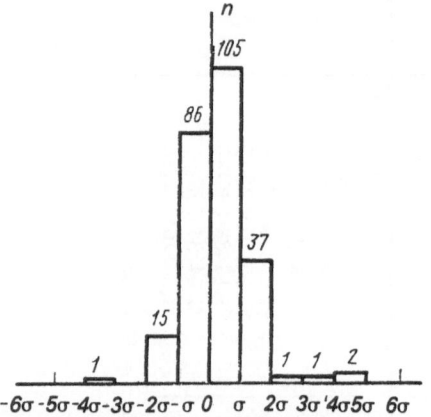

**FIG. 10.** Histogram of deviation of the values of compressibility for Freon-21, calculated by Eq. (0.15), from the experimental data [2.10].

**TABLE 18.** Comparison of Calculated Results with Experimental Data [2.16] for Speed of Sound for Freon-21

| $T$, K | $p$, MPa | $\omega_{exp}$, m/s | $\omega_{calc}$, m/s | $\Delta\omega$, m/s | $\delta w$, % |
|--------|----------|---------|---------|---------|---------|
| 463.16 | 0.49  | 200.0 | 200.71 | —0.71 | —0.34 |
| 463.16 | 0.98  | 194.9 | 196.05 | —1.15 | —0.58 |
| 463.16 | 1.47  | 189.4 | 191.17 | —1.17 | —0.61 |
| 463.16 | 1.96  | 183.6 | 186.01 | —2.41 | —1.30 |
| 463.16 | 2.45  | 177.4 | 180.60 | —3.20 | —1.77 |
| 463.16 | 2.94  | 170.7 | 174.81 | —4.11 | —2.36 |
| 451.66 | 0.49  | 197.3 | 197.86 | —0.56 | —0.28 |
| 451.66 | 0.98  | 191.8 | 192.70 | —0.90 | —0.47 |
| 451.66 | 1.47  | 185.9 | 187.20 | —1.30 | —0.69 |
| 451.66 | 1.96  | 179.7 | 181.35 | —1.60 | —0.88 |
| 451.66 | 2.45  | 173.1 | 175.07 | —1.97 | —1.12 |
| 451.66 | 2.94  | 166.0 | 168.18 | —2.18 | —1.29 |
| 303.46 | 0.049 | 166.8 | 166.95 | —0.15 | —0.09 |
| 303.46 | 0.098 | 165.1 | 165.54 | —0.44 | —0.27 |
| 303.46 | 0.147 | 163.2 | 164.11 | —0.91 | —0.55 |

## 2.3. A REVIEW OF PUBLISHED DATA ABOUT VISCOSITY AND THERMAL CONDUCTIVITY

Experimental studies of viscosity of liquid and gaseous Freon-21 at high pressures and in a wide temperature interval were carried out recently (Table 20). Therefore, handbooks published in the 1970s [0.8, 0.9, 0.10, 0.34] include limited and sometimes inaccurate information about viscosity of Freon-21 at atmospheric pressure $\eta_T$ and on the saturation line $\eta_l$. Reference [0.10] contains only estimated data about $\eta_T$ and $\eta_l$ based on the calculated data in [0.7, 2.6]. Tables in [0.8, 0.9] are mostly based on the experimental results of Makita [2.32], which proved to be unreliable.

**TABLE 19.** Comparison of Calculated Results with Experimental Data [2.5] for Heat Capacity at Constant Pressure for Freon-21

| $T$, K | $p$, MPa | $c_p^{exp}$, kJ/kg | $c_p^{calc}$, kJ/kg | $\delta c_p$, % | $T$, K | $p$, MPa | $c_p^{exp}$, kJ/kg | $c_p^{calc}$, kJ/kg | $\delta c_p$, % |
|--------|----------|-------|-------|-------|--------|----------|-------|-------|-------|
| 303.05 | 0.098 | 0.614 | 0.608 | 0.98  | 411.75 | 0.50 | 0.720 | 0.722 | —0.3 |
| 318.65 | 0.098 | 0.626 | 0.626 | 0     | 446.65 | 0.50 | 0.740 | 0.735 | 0.7  |
| 337.55 | 0.098 | 0.641 | 0.643 | —0.3  | 436.55 | 2.0  | 0.848 | 0.845 | 0.4  |
| 416.05 | 0.60  | 0.727 | 0.730 | —0.4  | 444.75 | 2.0  | 0.835 | 0.832 | 0,4  |
| 446.85 | 0.60  | 0.745 | 0.740 | 0.7   | 452.85 | 2.0  | 0.831 | 0.822 | 1    |
| 390.15 | 0.50  | 0.712 | 0.717 | —0,7  |        |      |       |       |      |

**TABLE 20.** Experimental Studies of Viscosity of Freon-21

| Year | Authors | Temperature, K | Pressure, MPa | Phase[a] | Number of experimental points | Method[b] | Reference |
|------|---------|----------------|---------------|----------|-------------------------------|-----------|-----------|
| 1939 | Benning and Markwood | 274—353 | 0.03—0.1 | G | 7 | RB | [2.35] |
|      |          | 243—333 | $p_s$ | L | 4 | RB | [2.35] |
| 1954 | Makita | 298—423 | 0.1—0.6 | G | 10 | RB | [2.32] |
| 1955 | Wellman | 303—362 | 0.1 | G | 4 | RB |  |
| 1956 | Kinser | 208—248 | $p_s$ | L | 6 | RB | [2.38] |
| 1958 | McCullum | 363—464 | 0.1 | G | 4 | RB |  |
| 1959 | Tsui | 363—423 | 0.1 | G | 3 | RB |  |
| 1959 | Kamien and Witzell | 303—363 | 0.1—0.6 | G | 12 | RB | [2.30] |
| 1970 | Phillips and Murphy | 208—347 | $p_s$ | L | 11 | Ca | [1.50] |
| 1975 | Geller et al. | 273—434 | 1—59 | L | 65 | Ca | [2.4] |
| 1975 | Raskazov et al. | 273—473 | 0.1—20 | G, L | 53 | Ca | [2.13] |

[a] G, Gas; L, liquid.
[b] Ca, Capillary; RB, rolling ball.

At the present time, reliable equations and experimentally established tables of viscosity for Freon-21 can be obtained in the region $T = 230–473$ K and $p = 0.1–60$ MPa.

It is obvious from Table 20 that the method of a rolling ball had been used for a long time to investigate the viscosity of Freon-21 at low pressures in the gas phase and on the saturation line. This method was criticized more than once in [1.5, 1.6]. At the same time, a sizable part of the measured values of $\eta_T$ in Refs. [2.30, 2.38, 3.35] agree sufficiently well with each other and with the measurements made using the capillary method [2.13]. Only the data of Makita [2.32] are an exception, since they differ by 8–10%. The data on other substances obtained by the same author were also found to be as inaccurate [2.3].

The measured values of $\eta_l$ [2.38, 3.35], which were obtained by the rolling-ball method, cannot be corroborated by later experimental data.

The results of Benning and Markwood [3.35] about the viscosity of liquid Freon-21 near the saturation line are too high (by up to 20%) in comparison to the new experimental data in [2.4, 2.13]. The experimental values of Kinser [2.38], on the contrary, are lower (by 10%) relative to the data of Phillips and Murphy [1.50]. Thus, the results of measurements of viscosity of Freon-21 in the liquid phase near the saturation line obtained by the rolling-ball method are apparently not reliable and will not be considered further.

The measurements of Phillips and Murphy [1.49, 1.50] were made with the method of a capillary viscometer with "a floating level" (without pressure unloading) at Reynolds numbers from 50 to 700; the results occupy an intermediate position between the data in [3.35] and [2.38]. However, comparing these results with new experimental values [2.4, 2.13] obtained using capillary viscometers of advanced design indicates that discrepancy rises with increasing temperature and reaches 12% at 347 K. It is pertinent to mention that in Ref. [1.50], the information necessary to analyze the measurements in detail is absent. Therefore, it is not possible to give a simple explanation for such a considerable deviation.

Geller et al. [2.4] used capillary viscometers designed by I. F. Golubev (variant V). All the measurements were made at Reynolds numbers not exceeding 800. "Link up" values for viscosity obtained by various viscometers was accomplished on two or three isotherms with deviation not exceeding the experimental error (according to the authors, ±1.5%). Freon-21, as shown by chromatography, contained less than 0.1% impurities. The ability to maintain the purity of Freon-21 upon completion of the experiments is a sign of the absence of chemical decomposition.

D. C. Raskazov et al. [2.13] used a modified Rankine viscometer with a closed-contour capillary. The distinctive feature of this viscometer was the removal of mercury, which creates the pressure in the capillary, to the region of room temperature and fixing the falling time of the mercury droplets by the noncontact method using high-quality resonance contour. The maximum relative error of experimental data, according to the author's evaluation did not exceed 2–2.3%. The content of impurities in Freon-21 amounted to 0.13%.

Analysis of the results in [2.4, 2.13] indicated that the experimental data agree with each other with deviation in the region of combined error of the considered investigations. There exists, therefore, adequate agreement of independent measurements for the gaseous phase of Freon-21 at atmospheric pressure as well as for the liquid phase. For the determination of viscosity of Freon-21 at atmospheric pressure, Eq. (1.18) is recommended with the following values for the constants:

$$\eta_{T_{cr}} = 16.7 \ \mu Pa \cdot s$$

$$a_0 = 0.73904; \qquad\qquad a_2 = -0.01239;$$
$$a_1 = 0.12559; \qquad\qquad a_3 = -0.00157.$$

When obtaining this equation, the bulk of initial $\eta_T$, $T$ data sets included 32 experimental values of $\eta_T$ from Refs. [2.13, 2.30, 2.38, 3.35] in the temperature range 273–473 K. To calculate the pressure dependence of viscosity for gaseous and liquid Freon-21, the generalized equation (0.35) is recommended in the interval $\omega = 0$–1.9. At $\omega \geqslant 1.9$ the generalized equation (0.41), the coefficients of which are given in Table 4, is recommended. The viscosity of Freon-21 at the reference point ($\tau = 0.7$ and $\pi = 0.7$) is found from the data in [2.4, 2.13] and is equal to $\eta_{0.7} = 272$ MPa $\cdot$ s.

Note that the coefficients $b_{\eta, i}$, $\alpha_{\eta, i}$, and $\beta_{\eta, i}$ were found on the basis of analysis of a highly representative aggregate of concordant experimental data for Freons-21, -22, and -23. The initial set of data included about 120 experimental values of viscosity for Freon-21 in the interval $T = 273$–473 K and $p = 0.1$–59 MPa, obtained in Refs. [2.4, 2.13, 2.30]. Shown in Figs. 11 and 12 are the deviations of experimental data from calculated data using these equations.

The only information existing reference books [0.8, 0.9, 0.10, 0.34] contain are thermal conductivity of Freon-21 at atmospheric pressure and near the saturation line. For example, their recommended values of $\lambda_l$ were taken from a theoretical work published in 1965 [2.14].

Up to the present time, sufficient experimental information has been accumulated dealing with thermal conductivity of Freon-21 in the gaseous and liquid phase at pressures up to 60 MPa in the range $T = 150$–470 K (Table 21).

To investigate the thermal conductivity of liquid freons, Markwood and Benning [2.33] used the relative version of the coaxial-cylinder method. In their calibration runs, Bridgeman's data on $\lambda$ for toluene and ethyl alcohol were adopted as a basis. These data turned out to be higher by 15–25% than the more reliable data of other researchers. This was due to the presence of critical errors in the method used which were related to the considerable outflow of heat. Markwood and Benning measured thermal conductivity of gaseous Freon-21 using the relative method of heated filament. Values of thermal conductivity of air and carbon dioxide were taken as reference data. Their values of $\lambda_T$, measured at $T = 303$–363 K, agree somewhat better with other gas data than do their values of $\lambda_l$ with others.

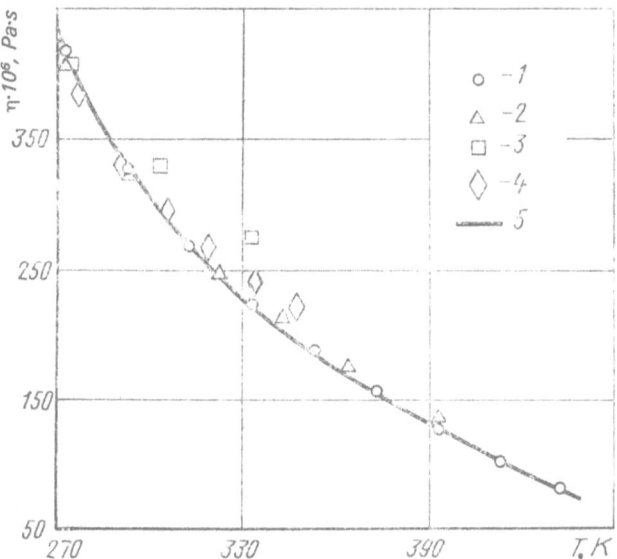

**FIG. 11.** Viscosity of liquid Freon-21 along the saturation line. Experimental data: 1, Geller and others [2.4]; 2, Raskazov and others [2.13]; 3, Benning and Markwood [3.35]; 4, Phillips and Murphy [1.50]. Calculated data: 5, present work.

In Ref. [2.35] measurements of $\lambda_l$ of Freon-21 were made using the method of a flat layer plate heated from the top. The natural convection in liquids with a positive coefficient of expansion was thus excluded. This method was used by

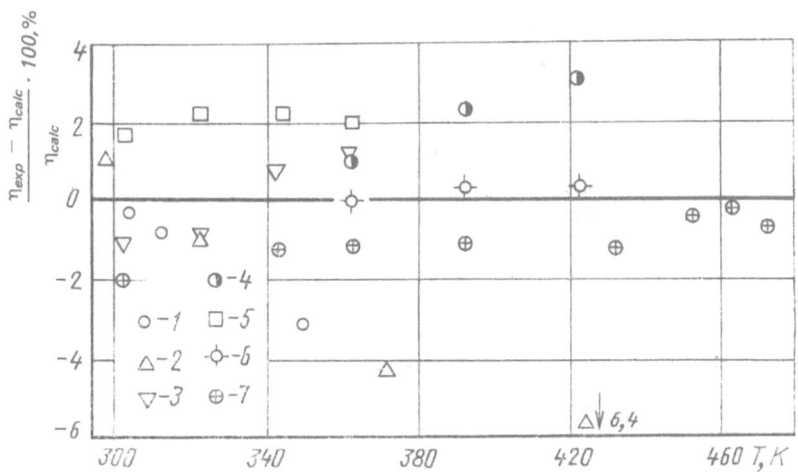

**FIG. 12.** Deviation of experimental values of viscosity of Freon-21 at atmospheric pressure from the values adopted in the present work. 1, Benning and Markwood [3.35]; 2, Makita [2.32]; 3, Wellman [2.38]; 4, McCullum [2.38]; 5, Kamien and Witzell [2.30]; 6, Tsui [2.38]; 7, Geller and others [2.4].

**TABLE 21.** Experimental Studies on Thermal Conductivity of Freon-21

| Year | Authors | Temperature, K | Pressure, MPa | Phase[a] | Number of experimental points | Method[b] | Reference |
|------|---------|----------------|---------------|----------|-------------------------------|-----------|-----------|
| 1943 | Markwood ard Benning | 303—363 | 0.1 | G | 2 | H | [2.33] |
|      |         | 273—348 | $\sim p_s$ | L | 3 | CC | [2.33] |
| 1959 | Powell and Challoner | 253—293 | $\sim p_s$ | L | 2 | F | [2.35] |
| 1964 | Djalalian | 215—273 | $\sim p_s$ | L | 7 | H | [1.29] |
| 1964 | Masia et al. | 278—408 | 0.003—0.12 | G | 8 | H | [2.34] |
| 1966 | Powell et al. | 286 | $\sim p_s$ | L | 1 | TC | [2.36] |
| 1967 | Tauscher | 148—323 | $\sim p_s$ | L | 13 | NH | [1.63] |
| 1970 | Gruzdev and Shestova | 325—465 | 0.16—4.5 | G | 51 | CC | [2.7] |
|      |         | 305—400 | 0.5—5.5 | L | 13 | CC | [2.7] |
| 1974 | Geller and Feredri | 210—432 | 5—59 | L | 64 | H | [2.5] |

[a] G, Gas; L, liquid.
[b] CC, Coaxial cylinder; F, flat plate; H, heated filament; NH, nonstationary heated filament; TC, thermal comparator.

the researchers earlier [2.28] when determining thermal conductivity at atmospheric pressure for seven technically important liquids including toluene.

According to the authors' evaluation, the error in the experimental data is equal to ±1%. However, the experimental data of Powell and Challoner for the thermal conductivity of toluene differ, in fact, by 3–4% from the recommended values in Ref. [2.2]. The results of measurements of liquid Freon-21 are higher by 8–10% in comparison with the data of other researchers.

Powell and coworkers [2.36] consequently used the relative method of thermal comparator to measure $\lambda_l$. The values of thermal conductivity of toluene and carbon tetrachloride which were obtained earlier [2.28] served as reference values. The error was estimated to be ±5%. The thermal conductivity of Freon-21 at a temperature of 286 K [2.36] also differs from the values given by other researchers.

The data of Masia et al. [2.34] obtained by the heated-filament method are sufficiently reliable and were used earlier when compiling reference tables in [2.31] and working out the generalized equation (1.21).

Tauscher [1.63] used the comparative version of the nonfixed method of heated platinum filament. The diameter of the thread at both probes was 20 μm, and the length was 50 mm. Highly viscous silicone oil was used as the reference liquid, the λ of which was determined during control tests for toluene. Control tests were conducted after each series of measurements. Thermal conductivity values of toluene recommended in Ziebland's work [2.40] were taken as reference points. The scatter in Tauscher's experimental points, as shown by the results of the control tests with toluene, is about 3–4%. At the same time, the experimental data for thermal conductivity of Freon-21 agree fairly well with the results of much later investigations [2.5, 2.7].

Analysis of the work of V. A. Gruzdev and A. I. Shestova [2.7, 2.8] shows that their data is sufficiently reliable. This is evidenced by the agreement (deviation is within the limits of the sum of the errors in the compared results) with the data in [2.33, 2.34] regarding thermal conductivity of gaseous Freon-21 at atmospheric pressure and with the results in [1.63, 2.5] for liquid Freon-21 along the saturation line.

V. Z. Geller and V. G. Peredri [2.5] used a modified version of the stationary heated-filament method. A thin-wall platinum capillary was adopted as an external resistance thermometer; this allowed an increase in the accuracy of determination of thermal conductivity. The absence of convection was ensured by measurements at two to four different temperature gradients in the layer (Rayleigh number did not exceed 1500). When computing, corrections were introduced to compensate for the eccentricity of the filament, heat outflow from the ends, and the change in geometric sizes of the measuring cell. In all, the corrections did not exceed 0.3–0.5%. Correction for radiative heat losses were not introduced due to the absence of infrared absorption in the spectrum of liquid Freon-21. This, however, as shown by the calculations did not noticeably distort the results of the measurements. The tests were carried out at a gap of 0.5 mm.

Purity of Freon-21, tested by chromatography, amounted to 99.9%. The error of experimental data was evaluated by the authors to be $\pm 1.2\%$. The results of measurements [2.5] near the saturation curve are in good agreement with the data in [1.63, 2.7].

Due to the aforementioned reasons, it is expedient to exclude data in [2.33, 2.35, 2.36] from the analysis. The remaining results of independent measurements agree well, although the desired high level of agreement of experimental data has not yet been reached.

To determine the thermal conductivity of Freon-21 at atmospheric pressure, Eq. (1.20) is recommended with the following values of coefficients:

$$a_0 = -6.378 \cdot 10^{-3}; \ a_1 = 5.135 \cdot 10^{-5}$$

where $\lambda$ is in W/(m $\cdot$ K). These values of the coefficients were obtained as a result of analyzing 30 values of $\lambda_T$ in the range of $T = 250-465$ K from the data in [2.7, 2.33, 2.34].

The temperature dependence of thermal conductivity of liquid Freon-21 along the saturation curve is described by Eqs. (0.46), the coefficients of which are equal to

$$
\begin{aligned}
c_1 &= -0.22593 & c_4 &= 4.03891; \\
c_2 &= 2.87771 & c_5 &= -0.98069; \\
c_3 &= -5.26325 & c_6 &= -0.00627; \\
\lambda_{cr} &= 373 \cdot 10^{-4} \ \text{W/(m} \cdot \text{K)}
\end{aligned}
$$

These coefficients are found as a result of analyzing 40 values of $\lambda_s$ in the interval $T = 148-432$ K using data in [1.29, 1.63, 2.5, 2.7].

To calculate the dependence of thermal conductivity on pressure, a system of equations agreeing with each other are recommended: in the interval $\omega = 0-1.9$, Eq. (0.34) with individual coefficients; at $\omega \geq 1.9$, the generalized equation (0.45).

The numerical values of the coefficients $b_{\lambda, i}$ of equation (0.34) are found from data in [2.7] in the interval of density values $10-230$ kg/m$^3$ and equal to

$$
\begin{aligned}
b_1 &= 0.026504; & b_3 &= -0.041032; \\
b_2 &= 0.038283; & b_4 &= 0.026114.
\end{aligned}
$$

where $\lambda \cdot 10^4$ W/(m $\cdot$ K).

The coefficients $\alpha_{\lambda, ij}$ of the generalized equation (0.45) are given in Table 5 and were obtained from measurements of $\lambda$ for Freons-21, -22, and -23 (272 points) in the range $\tau = 0.3-1.0$ and $\pi = 1-12$. The initial data included 51 values of $\lambda$ for Freon-21 in the interval $T = 210-432$ K and $p = 5-59$ MPa [2.5].

Figures 13 and 14 show the deviation of experimental data from the calculated results using these equations.

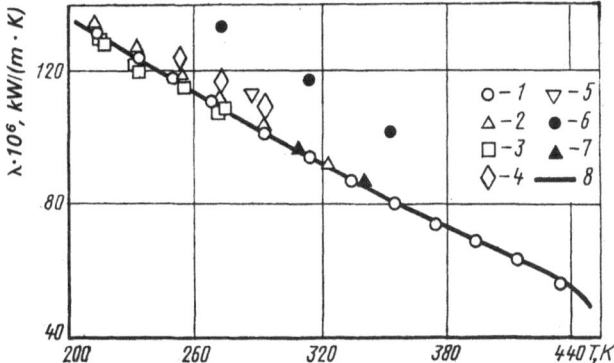

**FIG. 13.** Thermal conductivity of liquid Freon-21 along the saturation line. Experimental data: 1, Geller and Peredri [2.5]; 2, Tauscher [1.63]; 3, Djalalian [1.29]; 4, Powell and Challoner [2.35]; 5, Powell et al. [2.36]; 6, Markwood and Benning [2.33]; 7, Gruzdev and Shestova [2.7]. Calculated data: 8, present work.

## 2.4. TABLES OF
## THERMOPHYSICAL PROPERTIES

The thermodynamic tables for gaseous and liquid Freon-21 were computed from the equations indicated in Sec. 2.2 and cover the region $T = 303-473$ K and $p = 0.01-20$ MPa (Tables 22 and 23). The system of equations given in Sec. 2.3 for the calculation of viscosity and thermal conductivity of Freon-21 is applicable to all the experimentally investigated regions of state. But the accurate values of thermodynamic functions ($\rho$ and $c_p$), which are essential for the determination of $v$, $a$, and Pr are available only at pressures up to 20 MPa and at temperatures higher than $T_{NBP}$. Therefore, the recommended tables of transport properties (Tables 24 and 25) cover the same region of state as in thermodynamic tables.

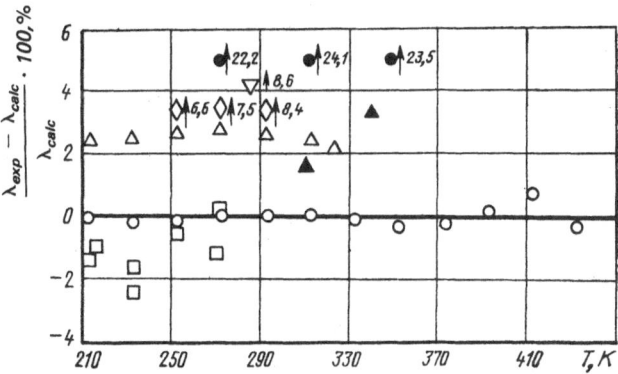

**FIG. 14.** Deviation of experimental values of thermal conductivity of liquid Freon-21 along the saturation line from the adopted values in the present work. Experimental data: 1, Geller and Peredri [2.5]; 2, Tauscher [1.63]; 3, Djalalian [1.29]; 4, Powell and Challoner [2.35]; 5, Powell et al. [2.36]; 6, Markwood and Benning [2.33]; 7, Gruzdev and Shestova [2.7]. Calculated data: 8, present work.

**TABLE 22.** Thermodynamic Properties of Freon-21 on the Saturation Line

| $T$ | $p$ | $\rho'$ | $\rho''$ | $h'$ | $h''$ | $s'$ | $s''$ | $c_p'$ | $c_p''$ | $r$ | $\sigma \cdot 10^3$ |
|---|---|---|---|---|---|---|---|---|---|---|---|
| 303.15 | 0.2150 | 1350.4 | 9.32 | 283.9 | 498.3 | 2.071 | 2.785 | 0.981 | 0.639 | 214.4 | 16.99 |
| 305.15 | 0.2296 | 1345.7 | 9.91 | 285.8 | 499.2 | 2.077 | 2.784 | 0.978 | 0.643 | 213.4 | 16.70 |
| 307.15 | 0.2448 | 1340.9 | 10.53 | 287.8 | 500.2 | 2.084 | 2.782 | 0.976 | 0.647 | 212.4 | 16.42 |
| 309.15 | 0.2607 | 1336.1 | 11.17 | 289.7 | 501.1 | 2.091 | 2.781 | 0.974 | 0.651 | 211.4 | 16.14 |
| 311.15 | 0.2774 | 1331.2 | 11.85 | 291.7 | 502.1 | 2.097 | 2.779 | 0.973 | 0.655 | 210.4 | 15.85 |
| 313.15 | 0.2948 | 1326.3 | 12.55 | 293.7 | 503.0 | 2.103 | 2.778 | 0.972 | 0.659 | 209.3 | 15.57 |
| 315.15 | 0.3130 | 1321.4 | 13.29 | 295.6 | 503.9 | 2.110 | 2.776 | 0.971 | 0.663 | 208.3 | 15.29 |
| 317.15 | 0.3321 | 1316.4 | 14.05 | 297.6 | 504.8 | 2.116 | 2.775 | 0.971 | 0.667 | 207.2 | 15.01 |
| 319.15 | 0.3520 | 1311.3 | 14.85 | 299.5 | 505.7 | 2.122 | 2.774 | 0.971 | 0.671 | 206.2 | 14.73 |
| 321.15 | 0.3728 | 1306.2 | 15.69 | 301.5 | 506.7 | 2.129 | 2.773 | 0.972 | 0.675 | 205.2 | 14.45 |
| 323.15 | 0.3945 | 1301.1 | 16.56 | 303.4 | 507.6 | 2.135 | 2.771 | 0.973 | 0.680 | 204.2 | 14.18 |
| 325.15 | 0.4172 | 1295.9 | 17.47 | 305.4 | 508.4 | 2.141 | 2.770 | 0.974 | 0.684 | 203.0 | 13.90 |
| 327.15 | 0.4408 | 1290.7 | 18.42 | 307.3 | 509.3 | 2.147 | 2.769 | 0.975 | 0.689 | 202.0 | 13.63 |
| 329.15 | 0.4655 | 1285.4 | 19.41 | 309.3 | 510.2 | 2.153 | 2.768 | 0.977 | 0.693 | 200.9 | 13.36 |
| 331.15 | 0.4912 | 1280.1 | 20.44 | 311.2 | 511.1 | 2.159 | 2.766 | 0.979 | 0.698 | 199.9 | 13.08 |
| 333.15 | 0.5180 | 1274.7 | 21.51 | 313.2 | 511.9 | 2.165 | 2.765 | 0.982 | 0.703 | 198.7 | 12.81 |
| 335.15 | 0.5459 | 1269.3 | 22.63 | 315.2 | 512.8 | 2.171 | 2.764 | 0.984 | 0.708 | 197.6 | 12.54 |
| 337.15 | 0.5749 | 1263.8 | 23.80 | 317.2 | 513.6 | 2.177 | 2.763 | 0.987 | 0.713 | 196.4 | 12.28 |
| 339.15 | 0.6051 | 1258.3 | 25.02 | 319.1 | 514.4 | 2.183 | 2.762 | 0.990 | 0.718 | 195.3 | 12.01 |
| 341.15 | 0.6365 | 1252.7 | 26.28 | 321.1 | 515.2 | 2.188 | 2.761 | 0.994 | 0.723 | 194.1 | 11.74 |
| 343.15 | 0.6691 | 1247.1 | 27.60 | 323.1 | 516.0 | 2.194 | 2.760 | 0.998 | 0.728 | 192.9 | 11.48 |
| 345.15 | 0.7030 | 1241.4 | 28.97 | 325.1 | 516.8 | 2.200 | 2.758 | 1.001 | 0.734 | 191.7 | 11.21 |
| 347.15 | 0.7381 | 1235.7 | 30.39 | 327.2 | 517.6 | 2.206 | 2.757 | 1.006 | 0.740 | 190.4 | 10.95 |

**TABLE 22.** Thermodynamic Properties of Freon-21 on the Saturation Line (*Continued*)

| T | p | ρ' | ρ'' | h' | h'' | s' | s'' | $c'_p$ | $c''_p$ | r | $\sigma \cdot 10^3$ |
|---|---|----|-----|----|-----|----|-----|--------|---------|---|---------------------|
| 349.15 | 0.7745 | 1229.9 | 31.88 | 329.2 | 518.4 | 2.212 | 2.756 | 1.010 | 0.745 | 189.2 | 10.69 |
| 351.15 | 0.8123 | 1224.1 | 33.42 | 331.2 | 519.2 | 2.217 | 2.755 | 1.015 | 0.751 | 188.0 | 10.43 |
| 353.15 | 0.8514 | 1218.1 | 35.02 | 333.2 | 519.9 | 2.223 | 2.754 | 1.020 | 0.757 | 186.7 | 10.17 |
| 355.15 | 0.8919 | 1212.2 | 36.68 | 335.3 | 520.6 | 2.229 | 2.753 | 1.025 | 0.764 | 185.3 | 9.91 |
| 357.15 | 0.9338 | 1206.1 | 38.41 | 337.3 | 521.4 | 2.235 | 2.753 | 1.030 | 0.770 | 184.1 | 9.66 |
| 359.15 | 0.9772 | 1200.0 | 40.21 | 339.4 | 522.1 | 2.240 | 2.751 | 1.036 | 0.777 | 182.7 | 9.40 |
| 361.15 | 1.0220 | 1193.8 | 42.08 | 341.5 | 522.7 | 2.246 | 2.750 | 1.042 | 0.784 | 181.2 | 9.15 |
| 363.15 | 1.0684 | 1187.6 | 44.02 | 343.6 | 523.4 | 2.252 | 2.749 | 1.048 | 0.791 | 179.8 | 8.90 |
| 365.15 | 1.1162 | 1181.3 | 46.03 | 345.7 | 524.1 | 2.257 | 2.748 | 1.054 | 0.798 | 178.4 | 8.65 |
| 367.15 | 1.1657 | 1174.9 | 48.12 | 347.8 | 524.7 | 2.263 | 2.747 | 1.061 | 0.806 | 176.9 | 8.40 |
| 369.15 | 1.2167 | 1168.4 | 50.29 | 349.9 | 525.4 | 2.269 | 2.745 | 1.068 | 0.814 | 175.5 | 8.15 |
| 371.15 | 1.2693 | 1161.8 | 52.55 | 352.0 | 526.0 | 2.274 | 2.744 | 1.075 | 0.822 | 174.0 | 7.90 |
| 373.15 | 1.3236 | 1155.2 | 54.89 | 354.2 | 526.6 | 2.280 | 2.743 | 1.083 | 0.831 | 172.4 | 7.66 |
| 375.15 | 1.3795 | 1148.5 | 57.32 | 356.4 | 527.1 | 2.286 | 2.742 | 1.091 | 0.840 | 170.7 | 7.42 |
| 377.15 | 1.4372 | 1141.7 | 59.85 | 358.5 | 527.7 | 2.291 | 2.741 | 1.099 | 0.849 | 169.2 | 7.17 |
| 379.15 | 1.4966 | 1134.7 | 62.47 | 360.7 | 528.2 | 2.297 | 2.740 | 1.108 | 0.859 | 167.5 | 6.93 |
| 381.15 | 1.5577 | 1127.7 | 65.20 | 363.0 | 528.8 | 2.303 | 2.739 | 1.117 | 0.869 | 165.8 | 6.70 |
| 383.15 | 1.6207 | 1120.6 | 68.03 | 365.2 | 529.3 | 2.308 | 2.737 | 1.127 | 0.880 | 164.1 | 6.46 |
| 385.15 | 1.6855 | 1113.4 | 70.98 | 367.4 | 529.7 | 2.314 | 2.736 | 1.137 | 0.891 | 162.3 | 6.22 |
| 387.15 | 1.7522 | 1106.0 | 74.04 | 369.7 | 530.2 | 2.320 | 2.735 | 1.148 | 0.903 | 160.5 | 5.99 |
| 389.15 | 1.8207 | 1098.5 | 77.22 | 372.0 | 530.6 | 2.325 | 2.734 | 1.159 | 0.915 | 158.6 | 5.76 |
| 391.15 | 1.8912 | 1091.0 | 80.54 | 374.3 | 531.0 | 2.331 | 2.733 | 1.171 | 0.928 | 156.7 | 5.53 |
| 393.15 | 1.9637 | 1083.2 | 83.98 | 376.6 | 531.4 | 2.337 | 2.731 | 1.183 | 0.942 | 154.8 | 5.30 |
| 395.15 | 2.038 | 1075.4 | 87.58 | 378.9 | 531.8 | 2.343 | 2.730 | 1.197 | 0.957 | 152.9 | 5.08 |
| 397.15 | 2.115 | 1067.3 | 91.32 | 381.3 | 532.1 | 2.348 | 2.729 | 1.211 | 0.973 | 150.8 | 4.85 |

**TABLE 22.** Thermodynamic Properties of Freon-21 on the Saturation Line (*Continued*)

| $T$ | $p$ | $\rho'$ | $\rho''$ | $h'$ | $h''$ | $s'$ | $s''$ | $c_p'$ | $c_p''$ | $r$ | $\sigma \cdot 10^3$ |
|---|---|---|---|---|---|---|---|---|---|---|---|
| 399.15 | 2.193 | 1059.2 | 95.22 | 383.6 | 532.4 | 2.354 | 2.727 | 1.226 | 0.989 | 148.8 | 4.63 |
| 401.15 | 2.274 | 1050.8 | 99.30 | 386.1 | 532.6 | 2.360 | 2.726 | 1.242 | 1.007 | 146.5 | 4.41 |
| 403.15 | 2.357 | 1042.3 | 103.55 | 388.5 | 532.9 | 2.366 | 2.724 | 1.260 | 1.027 | 144.4 | 4.20 |
| 405.15 | 2.442 | 1033.5 | 108.00 | 390.9 | 533.0 | 2.372 | 2.723 | 1.278 | 1.047 | 142.1 | 3.98 |
| 407.15 | 2.530 | 1024.6 | 112.65 | 393.4 | 533.2 | 2.378 | 2.721 | 1.298 | 1.070 | 139.8 | 3.77 |
| 409.15 | 2.620 | 1015.4 | 117.53 | 395.9 | 533.3 | 2.383 | 2.719 | 1.320 | 1.094 | 137.4 | 3.56 |
| 411.15 | 2.712 | 1006.1 | 122.64 | 398.5 | 533.4 | 2.389 | 2.718 | 1.344 | 1.121 | 134.9 | 3.35 |
| 413.15 | 2.806 | 996.4 | 128.01 | 401.1 | 533.4 | 2.395 | 2.716 | 1.370 | 1.150 | 132.3 | 3.14 |
| 415.15 | 2.903 | 986.5 | 133.66 | 403.7 | 533.3 | 2.402 | 2.714 | 1.398 | 1.182 | 129.6 | 2.94 |
| 417.15 | 3.003 | 976.2 | 139.62 | 406.4 | 533.2 | 2.408 | 2.712 | 1.430 | 1.217 | 126.8 | 2.74 |
| 419.15 | 3.105 | 965.7 | 145.91 | 409.1 | 533.1 | 2.414 | 2.710 | 1.465 | 1.257 | 124.0 | 2.54 |
| 421.15 | 3.210 | 954.7 | 152.57 | 411.8 | 532.9 | 2.420 | 2.708 | 1.504 | 1.301 | 121.1 | 2.35 |
| 423.15 | 3.317 | 943.4 | 159.63 | 414.6 | 532.6 | 2.427 | 2.705 | 1.549 | 1.352 | 118.0 | 2.16 |
| 425.15 | 3.428 | 931.6 | 167.15 | 417.5 | 532.2 | 2.433 | 2.703 | 1.600 | 1.409 | 114.7 | 1.97 |
| 427.15 | 3.541 | 919.3 | 175.18 | 420.4 | 531.7 | 2.440 | 2.700 | 1.658 | 1.476 | 111.3 | 1.79 |
| 429.15 | 3.657 | 906.4 | 183.79 | 423.4 | 531.2 | 2.446 | 2.697 | 1.726 | 1.553 | 107.8 | 1.61 |
| 431.15 | 3.776 | 892.8 | 193.05 | 426.5 | 530.5 | 2.453 | 2.694 | 1.806 | 1.645 | 104.0 | 1.43 |
| 433.15 | 3.898 | 878.6 | 203.08 | 429.7 | 529.7 | 2.460 | 2.691 | 1.902 | 1.756 | 100.0 | 1.26 |

**TABLE 23.** Thermodynamic Properties of Freon-21 in the Single-Phase Region

| $T = 303.15$ K | | | | | | | | |
|---|---|---|---|---|---|---|---|---|
| $p$ | $\rho$ | $z$ | $h$ | $s$ | $c_p$ | $w$ | $\mu$ | $\alpha \cdot 10^3$ |
| 0.01 | 0.41 | 0.9975 | 502.8 | 3.044 | 0.600 | 168.0 | 34.90 | 3.327 |
| 0.02 | 0.82 | 0.9949 | 502.6 | 2.987 | 0.601 | 167.6 | 34.95 | 3.356 |
| 0.03 | 1.23 | 0.9923 | 502.4 | 2.954 | 0.603 | 167.3 | 35.01 | 3.385 |
| 0.04 | 1.65 | 0.9897 | 502.1 | 2.930 | 0.605 | 167.0 | 35.07 | 3.414 |
| 0.05 | 2.07 | 0.9871 | 501.9 | 2.912 | 0.607 | 166.7 | 35.13 | 3.444 |
| 0.1 | 4.19 | 0.9740 | 500.9 | 2.853 | 0.616 | 165.1 | 35.43 | 3.600 |
| 0.2 | 8.63 | 0.9468 | 498.6 | 2.792 | 0.636 | 161.7 | 36.10 | 3.952 |
| 0.3 | 1350.66 | 0.0091 | 283.9 | 2.071 | 0.981 | 716.2 | −0.35 | 1.758 |
| 0.4 | 1350.95 | 0.0121 | 283.9 | 2.070 | 0.980 | 717.0 | −0.35 | 1.755 |
| 0.5 | 1351.24 | 0.0151 | 284.0 | 2.070 | 0.980 | 717.9 | −0.35 | 1.753 |
| 0.6 | 1351.53 | 0.0181 | 284.0 | 2.070 | 0.980 | 718.8 | −0.35 | 1.751 |
| 0.7 | 1351.82 | 0.0211 | 284.0 | 2.070 | 0.979 | 719.6 | −0.36 | 1.748 |
| 0.8 | 1352.11 | 0.0242 | 284.1 | 2.070 | 0.979 | 720.5 | −0.36 | 1.746 |
| 0.9 | 1352.39 | 0.0272 | 284.1 | 2.070 | 0.979 | 721.4 | −0.36 | 1.744 |
| 1.0 | 1352.68 | 0.0302 | 284.1 | 2.070 | 0.978 | 722.2 | −0.36 | 1.741 |
| 1.5 | 1354.10 | 0.0452 | 284.3 | 2.069 | 0.977 | 726.5 | −0.36 | 1.730 |
| 2.0 | 1355.50 | 0.0603 | 284.5 | 2.068 | 0.975 | 730.8 | −0.36 | 1.719 |
| 2.5 | 1356.89 | 0.0752 | 284.7 | 2.068 | 0.974 | 735.0 | −0.37 | 1.708 |
| 3.0 | 1358.26 | 0.0902 | 284.8 | 2.067 | 0.972 | 739.2 | −0.37 | 1.698 |
| 3.5 | 1359.62 | 0.1051 | 285.0 | 2.067 | 0.971 | 743.3 | −0.37 | 1.688 |
| 4.0 | 1360.96 | 0.1200 | 285.2 | 2.066 | 0.969 | 747.4 | −0.37 | 1.678 |
| 4.5 | 1362.29 | 0.1349 | 285.4 | 2.065 | 0.968 | 751.5 | −0.38 | 1.668 |
| 5.0 | 1363.60 | 0.1497 | 285.6 | 2.065 | 0.966 | 755.6 | −0.38 | 1.658 |
| 5.5 | 1364.90 | 0.1645 | 285.7 | 2.064 | 0.965 | 759.6 | −0.38 | 1.649 |
| 6.0 | 1366.19 | 0.1793 | 285.9 | 2.063 | 0.963 | 763.6 | −0.38 | 1.640 |
| 6.5 | 1367.47 | 0.1941 | 286.1 | 2.063 | 0.962 | 767.6 | −0.38 | 1.631 |
| 7.0 | 1368.73 | 0.2088 | 286.3 | 2.062 | 0.961 | 771.5 | −0.39 | 1.622 |
| 7.5 | 1369.98 | 0.2236 | 286.5 | 2.062 | 0.959 | 775.4 | −0.39 | 1.614 |
| 8.0 | 1371.21 | 0.2382 | 286.7 | 2.061 | 0.958 | 779.3 | −0.39 | 1.605 |
| 8.5 | 1372.44 | 0.2529 | 286.8 | 2.061 | 0.957 | 783.1 | −0.39 | 1.597 |
| 9.0 | 1373.65 | 0.2675 | 287.0 | 2.060 | 0.956 | 786.9 | −0.40 | 1.589 |
| 9.5 | 1374.85 | 0.2822 | 287.2 | 2.059 | 0.954 | 790.7 | −0.40 | 1.581 |
| 10.0 | 1376.05 | 0.2968 | 287.4 | 2.059 | 0.953 | 794.5 | −0.40 | 1.573 |
| 11.0 | 1378.39 | 0.3259 | 287.8 | 2.058 | 0.951 | 802.0 | −0.40 | 1.558 |
| 12.0 | 1380.70 | 0.3549 | 288.2 | 2.057 | 0.949 | 809.4 | −0.41 | 1.544 |
| 13.0 | 1382.97 | 0.3838 | 288.6 | 2.055 | 0.946 | 816.7 | −0.41 | 1.529 |
| 14.0 | 1385.20 | 0.4127 | 289.0 | 2.054 | 0.944 | 823.9 | −0.41 | 1.516 |
| 15.0 | 1387.39 | 0.4415 | 289.3 | 2.053 | 0.942 | 831.0 | −0.42 | 1.503 |
| 16.0 | 1389.55 | 0.4702 | 289.7 | 2.052 | 0.940 | 838.0 | −0.42 | 1.490 |
| 17.0 | 1391.67 | 0.4988 | 290.1 | 2.051 | 0.938 | 845.0 | −0.42 | 1.478 |
| 18.0 | 1393.76 | 0.5274 | 290.5 | 2.050 | 0.936 | 851.8 | −0.43 | 1.466 |
| 19.0 | 1395.82 | 0.5558 | 298.9 | 2.049 | 0.934 | 858.6 | −0.43 | 1.455 |

**TABLE 23.** Thermodynamic Properties of Freon-21 in the Single-Phase Region (*Continued*)

$$T = 313.15 \text{ K}$$

| $p$ | $\rho$ | $z$ | $h$ | $s$ | $c_p$ | $w$ | $\mu$ | $\alpha \cdot 10^3$ |
|---|---|---|---|---|---|---|---|---|
| 0.01 | 0.40 | 0.9977 | 508.8 | 3.064 | 0.609 | 170.5 | 31.75 | 3.218 |
| 0.02 | 0.79 | 0.9954 | 508.6 | 3.008 | 0.610 | 170.2 | 31.80 | 3.243 |
| 0.03 | 1.19 | 0.9931 | 508.5 | 2.975 | 0.612 | 170.0 | 31.85 | 3.268 |
| 0.04 | 1.60 | 0.9908 | 508.3 | 2.951 | 0.614 | 169.7 | 31.90 | 3.293 |
| 0.05 | 2.00 | 0.9885 | 508.1 | 2.933 | 0.615 | 169.4 | 31.94 | 3.319 |
| 0.1 | 4.05 | 0.9767 | 507.1 | 2.874 | 0.623 | 167.9 | 32.20 | 3.453 |
| 0.2 | 8.30 | 0.9525 | 505.0 | 2.814 | 0.640 | 164.9 | 32.76 | 3.749 |
| 0.3 | 1326.32 | 0.0889 | 293.7 | 2.103 | 0.972 | 696.4 | —0.32 | 1.881 |
| 0.4 | 1326.64 | 0.0119 | 293.7 | 2.103 | 0.971 | 697.4 | —0.32 | 1.878 |
| 0.5 | 1326.96 | 0.0149 | 293.7 | 2.103 | 0.971 | 698.3 | —0.32 | 1.875 |
| 0.6 | 1327.28 | 0.0179 | 293.7 | 2.103 | 0.970 | 699.2 | —0.32 | 1.872 |
| 0.7 | 1327.60 | 0.0208 | 293.8 | 2.103 | 0.970 | 700.1 | —0.32 | 1.869 |
| 0.8 | 1327.91 | 0.0238 | 293.8 | 2.103 | 0.970 | 701.1 | —0.32 | 1.866 |
| 0.9 | 1328.23 | 0.0268 | 293.8 | 2.103 | 0.969 | 702.0 | —0.32 | 1.863 |
| 1.0 | 1328.54 | 0.0298 | 293.9 | 2.102 | 0.969 | 702.9 | —0.32 | 1.860 |
| 1.5 | 1330.11 | 0.0446 | 294.0 | 2.102 | 0.967 | 707.5 | —0.33 | 1.847 |
| 2.0 | 1331.65 | 0.0594 | 294.2 | 2.101 | 0.965 | 712.0 | —0.33 | 1.833 |
| 2.5 | 1333.17 | 0.0741 | 294.3 | 2.100 | 0.964 | 716.5 | —0.34 | 1.820 |
| 3.0 | 1334.67 | 0.0889 | 294.5 | 2.100 | 0.962 | 720.9 | —0.34 | 1.807 |
| 3.5 | 1336.16 | 0.1035 | 294.7 | 2.099 | 0.960 | 725.3 | —0.34 | 1.795 |
| 4.0 | 1337.62 | 0.1182 | 294.8 | 2.098 | 0.958 | 729.7 | —0.34 | 1.783 |
| 4.5 | 1339.07 | 0.1328 | 295.0 | 2.098 | 0.957 | 734.0 | —0.35 | 1.771 |
| 5.0 | 1340.51 | 0.1474 | 295.2 | 2.097 | 0.955 | 738.3 | —0.35 | 1.759 |
| 5.5 | 1341.92 | 0.1620 | 295.3 | 2.096 | 0.954 | 742.6 | —0.35 | 1.748 |
| 6.0 | 1343.32 | 0.1766 | 295.5 | 2.096 | 0.952 | 746.8 | —0.36 | 1.737 |
| 6.5 | 1344.71 | 0.1911 | 295.7 | 2.095 | 0.951 | 751.0 | —0.36 | 1.726 |
| 7.0 | 1346.08 | 0.2056 | 295.9 | 2.094 | 0.949 | 755.1 | —0.36 | 1.716 |
| 7.5 | 1347.43 | 0.2200 | 296.0 | 2.094 | 0.948 | 759.2 | —0.36 | 1.705 |
| 8.0 | 1348.77 | 0.2345 | 296.2 | 2.093 | 0.946 | 765.3 | —0.37 | 1.695 |
| 8.5 | 1350.10 | 0.2489 | 296.4 | 2.093 | 0.945 | 767.4 | —0.37 | 1.665 |
| 9.0 | 1351.41 | 0.2633 | 296.5 | 2.092 | 0.944 | 771.4 | —0.37 | 1.676 |
| 9.5 | 1352.71 | 0.2776 | 296.7 | 2.091 | 0.942 | 775.4 | —0.38 | 1.666 |
| 10.0 | 1354.00 | 0.2920 | 296.9 | 2.091 | 0.941 | 779.3 | —0.38 | 1.657 |
| 11.0 | 1356.53 | 0.3206 | 297.3 | 2.089 | 0.938 | 787.2 | —0.38 | 1.639 |
| 12.0 | 1359.02 | 0.3491 | 297.6 | 2.088 | 0.936 | 794.9 | —0.39 | 1.622 |
| 13.0 | 1361.46 | 0.3775 | 298.0 | 2.087 | 0.934 | 802.6 | —0.39 | 1.606 |
| 14.0 | 1363.86 | 0.4058 | 298.3 | 2.086 | 0.931 | 810.1 | —0.40 | 1.590 |
| 15.0 | 1366.21 | 0.4340 | 298.7 | 2.085 | 0.929 | 817.5 | —0.40 | 1.575 |
| 16.0 | 1368.52 | 0.4622 | 299.1 | 2.084 | 0.927 | 824.9 | —0.40 | 1.560 |
| 17.0 | 1370.79 | 0.4902 | 299.5 | 2.082 | 0.925 | 832.1 | —0.41 | 1.546 |
| 18.0 | 1373.03 | 0.5182 | 299.8 | 2.081 | 0.923 | 839.3 | —0.41 | 1.532 |
| 19.0 | 1375.22 | 0.5462 | 300.2 | 2.080 | 0.921 | 846.4 | —0.41 | 1.519 |
| 20.0 | 1377.39 | 0.5740 | 300.6 | 2.079 | 0.919 | 853.4 | —0.42 | 1.506 |

**TABLE 23.** Thermodynamic Properties of Freon-21 in the Single-Phase Region (*Continued*)

$$T = 323,15 \ K$$

| $p$ | $\rho$ | $z$ | $h$ | $s$ | $c_p$ | $w$ | $\mu$ | $\alpha \cdot 10^3$ |
|------|--------|--------|-------|-------|-------|-------|--------|---------|
| 0.01 | 0.38 | 0.9979 | 515.0 | 3.084 | 0.618 | 173.1 | 29.02 | 3.116 |
| 0.02 | 0.77 | 0.9959 | 514.8 | 3.028 | 0.620 | 172.8 | 29.06 | 3.137 |
| 0.03 | 1.16 | 0.9938 | 514.6 | 2.995 | 0.621 | 172.5 | 29.10 | 3.159 |
| 0.04 | 1.55 | 0.9917 | 514.4 | 2.971 | 0.622 | 172.3 | 29.15 | 3.181 |
| 0.05 | 1.94 | 0.9896 | 514.3 | 2.953 | 0.624 | 172.0 | 29.19 | 3.204 |
| 0.1 | 3.91 | 0.9791 | 513.3 | 2.895 | 0.630 | 170.7 | 29.40 | 3.319 |
| 0.2 | 8.00 | 0.9574 | 511.5 | 2.835 | 0.645 | 167.9 | 29.87 | 3.572 |
| 0.3 | 12.29 | 0.9349 | 509.5 | 2.798 | 0.662 | 165.0 | 30.39 | 3.860 |
| 0.4 | 1301.11 | 0.0116 | 303.4 | 2.135 | 0.973 | 672.5 | −0.28 | 2.012 |
| 0.5 | 1301.46 | 0.0147 | 303.4 | 2.135 | 0.972 | 673.5 | −0.28 | 2.008 |
| 0.6 | 1301.82 | 0.0177 | 303.5 | 2.134 | 0.972 | 674.5 | −0.28 | 2.005 |
| 0.7 | 1302.17 | 0.0206 | 303.5 | 2.134 | 0.971 | 675.5 | −0.28 | 2.001 |
| 0.8 | 1302.52 | 0.0235 | 303.5 | 2.134 | 0.971 | 676.5 | −0.28 | 1.997 |
| 0.9 | 1302.87 | 0.0265 | 303.5 | 2.134 | 0.970 | 677.5 | −0.28 | 1.994 |
| 1.0 | 1303.22 | 0.0294 | 303.6 | 2.134 | 0.970 | 678.5 | −0.28 | 1.990 |
| 1.5 | 1304.96 | 0.0440 | 303.7 | 2.133 | 0.968 | 683.3 | −0.29 | 1.973 |
| 2.0 | 1306.67 | 0.0586 | 303.8 | 2.132 | 0.966 | 688.2 | −0.29 | 1.956 |
| 2.5 | 1308.35 | 0.0732 | 304.0 | 2.131 | 0.964 | 692.9 | −0.30 | 1.940 |
| 3.0 | 1310.01 | 0.0877 | 304.1 | 2.131 | 0.962 | 697.6 | −0.30 | 1.924 |
| 3.5 | 1311.65 | 0.1022 | 304.3 | 2.130 | 0.960 | 702.3 | −0.30 | 1.909 |
| 4.0 | 1313.27 | 0.1167 | 304.4 | 2.129 | 0.958 | 706.9 | −0.31 | 1.894 |
| 4.5 | 1314.86 | 0.1311 | 304.6 | 2.129 | 0.956 | 711.5 | −0.31 | 1.880 |
| 5.0 | 1316.44 | 0.1455 | 304.7 | 2.128 | 0.954 | 716.1 | −0.32 | 1.866 |
| 5.5 | 1317.99 | 0.1599 | 304.9 | 2.127 | 0.952 | 720.6 | −0.32 | 1.852 |
| 6.0 | 1319.53 | 0.1742 | 305.0 | 2.126 | 0.951 | 725.0 | −0.32 | 1.839 |
| 6.5 | 1321.04 | 0.1885 | 305.2 | 2.126 | 0.949 | 729.4 | −0.33 | 1.826 |
| 7.0 | 1322.54 | 0.2028 | 305.3 | 2.125 | 0.947 | 733.8 | −0.33 | 1.813 |
| 7.5 | 1324.02 | 0.2170 | 305.5 | 2.124 | 0.946 | 738.1 | −0.33 | 1.801 |
| 8.0 | 1325.49 | 0.2312 | 305.7 | 2.124 | 0.944 | 742.5 | −0.34 | 1.789 |
| 8.5 | 1326.93 | 0.2454 | 305.8 | 2.123 | 0.942 | 746.7 | −0.34 | 1.777 |
| 9.0 | 1328.36 | 0.2595 | 306.0 | 2.122 | 0.941 | 750.9 | −0.34 | 1.766 |
| 9.5 | 1329.78 | 0.2737 | 306.1 | 2.122 | 0.939 | 755.1 | −0.35 | 1.754 |
| 10.0 | 1331.17 | 0.2878 | 306.3 | 2.121 | 0.938 | 759.3 | −0.35 | 1.744 |
| 11.0 | 1333.93 | 0.3159 | 306.6 | 2.120 | 0.935 | 767.5 | −0.36 | 1.722 |
| 12.0 | 1336.62 | 0.3439 | 307.0 | 2.118 | 0.932 | 775.6 | −0.36 | 1.702 |
| 13.0 | 1339.26 | 0.3718 | 307.3 | 2.117 | 0.930 | 783.6 | −0.37 | 1.683 |
| 14.0 | 1341.85 | 0.3997 | 307.6 | 2.116 | 0.927 | 791.5 | −0.37 | 1.664 |
| 15.0 | 1344.38 | 0.4274 | 308.0 | 2.115 | 0.925 | 799.3 | −0.38 | 1.646 |
| 16.0 | 1346.87 | 0.4551 | 308.3 | 2.114 | 0.923 | 806.9 | −0.38 | 1.629 |
| 17.0 | 1349.32 | 0.4826 | 308.7 | 2.112 | 0.920 | 814.5 | −0.39 | 1.613 |
| 18.0 | 1351.71 | 0.5101 | 309.1 | 2.111 | 0.918 | 821.9 | −0.39 | 1.597 |
| 19.0 | 1354.07 | 0.5375 | 309.4 | 2.110 | 0.916 | 829.3 | −0.39 | 1.582 |
| 20.0 | 1356.39 | 0.5648 | 309.8 | 2.109 | 0.914 | 836.6 | −0.40 | 1.567 |

**TABLE 23.** Thermodynamic Properties of Freon-21 in the Single-Phase Region (*Continued*)

$$T = 333.15 \ K$$

| $p$ | $\rho$ | $z$ | $h$ | $s$ | $c_p$ | $w$ | $\mu$ | $a \cdot 10^3$ |
|---|---|---|---|---|---|---|---|---|
| 0.01 | 0.37 | 0.9981 | 521.2 | 3.104 | 0.627 | 175.5 | 26.65 | 3.020 |
| 0.02 | 0.75 | 0.9963 | 521.1 | 3.048 | 0.629 | 175.3 | 26.68 | 3.039 |
| 0.03 | 1.12 | 0.9944 | 520.9 | 3.015 | 0.630 | 175.1 | 26.72 | 3.058 |
| 0.04 | 1.50 | 0.9925 | 520.7 | 2.991 | 0.631 | 174.8 | 26.75 | 3.078 |
| 0.05 | 1.88 | 0.9906 | 520.6 | 2.973 | 0.632 | 174.6 | 26.79 | 3.097 |
| 0.1 | 3.79 | 0.9811 | 519.7 | 2.915 | 0.638 | 173.4 | 26.97 | 3.197 |
| 0.2 | 7.73 | 0.9617 | 517.9 | 2.855 | 0.651 | 170.8 | 27.36 | 3.415 |
| 0.3 | 11.84 | 0.9415 | 516.1 | 2.819 | 0.666 | 168.2 | 27.79 | 3.659 |
| 0.4 | 16.15 | 0.9206 | 514.2 | 2.791 | 0.681 | 165.4 | 28.26 | 3.935 |
| 0.5 | 20.67 | 0.8988 | 512.3 | 2.769 | 0.699 | 162.5 | 28.80 | 4.251 |
| 0.6 | 1275.05 | 0.0175 | 313.2 | 2.165 | 0.981 | 645.2 | —0.22 | 2.155 |
| 0.7 | 1275.44 | 0.0204 | 313.3 | 2.165 | 0.981 | 646.3 | —0.23 | 2.150 |
| 0.8 | 1275.84 | 0.0233 | 313.3 | 2.164 | 0.980 | 647.3 | —0.23 | 2.146 |
| 0.9 | 1276.23 | 0.0262 | 313.3 | 2.164 | 0.980 | 648.4 | —0.23 | 2.141 |
| 1.0 | 1276.63 | 0.0291 | 313.3 | 2.164 | 0.979 | 649.5 | —0.23 | 2.137 |
| 1.5 | 1278.57 | 0.0436 | 313.4 | 2.163 | 0.976 | 654.7 | —0.24 | 2.115 |
| 2.0 | 1280.49 | 0.0580 | 313.5 | 2.162 | 0.974 | 659.8 | —0.24 | 2.094 |
| 2.5 | 1282.37 | 0.0724 | 313.7 | 2.162 | 0.971 | 664.9 | —0.25 | 2.074 |
| 3.0 | 1284.22 | 0.0868 | 313.8 | 2.161 | 0.969 | 669.9 | —0.25 | 2.054 |
| 3.5 | 1286.05 | 0.1011 | 313.9 | 2.160 | 0.967 | 674.9 | —0.26 | 2.035 |
| 4.0 | 1287.85 | 0.1154 | 314.0 | 2.159 | 0.965 | 679.8 | —0.26 | 2.017 |
| 4.5 | 1289.62 | 0.1297 | 314.2 | 2.158 | 0.962 | 684.7 | —0.27 | 1.999 |
| 5.0 | 1291.36 | 0.1439 | 314.3 | 2.158 | 0.960 | 689.5 | —0.27 | 1.982 |
| 5.5 | 1293.09 | 0.1580 | 314.4 | 2.157 | 0.958 | 694.3 | —0.28 | 1.965 |
| 6.0 | 1294.78 | 0.1722 | 314.6 | 2.156 | 0.956 | 699.0 | —0.28 | 1.949 |
| 6.5 | 1296.46 | 0.1863 | 314.7 | 2.155 | 0.954 | 703.6 | —0.29 | 1.933 |
| 7.0 | 1298.11 | 0.2004 | 314.8 | 2.155 | 0.952 | 708.3 | —0.29 | 1.918 |
| 7.5 | 1299.74 | 0.2144 | 315.0 | 2.154 | 0.951 | 712.8 | —0.30 | 1.903 |
| 8.0 | 1301.34 | 0.2284 | 315.1 | 2.153 | 0.949 | 717.4 | —0.30 | 1.889 |
| 8.5 | 1302.93 | 0.2424 | 315.3 | 2.152 | 0.947 | 721.9 | —0.30 | 1.875 |
| 9.0 | 1304.50 | 0.2564 | 315.4 | 2.152 | 0.945 | 726.3 | —0.31 | 1.861 |
| 9.5 | 1306.05 | 0.2703 | 315.6 | 2.151 | 0.944 | 730.7 | —0.31 | 1.847 |
| 10.0 | 1307.58 | 0.2842 | 315.7 | 2.150 | 0.942 | 735.1 | —0.32 | 1.834 |
| 11.0 | 1310.58 | 0.3119 | 316.0 | 2.149 | 0.939 | 743.7 | —0.32 | 1.809 |
| 12.0 | 1313.52 | 0.3395 | 316.3 | 2.148 | 0.936 | 752.2 | —0.33 | 1.785 |
| 13.0 | 1316.39 | 0.3670 | 316.6 | 2.146 | 0.933 | 760.5 | —0.34 | 1.763 |
| 14.0 | 1319.20 | 0.3943 | 316.9 | 2.145 | 0.930 | 768.7 | —0.34 | 1.741 |
| 15.0 | 1321.95 | 0.4216 | 317.3 | 2.144 | 0.927 | 776.8 | —0.35 | 1.720 |
| 16.0 | 1324.64 | 0.4488 | 317.6 | 2.142 | 0.925 | 784.8 | —0.35 | 1.700 |
| 17.0 | 1327.28 | 0.4759 | 317.9 | 2.141 | 0.922 | 792.6 | —0.36 | 1.681 |
| 18.0 | 1329.86 | 0.5029 | 318.2 | 2.140 | 0.920 | 800.4 | —0.36 | 1.663 |
| 19.0 | 1332.40 | 0.5299 | 318.6 | 2.138 | 0.917 | 808.0 | —0.37 | 1.645 |
| 20.0 | 1334.89 | 0.5567 | 318.9 | 2.137 | 0.915 | 815.6 | —0.38 | 1.628 |

**TABLE 23.** Thermodynamic Properties of Freon-21 in the Single-Phase Region (*Continued*)

$$T = 343.15 \text{ K}$$

| $p$ | $\rho$ | $z$ | $h$ | $s$ | $c_p$ | $w$ | $\mu$ | $\alpha \cdot 10^3$ |
|---|---|---|---|---|---|---|---|---|
| 0.01 | 0.36 | 0.9983 | 527.5 | 3.123 | 0.636 | 178.0 | 24.56 | 2.931 |
| 0.02 | 0.72 | 0.9966 | 527.4 | 3.067 | 0.637 | 177.8 | 24.59 | 2.947 |
| 0.03 | 1.09 | 0.9949 | 527.2 | 3.034 | 0.638 | 177.6 | 24.62 | 2.964 |
| 0.04 | 1.45 | 0.9932 | 527.1 | 3.010 | 0.639 | 177.3 | 24.65 | 2.981 |
| 0.05 | 1.82 | 0.9915 | 526.9 | 2.992 | 0.640 | 177.1 | 24.68 | 2.998 |
| 0.1 | 3.67 | 0.9829 | 526.1 | 2.934 | 0.646 | 176.0 | 24.83 | 3.086 |
| 0.2 | 7.47 | 0.9654 | 524.5 | 2.875 | 0.657 | 173.7 | 25.16 | 3.274 |
| 0.3 | 11.42 | 0.9473 | 522.8 | 2.839 | 0.670 | 171.3 | 25.52 | 3.483 |
| 0.4 | 15.54 | 0.9286 | 521.1 | 2.812 | 0.683 | 168.8 | 25.91 | 3.716 |
| 0.5 | 19.84 | 0.9092 | 519.3 | 2.790 | 0.699 | 166.2 | 26.34 | 3.978 |
| 0.6 | 24.35 | 0.8891 | 517.4 | 2.771 | 0.715 | 163.4 | 26.81 | 4.275 |
| 0.7 | 1247.25 | 0.0202 | 323.1 | 2.194 | 0.997 | 613.1 | —0.16 | 2.325 |
| 0.8 | 1247.70 | 0.0231 | 323.2 | 2.194 | 0.997 | 614.2 | —0.16 | 2.319 |
| 0.9 | 1248.15 | 0.0260 | 323.2 | 2.194 | 0.996 | 615.3 | —0.17 | 2.313 |
| 1.0 | 1248.60 | 0.0289 | 323.2 | 2.194 | 0.995 | 616.5 | —0.17 | 2.308 |
| 1.5 | 1250.81 | 0.0433 | 323.3 | 2.193 | 0.992 | 622.1 | —0.18 | 2.280 |
| 2.0 | 1252.98 | 0.0576 | 323.4 | 2.192 | 0.989 | 627.6 | —0.18 | 2.253 |
| 2.5 | 1255.11 | 0.0719 | 323.5 | 2.191 | 0.986 | 633.1 | —0.19 | 2.227 |
| 3.0 | 1257.20 | 0.0861 | 323.5 | 2.190 | 0.983 | 638.5 | —0.20 | 2.202 |
| 3.5 | 1259.26 | 0.1003 | 323.6 | 2.189 | 0.980 | 643.8 | —0.20 | 2.178 |
| 4.0 | 1261.28 | 0.1144 | 323.7 | 2.188 | 0.978 | 649.0 | —0.21 | 2.155 |
| 4.5 | 1263.26 | 0.1285 | 323.9 | 2.187 | 0.975 | 654.2 | —0.22 | 2.133 |
| 5.0 | 1265.22 | 0.1426 | 324.0 | 2.187 | 0.972 | 659.3 | —0.22 | 2.112 |
| 5.5 | 1267.14 | 0.1566 | 324.1 | 2.186 | 0.970 | 664.4 | —0.23 | 2.091 |
| 6.0 | 1269.03 | 0.1706 | 324.2 | 2.185 | 0.969 | 669.4 | —0.24 | 2.071 |
| 6.5 | 1270.89 | 0.1845 | 324.3 | 2.184 | 0.965 | 674.3 | —0.24 | 2.052 |
| 7.0 | 1272.73 | 0.1984 | 324.4 | 2.183 | 0.963 | 679.2 | —0.25 | 2.033 |
| 7.5 | 1274.53 | 0.2123 | 324.5 | 2.183 | 0.961 | 684.0 | —0.25 | 2.015 |
| 8.0 | 1276.31 | 0.2261 | 324.7 | 2.182 | 0.959 | 688.8 | —0.26 | 1.998 |
| 8.5 | 1278.07 | 0.2399 | 324.8 | 2.181 | 0.957 | 693.5 | —0.26 | 1.981 |
| 9.0 | 1279.80 | 0.2537 | 324.9 | 2.180 | 0.955 | 698.2 | —0.27 | 1.964 |
| 9.5 | 1281.51 | 0.2674 | 325.0 | 2.179 | 0.953 | 702.9 | —0.27 | 1.948 |
| 10.0 | 1283.19 | 0.2811 | 325.2 | 2.179 | 0.951 | 707.5 | —0.28 | 1.932 |
| 11.0 | 1286.49 | 0.3085 | 325.4 | 2.177 | 0.947 | 716.5 | —0.28 | 1.902 |
| 12.0 | 1289.71 | 0.3357 | 325.7 | 2.176 | 0.944 | 725.4 | —0.29 | 1.874 |
| 13.0 | 1292.85 | 0.3627 | 326.0 | 2.174 | 0.941 | 734.1 | —0.30 | 1.847 |
| 14.0 | 1295.91 | 0.3897 | 326.3 | 2.173 | 0.937 | 742.7 | —0.31 | 1.822 |
| 15.0 | 1298.91 | 0.4166 | 326.6 | 2.171 | 0.934 | 751.1 | —0.32 | 1.797 |
| 16.0 | 1301.83 | 0.4434 | 326.9 | 2.176 | 0.932 | 759.4 | —0.32 | 1.774 |
| 17.0 | 1304.69 | 0.4701 | 327.2 | 2.169 | 0.929 | 767.5 | —0.33 | 1.752 |
| 18.0 | 1307.49 | 0.4966 | 327.5 | 2.167 | 0.926 | 775.6 | —0.34 | 1.731 |
| 19.0 | 1310.24 | 0.5231 | 327.8 | 2.166 | 0.924 | 783.5 | —0.34 | 1.710 |
| 20.0 | 1312.93 | 0.5495 | 328.1 | 2.165 | 0.921 | 791.3 | —0.35 | 1.691 |

**TABLE 23.** Thermodynamic Properties of Freon-21 in the Single-Phase Region (*Continued*)

$$T = 353.15 \text{ K}$$

| $p$ | $\rho$ | $z$ | $h$ | $s$ | $c_p$ | $w$ | $\mu$ | $\alpha \cdot 10^3$ |
|------|--------|--------|--------|--------|--------|--------|--------|--------|
| 0.01 | 0.35 | 0.9985 | 534.0 | 3.142 | 0.645 | 180.4 | 22.72 | 2.846 |
| 0.02 | 0.70 | 0.9969 | 533.8 | 3.086 | 0.646 | 180.2 | 22.74 | 2.861 |
| 0.03 | 1.06 | 0.9954 | 533.7 | 3.053 | 0.647 | 180.0 | 22.77 | 2.876 |
| 0.04 | 1.41 | 0.9939 | 533.5 | 3.029 | 0.648 | 179.8 | 22.80 | 2.891 |
| 0.05 | 1.77 | 0.9923 | 533.4 | 3.011 | 0.649 | 179.6 | 22.82 | 2.906 |
| 0.1 | 3.56 | 0.9845 | 532.6 | 2.953 | 0.654 | 178.6 | 22.95 | 2.983 |
| 0.2 | 7.24 | 0.9687 | 531.1 | 2.894 | 0.664 | 176.5 | 23.23 | 3.148 |
| 0.3 | 11.04 | 0.9523 | 529.5 | 2.858 | 0.675 | 174.3 | 23.53 | 3.328 |
| 0.4 | 14.99 | 0.9355 | 527.9 | 2.832 | 0.687 | 172.0 | 23.85 | 3.526 |
| 0.5 | 19.09 | 0.9182 | 526.3 | 2.810 | 0.700 | 169.6 | 24.20 | 3.746 |
| 0.6 | 23.36 | 0.9003 | 524.5 | 2.792 | 0.714 | 167.2 | 24.57 | 3.992 |
| 0.7 | 27.83 | 0.8818 | 522.7 | 2.776 | 0.730 | 164.6 | 24.99 | 4.269 |
| 0.8 | 32.51 | 0.8625 | 520.9 | 2.761 | 0.747 | 161.9 | 25.45 | 4.583 |
| 0.9 | 1218.38 | 0.0259 | 333.2 | 2.223 | 1.019 | 578.8 | —0.09 | 2.521 |
| 1.0 | 1218.90 | 0.0288 | 333.2 | 2.223 | 1.018 | 580.1 | —0.09 | 2.514 |
| 1.5 | 1221.45 | 0.0430 | 333.3 | 2.222 | 1.014 | 586.2 | —0.10 | 2.477 |
| 2.0 | 1223.95 | 0.0573 | 333.4 | 2.221 | 1.010 | 592.1 | —0.11 | 2.442 |
| 2.5 | 1226.39 | 0.0715 | 333.4 | 2.220 | 1.006 | 598.0 | —0.12 | 2.408 |
| 3.0 | 1228.78 | 0.0856 | 333.5 | 2.219 | 1.003 | 603.8 | —0.13 | 2.376 |
| 3.5 | 1231.12 | 0.0997 | 333.5 | 2.218 | 0.999 | 609.5 | —0.14 | 2.346 |
| 4.0 | 1233.42 | 0.1137 | 333.6 | 2.217 | 0.996 | 615.2 | —0.15 | 2.316 |
| 4.5 | 1235.67 | 0.1277 | 333.7 | 2.216 | 0.993 | 620.7 | —0.16 | 2.288 |
| 5.0 | 1237.88 | 0.1416 | 333.8 | 2.215 | 0.990 | 626.2 | —0.16 | 2.261 |
| 5.5 | 1240.04 | 0.1555 | 333.8 | 2.214 | 0.987 | 631.6 | —0.17 | 2.235 |
| 6.0 | 1242.17 | 0.1693 | 333.9 | 2.213 | 0.984 | 636.9 | —0.18 | 2.210 |
| 6.5 | 1244.26 | 0.1831 | 334.0 | 2.212 | 0.981 | 642.1 | —0.19 | 2.186 |
| 7.0 | 1246.32 | 0.1969 | 334.1 | 2.211 | 0.978 | 647.3 | —0.19 | 2.163 |
| 7.5 | 1248.34 | 0.2106 | 334.2 | 2.211 | 0.976 | 652.4 | —0.20 | 2.141 |
| 8.0 | 1250.33 | 0.2243 | 334.3 | 2.210 | 0.973 | 657.5 | —0.21 | 2.119 |
| 8.5 | 1252.28 | 0.2379 | 334.4 | 2.209 | 0.971 | 662.5 | —0.21 | 2.098 |
| 9.0 | 1254.21 | 0.2515 | 334.5 | 2.208 | 0.969 | 667.4 | —0.22 | 2.078 |
| 9.5 | 1256.10 | 0.2651 | 334.6 | 2.207 | 0.966 | 672.3 | —0.22 | 2.059 |
| 10.0 | 1257.97 | 0.2787 | 334.7 | 2.206 | 0.964 | 677.2 | —0.23 | 2.040 |
| 11.0 | 1261.62 | 0.3056 | 335.0 | 2.205 | 0.960 | 686.7 | —0.24 | 2.004 |
| 12.0 | 1265.17 | 0.3325 | 335.2 | 2.203 | 0.956 | 696.0 | —0.25 | 1.970 |
| 13.0 | 1268.62 | 0.3592 | 335.4 | 2.202 | 0.952 | 705.1 | —0.26 | 1.938 |
| 14.0 | 1271.98 | 0.3858 | 335.7 | 2.200 | 0.948 | 714.0 | —0.27 | 1.908 |
| 15.0 | 1275.25 | 0.4123 | 336.0 | 2.199 | 0.945 | 722.8 | —0.28 | 1.879 |
| 16.0 | 1278.45 | 0.4387 | 336.2 | 2.197 | 0.942 | 731.5 | —0.29 | 1.852 |
| 17.0 | 1281.56 | 0.4650 | 336.5 | 2.196 | 0.939 | 739.9 | —0.30 | 1.826 |
| 18.0 | 1284.61 | 0.4912 | 336.8 | 2.194 | 0.936 | 748.3 | —0.30 | 1.802 |
| 19.0 | 1287.59 | 0.5173 | 337.1 | 2.193 | 0.933 | 756.5 | —0.32 | 1.778 |
| 20.0 | 1290.58 | 0.5433 | 337.4 | 2.192 | 0.930 | 764.6 | —0.32 | 1.756 |

**TABLE 23.** Thermodynamic Properties of Freon-21 in the Single-Phase Region (*Continued*)

$$T = 363.15 \text{ K}$$

| $p$ | $\rho$ | $z$ | $h$ | $s$ | $c_p$ | $w$ | $\mu$ | $\alpha \cdot 10^3$ |
|---|---|---|---|---|---|---|---|---|
| 0.01 | 0.34 | 0.9986 | 540.4 | 3.160 | 0.654 | 182.8 | 21.09 | 2.767 |
| 0.02 | 0.68 | 0.9972 | 540.3 | 3.184 | 0.654 | 182.6 | 21.11 | 2.780 |
| 0.03 | 1.03 | 0.9958 | 540.2 | 3.071 | 0.655 | 182.4 | 21.13 | 2.793 |
| 0.04 | 1.37 | 0.9944 | 540.0 | 3.047 | 0.656 | 182.2 | 21.15 | 2.806 |
| 0.05 | 1.72 | 0.9930 | 539.9 | 3.029 | 0.657 | 182.0 | 21.18 | 2.819 |
| 0.1 | 3.46 | 0.9859 | 539.2 | 2.972 | 0.661 | 181.1 | 21.29 | 2.888 |
| 0.2 | 7.02 | 0.9715 | 537.8 | 2.913 | 0.670 | 179.2 | 21.52 | 3.032 |
| 0.3 | 10.69 | 0.9568 | 536.3 | 2.877 | 0.680 | 177.2 | 21.77 | 3.189 |
| 0.4 | 14.48 | 0.9416 | 534.8 | 2.851 | 0.690 | 175.1 | 22.03 | 3.360 |
| 0.5 | 18.40 | 0.9261 | 533.3 | 2.830 | 0.702 | 172.9 | 22.32 | 3.547 |
| 0.6 | 22.47 | 0.9101 | 531.7 | 2.812 | 0.714 | 170.7 | 22.63 | 3.753 |
| 0.7 | 26.70 | 0.8936 | 530.0 | 2.796 | 0.727 | 168.4 | 22.96 | 3.982 |
| 0.8 | 31.11 | 0.8765 | 528.4 | 2.782 | 0.742 | 166.0 | 23.32 | 4.237 |
| 0.9 | 35.72 | 0.8588 | 526.6 | 2.769 | 0.759 | 163.5 | 23.72 | 4.524 |
| 1.0 | 40.56 | 0.8404 | 524.7 | 2.757 | 0.777 | 160.9 | 24.15 | 4.850 |
| 1.5 | 1190.17 | 0.0430 | 343.6 | 2.251 | 1.043 | 547.2 | —0.01 | 2.722 |
| 2.0 | 1193.09 | 0.0571 | 343.6 | 2.249 | 1.038 | 553.8 | —0.02 | 2.674 |
| 2.5 | 1195.94 | 0.0713 | 343.6 | 2.248 | 1.033 | 560.2 | —0.04 | 2.629 |
| 3.0 | 1198.71 | 0.0853 | 343.6 | 2.247 | 1.028 | 566.5 | —0.05 | 2.587 |
| 3.5 | 1201.42 | 0.0993 | 343.6 | 2.246 | 1.024 | 572.7 | —0.06 | 2.546 |
| 4.0 | 1204.06 | 0.1132 | 343.7 | 2.245 | 1.020 | 578.8 | —0.07 | 2.508 |
| 4.5 | 1206.64 | 0.1271 | 343.7 | 2.244 | 1.015 | 584.7 | —0.08 | 2.472 |
| 5.0 | 1209.17 | 0.1410 | 343.8 | 2.243 | 1.012 | 590.6 | —0.09 | 2.437 |
| 5.5 | 1211.64 | 0.1547 | 343.8 | 2.242 | 1.008 | 596.4 | —0.10 | 2.404 |
| 6.0 | 1214.06 | 0.1685 | 343.9 | 2.241 | 1.004 | 602.1 | —0.11 | 2.372 |
| 6.5 | 1216.44 | 0.1821 | 343.9 | 2.240 | 1.001 | 607.7 | —0.12 | 2.342 |
| 7.0 | 1218.76 | 0.1958 | 344.0 | 2.239 | 0.998 | 613.2 | —0.13 | 2.312 |
| 7.5 | 1221.04 | 0.2094 | 344.1 | 2.238 | 0.995 | 618.7 | —0.14 | 2.285 |
| 8.0 | 1223.28 | 0.2229 | 344.1 | 2.237 | 0.992 | 624.1 | —0.15 | 2.258 |
| 8.5 | 1225.48 | 0.2364 | 344.2 | 2.236 | 0.989 | 629.4 | —0.16 | 2.232 |
| 9.0 | 1227.63 | 0.2499 | 344.3 | 2.235 | 0.986 | 634.6 | —0.16 | 2.207 |
| 9.5 | 1229.76 | 0.2633 | 344.4 | 2.235 | 0.983 | 639.8 | —0.17 | 2.183 |
| 10.0 | 1231.84 | 0.2767 | 344.5 | 2.234 | 0.981 | 644.9 | —0.18 | 2.160 |
| 11.0 | 1235.91 | 0.3034 | 344.6 | 2.232 | 0.976 | 654.9 | —0.19 | 2.117 |
| 12.0 | 1239.84 | 0.3299 | 344.8 | 2.230 | 0.971 | 664.7 | —0.20 | 2.076 |
| 13.0 | 1243.66 | 0.3563 | 345.0 | 2.229 | 0.966 | 674.2 | —0.22 | 2.038 |
| 14.0 | 1247.36 | 0.3826 | 345.2 | 2.227 | 0.962 | 683.6 | —0.23 | 2.002 |
| 15.0 | 1250.97 | 0.4087 | 345.5 | 2.225 | 0.958 | 692.8 | —0.24 | 1.968 |
| 16.0 | 1254.47 | 0.4348 | 345.7 | 2.224 | 0.955 | 701.8 | —0.25 | 1.936 |
| 17.0 | 1257.88 | 0.4607 | 345.9 | 2.222 | 0.951 | 710.6 | —0.26 | 1.906 |
| 18.0 | 1261.20 | 0.4865 | 346.2 | 2.221 | 0.948 | 719.2 | —0.27 | 1.877 |
| 19.0 | 1264.45 | 0.5122 | 346.4 | 2.219 | 0.945 | 727.8 | —0.28 | 1.850 |
| 20.0 | 1267.61 | 0.5378 | 346.7 | 2.218 | 0.942 | 736.1 | —0.28 | 1.824 |

**TABLE 23.** Thermodynamic Properties of Freon-21 in the Single-Phase Region (*Continued*)

$$T = 373.15 \text{ K}$$

| $p$ | $\rho$ | $z$ | $h$ | $s$ | $c_p$ | $w$ | $\mu$ | $\alpha \cdot 10^3$ |
|------|--------|--------|-------|-------|-------|-------|--------|--------|
| 0.01 | 0.33 | 0.9987 | 547.0 | 3.178 | 0.662 | 185.1 | 19.64 | 2.691 |
| 0.02 | 0.67 | 0.9975 | 546.9 | 3.122 | 0.663 | 185.0 | 19.65 | 2.703 |
| 0.03 | 1.00 | 0.9962 | 546.8 | 3.089 | 0.663 | 184.8 | 19.67 | 2.715 |
| 0.04 | 1.33 | 0.9949 | 546.6 | 3.065 | 0.664 | 184.6 | 19.69 | 2.727 |
| 0.05 | 1.67 | 0.9936 | 546.5 | 3.047 | 0.665 | 184.4 | 19.71 | 2.739 |
| 0.1 | 3.36 | 0.9872 | 545.8 | 2.990 | 0.669 | 183.6 | 19.80 | 2.799 |
| 0.2 | 6.81 | 0.9741 | 544.5 | 2.931 | 0.677 | 181.8 | 20.00 | 2.927 |
| 0.3 | 10.36 | 0.9607 | 543.1 | 2.896 | 0.686 | 180.0 | 20.21 | 3.065 |
| 0.4 | 14.01 | 0.9470 | 541.7 | 2.870 | 0.695 | 178.1 | 20.43 | 3.213 |
| 0.5 | 17.78 | 0.9330 | 540.3 | 2.849 | 0.705 | 176.1 | 20.67 | 3.374 |
| 0.6 | 21.67 | 0.9186 | 538.8 | 2.832 | 0.715 | 174.1 | 20.92 | 3.549 |
| 0.7 | 25.69 | 0.9038 | 537.3 | 2.816 | 0.727 | 172.0 | 21.19 | 3.740 |
| 0.8 | 29.87 | 0.8886 | 535.7 | 2.802 | 0.739 | 169.9 | 21.48 | 3.951 |
| 0.9 | 34.20 | 0.8729 | 534.1 | 2.790 | 0.753 | 167.7 | 21.79 | 4.184 |
| 1.0 | 38.72 | 0.8567 | 532.4 | 2.778 | 0.768 | 165.3 | 22.13 | 4.444 |
| 1.5 | 1156.46 | 0.0430 | 354.2 | 2.279 | 1.080 | 505.5 | —0.11 | 3.039 |
| 2.0 | 1158.96 | 0.0572 | 354.1 | 2.278 | 1.073 | 512.7 | —0.09 | 2.973 |
| 2.5 | 1163.34 | 0.0713 | 354.1 | 2.277 | 1.066 | 519.8 | —0.07 | 2.909 |
| 3.0 | 1166.62 | 0.0853 | 354.0 | 2.276 | 1.060 | 526.8 | —0.05 | 2.851 |
| 3.5 | 1169.81 | 0.0993 | 354.0 | 2.274 | 1.054 | 533.5 | —0.04 | 2.796 |
| 4.0 | 1172.90 | 0.1131 | 354.0 | 2.273 | 1.049 | 540.2 | —0.02 | 2.744 |
| 4.5 | 1175.92 | 0.1270 | 354.0 | 2.272 | 1.043 | 546.7 | —0.00 | 2.695 |
| 5.0 | 1178.85 | 0.1407 | 354.0 | 2.271 | 1.038 | 553.0 | —0.01 | 2.649 |
| 5.5 | 1181.71 | 0.1544 | 354.0 | 2.270 | 1.034 | 559.3 | —0.02 | 2.606 |
| 6.0 | 1184.50 | 0.1680 | 354.0 | 2.269 | 1.029 | 565.4 | —0.04 | 2.565 |
| 6.5 | 1187.22 | 0.1816 | 354.0 | 2.268 | 1.025 | 571.5 | —0.05 | 2.526 |
| 7.0 | 1189.88 | 0.1952 | 354.1 | 2.267 | 1.021 | 577.4 | —0.06 | 2.488 |
| 7.5 | 1192.49 | 0.2086 | 354.1 | 2.266 | 1.017 | 583.2 | —0.07 | 2.453 |
| 8.0 | 1195.03 | 0.2221 | 354.1 | 2.265 | 1.013 | 589.0 | —0.08 | 2.419 |
| 8.5 | 1197.53 | 0.2355 | 354.2 | 2.264 | 1.010 | 594.6 | —0.09 | 2.387 |
| 9.0 | 1199.97 | 0.2488 | 354.2 | 2.263 | 1.006 | 600.2 | —0.10 | 2.356 |
| 9.5 | 1202.36 | 0.2621 | 354.3 | 2.262 | 1.003 | 605.7 | —0.11 | 2.326 |
| 10.0 | 1204.71 | 0.2754 | 354.3 | 2.261 | 1.000 | 611.1 | —0.12 | 2.297 |
| 11.0 | 1209.27 | 0.3018 | 354.5 | 2.259 | 0.994 | 621.7 | —0.14 | 2.244 |
| 12.0 | 1213.67 | 0.3280 | 354.6 | 2.257 | 0.988 | 632.0 | —0.15 | 2.194 |
| 13.0 | 1217.91 | 0.3541 | 354.8 | 2.255 | 0.983 | 642.1 | —0.17 | 2.148 |
| 14.0 | 1222.02 | 0.3801 | 354.9 | 2.253 | 0.978 | 651.9 | —0.18 | 2.105 |
| 15.0 | 1226.00 | 0.4059 | 355.1 | 2.252 | 0.974 | 661.5 | —0.19 | 2.065 |
| 16.0 | 1229.86 | 0.4316 | 355.3 | 2.250 | 0.970 | 670.9 | —0.20 | 2.027 |
| 17.0 | 1233.61 | 0.4572 | 355.5 | 2.248 | 0.966 | 680.1 | —0.22 | 1.992 |
| 18.0 | 1237.26 | 0.4826 | 355.7 | 2.247 | 0.962 | 689.1 | —0.23 | 1.958 |
| 19.0 | 1240.80 | 0.5080 | 356.0 | 2.245 | 0.958 | 697.9 | —0.24 | 1.926 |
| 20.0 | 1244.26 | 0.5332 | 356.2 | 2.244 | 0.955 | 706.6 | —0.25 | 1.896 |

**TABLE 23.** Thermodynamic Properties of Freon-21 in the Single-Phase Region (*Continued*)

$$T = 383.15 \text{ K}$$

| $p$ | $\rho$ | $z$ | $h$ | $s$ | $c_p$ | $w$ | $\mu$ | $\alpha \cdot 10^3$ |
|---|---|---|---|---|---|---|---|---|
| 0.01 | 0.32 | 0.9988 | 553.7 | 3.196 | 0.670 | 187.5 | 18.33 | 2.620 |
| 0.02 | 0.65 | 0.9977 | 553.5 | 3.139 | 0.671 | 187.3 | 18.35 | 2.631 |
| 0.03 | 0.97 | 0.9965 | 553.4 | 3.106 | 0.671 | 187.1 | 18.36 | 2.641 |
| 0.04 | 1.30 | 0.9953 | 553.3 | 3.083 | 0.672 | 187.0 | 18.38 | 2.652 |
| 0.05 | 1.62 | 0.9942 | 553.2 | 3.065 | 0.673 | 186.8 | 18.40 | 2.662 |
| 0.1 | 3.27 | 0.9883 | 552.6 | 3.008 | 0.676 | 186.0 | 18.48 | 2.717 |
| 0.2 | 6.62 | 0.9764 | 551.3 | 2.949 | 0.684 | 184.4 | 18.64 | 2.830 |
| 0.3 | 10.05 | 0.9642 | 550.0 | 2.914 | 0.691 | 182.7 | 18.82 | 2.951 |
| 0.4 | 13.58 | 0.9518 | 548.7 | 2.888 | 0.700 | 181.0 | 19.00 | 3.081 |
| 0.5 | 17.20 | 0.9391 | 547.3 | 2.868 | 0.708 | 179.2 | 19.20 | 3.221 |
| 0.6 | 20.93 | 0.9261 | 546.0 | 2.850 | 0.718 | 177.4 | 19.41 | 3.371 |
| 0.7 | 24.78 | 0.9128 | 544.6 | 2.835 | 0.728 | 175.5 | 19.63 | 3.534 |
| 0.8 | 28.75 | 0.8992 | 543.1 | 2.822 | 0.739 | 173.5 | 19.86 | 3.710 |
| 0.9 | 32.85 | 0.8852 | 541.6 | 2.809 | 0.750 | 171.5 | 20.11 | 3.904 |
| 1.0 | 37.10 | 0.8708 | 540.1 | 2.798 | 0.763 | 169.5 | 20.38 | 4.116 |
| 1.5 | 61.24 | 0.7913 | 531.6 | 2.748 | 0.850 | 157.8 | 22.07 | 5.608 |
| 2.0 | 1123.85 | 0.0575 | 365.1 | 2.307 | 1.119 | 469.0 | 0.23 | 3.374 |
| 2.5 | 1127.99 | 0.0716 | 364.9 | 2.306 | 1.109 | 476.9 | 0.21 | 3.281 |
| 3.0 | 1131.97 | 0.0856 | 364.8 | 2.304 | 1.101 | 484.7 | 0.18 | 3.196 |
| 3.5 | 1135.81 | 0.0996 | 364.7 | 2.303 | 1.092 | 492.2 | 0.16 | 3.118 |
| 4.0 | 1139.51 | 0.1134 | 364.7 | 2.301 | 1.085 | 499.5 | 0.14 | 3.045 |
| 4.5 | 1143.10 | 0.1272 | 364.6 | 2.300 | 1.078 | 506.7 | 0.11 | 2.978 |
| 5.0 | 1146.56 | 0.1409 | 364.5 | 2.299 | 1.071 | 513.7 | 0.10 | 2.915 |
| 5.5 | 1149.93 | 0.1545 | 364.5 | 2.298 | 1.065 | 520.5 | 0.08 | 2.856 |
| 6.0 | 1153.19 | 0.1681 | 364.5 | 2.296 | 1.059 | 527.2 | 0.06 | 2.801 |
| 6.5 | 1156.37 | 0.1816 | 364.4 | 2.295 | 1.053 | 533.7 | 0.04 | 2.749 |
| 7.0 | 1159.45 | 0.1951 | 364.4 | 2.294 | 1.048 | 540.2 | 0.03 | 2.700 |
| 7.5 | 1162.46 | 0.2085 | 364.4 | 2.293 | 1.043 | 546.4 | 0.01 | 2.654 |
| 8.0 | 1165.39 | 0.2218 | 364.4 | 2.292 | 1.039 | 552.6 | 0.00 | 2.610 |
| 8.5 | 1168.25 | 0.2351 | 364.4 | 2.291 | 1.034 | 558.7 | —0.01 | 2.569 |
| 9.0 | 1171.04 | 0.2483 | 364.4 | 2.289 | 1.030 | 564.6 | —0.03 | 2.529 |
| 9.5 | 1173.77 | 0.2615 | 364.4 | 2.288 | 1.026 | 570.5 | —0.04 | 2.492 |
| 10.0 | 1176.43 | 0.2746 | 364.4 | 2.287 | 1.022 | 576.3 | —0.05 | 2.456 |
| 11.0 | 1181.68 | 0.3008 | 364.5 | 2.285 | 1.015 | 587.5 | —0.07 | 2.389 |
| 12.0 | 1186.55 | 0.3268 | 364.6 | 2.283 | 1.008 | 598.4 | —0.09 | 2.328 |
| 13.0 | 1191.31 | 0.3526 | 364.7 | 2.281 | 1.002 | 609.0 | —0.11 | 2.272 |
| 14.0 | 1195.89 | 0.3782 | 364.8 | 2.280 | 0.996 | 619.4 | —0.12 | 2.220 |
| 15.0 | 1200.32 | 0.4038 | 364.9 | 2.278 | 0.991 | 629.4 | —0.14 | 2.172 |
| 16.0 | 1204.59 | 0.4291 | 365.1 | 2.276 | 0.986 | 639.3 | —0.16 | 2.127 |
| 17.0 | 1208.73 | 0.4544 | 365.2 | 2.274 | 0.981 | 648.9 | —0.17 | 2.085 |
| 18.0 | 1212.74 | 0.4795 | 365.4 | 2.272 | 0.977 | 658.3 | —0.18 | 2.046 |
| 19.0 | 1216.63 | 0.5046 | 365.6 | 2.271 | 0.973 | 667.4 | —0.20 | 2.009 |
| 20.0 | 1220.41 | 0.5295 | 365.8 | 2.269 | 0.969 | 676.4 | —0.21 | 1.974 |

**TABLE 23.** Thermodynamic Properties of Freon-21 in the Single-Phase Region (*Continued*)

$$T = 393.15 \text{ K}$$

| $p$ | $\rho$ | $z$ | $h$ | $s$ | $c_p$ | $w$ | $\mu$ | $\alpha \cdot 10^3$ |
|---|---|---|---|---|---|---|---|---|
| 0.01 | 0.32 | 0.9989 | 560.4 | 3.213 | 0.678 | 189.7 | 17.16 | 2.553 |
| 0.02 | 0.63 | 0.9979 | 560.3 | 3.157 | 0.679 | 189.6 | 17.18 | 2.562 |
| 0.03 | 0.95 | 0.9968 | 560.2 | 3.124 | 0.679 | 189.5 | 17.19 | 2.572 |
| 0.04 | 1.26 | 0.9957 | 560.0 | 3.100 | 0.680 | 189.3 | 17.20 | 2.581 |
| 0.05 | 1.58 | 0.9947 | 559.9 | 3.082 | 0.681 | 189.2 | 17.22 | 2.591 |
| 0.1 | 3.18 | 0.9893 | 559.3 | 3.025 | 0.684 | 188.4 | 17.28 | 2.639 |
| 0.2 | 6.44 | 0.9784 | 558.2 | 2.967 | 0.690 | 186.9 | 17.43 | 2.741 |
| 0.3 | 9.77 | 0.9673 | 556.9 | 2.932 | 0.697 | 185.3 | 17.58 | 2.848 |
| 0.4 | 13.17 | 0.9560 | 555.7 | 2.906 | 0.705 | 183.8 | 17.73 | 2.962 |
| 0.5 | 16.67 | 0.9445 | 554.4 | 2.886 | 0.713 | 182.1 | 17.89 | 3.084 |
| 0.6 | 20.25 | 0.9327 | 553.1 | 2.869 | 0.721 | 180.5 | 18.07 | 3.214 |
| 0.7 | 23.94 | 0.9207 | 551.8 | 2.854 | 0.730 | 178.8 | 18.25 | 3.354 |
| 0.8 | 27.73 | 0.9085 | 550.5 | 2.841 | 0.739 | 177.0 | 18.44 | 3.504 |
| 0.9 | 31.63 | 0.8959 | 549.1 | 2.829 | 0.749 | 175.2 | 18.64 | 3.667 |
| 1.0 | 35.66 | 0.8831 | 547.7 | 2.817 | 0.760 | 173.4 | 18.85 | 3.843 |
| 1.5 | 58.08 | 0.8132 | 540.0 | 2.770 | 0.830 | 163.1 | 20.16 | 3.016 |
| 2.0 | 1083.62 | 0.0581 | 376.5 | 2.337 | 1.182 | 422.0 | 0.43 | 3.957 |
| 2.5 | 1088.89 | 0.0723 | 376.3 | 2.335 | 1.167 | 431.2 | 0.39 | 3.809 |
| 3.0 | 1093.90 | 0.0864 | 376.1 | 2.333 | 1.154 | 440.1 | 0.35 | 3.676 |
| 3.5 | 1098.67 | 0.1003 | 375.9 | 2.331 | 1.142 | 448.6 | 0.32 | 3.556 |
| 4.0 | 1103.23 | 0.1142 | 375.7 | 2.330 | 1.131 | 456.9 | 0.28 | 3.448 |
| 4.5 | 1107.60 | 0.1279 | 375.6 | 2.328 | 1.121 | 464.9 | 0.26 | 3.350 |
| 5.0 | 1111.80 | 0.1416 | 375.4 | 2.327 | 1.111 | 472.6 | 0.23 | 3.260 |
| 5.5 | 1115.84 | 0.1552 | 375.3 | 2.325 | 1.103 | 480.2 | 0.20 | 3.177 |
| 6.0 | 1119.74 | 0.1687 | 375.2 | 2.324 | 1.095 | 487.5 | 0.18 | 3.100 |
| 6.5 | 1123.50 | 0.1822 | 375.1 | 2.323 | 1.087 | 494.7 | 0.16 | 3.029 |
| 7.0 | 1127.14 | 0.1955 | 375.0 | 2.321 | 1.080 | 501.7 | 0.14 | 2.963 |
| 7.5 | 1130.67 | 0.2089 | 375.0 | 2.320 | 1.074 | 508.6 | 0.12 | 2.901 |
| 8.0 | 1134.09 | 0.2221 | 374.9 | 2.319 | 1.068 | 515.2 | 0.10 | 2.843 |
| 8.5 | 1137.41 | 0.2353 | 374.9 | 2.318 | 1.062 | 521.8 | 0.08 | 2.788 |
| 9.0 | 1140.64 | 0.2484 | 374.8 | 2.316 | 1.057 | 528.2 | 0.06 | 2.737 |
| 9.5 | 1143.79 | 0.2615 | 374.8 | 2.315 | 1.052 | 534.5 | 0.05 | 2.689 |
| 10.0 | 1146.85 | 0.2746 | 374.8 | 2.314 | 1.047 | 540.7 | 0.03 | 2.643 |
| 11.0 | 1152.74 | 0.3005 | 374.8 | 2.312 | 1.038 | 552.7 | 0.00 | 2.559 |
| 12.0 | 1158.37 | 0.3262 | 374.8 | 2.310 | 1.030 | 564.3 | −0.02 | 2.483 |
| 13.0 | 1163.74 | 0.3517 | 374.8 | 2.307 | 1.023 | 575.5 | −0.04 | 2.413 |
| 14.0 | 1168.89 | 0.3771 | 374.9 | 2.305 | 1.016 | 586.4 | −0.06 | 2.350 |
| 15.0 | 1173.84 | 0.4024 | 374.9 | 2.303 | 1.010 | 597.0 | −0.08 | 2.292 |
| 16.0 | 1178.60 | 0.4274 | 375.0 | 2.301 | 1.004 | 607.3 | −0.10 | 2.238 |
| 17.0 | 1183.19 | 0.4524 | 375.1 | 2.300 | 0.998 | 617.4 | −0.12 | 2.188 |
| 18.0 | 1187.62 | 0.4772 | 375.3 | 2.298 | 0.993 | 627.2 | −0.13 | 2.142 |
| 19.0 | 1191.91 | 0.5019 | 375.4 | 2.296 | 0.989 | 636.7 | −0.15 | 2.099 |
| 20.0 | 1196.06 | 0.5265 | 375.6 | 2.294 | 0.984 | 646.1 | −0.16 | 2.058 |

**TABLE 23.** Thermodynamic Properties of Freon-21 in the Single-Phase Region (*Continued*)

$$T = 403.15 \text{ K}$$

| $p$ | $\rho$ | $z$ | $h$ | $s$ | $c_p$ | $w$ | $\mu$ | $\alpha \cdot 10^3$ |
|---|---|---|---|---|---|---|---|---|
| 0.01 | 0.31 | 0.9990 | 567.2 | 3.230 | 0.686 | 192.0 | 16.10 | 2.489 |
| 0.02 | 0.62 | 0.9980 | 567.1 | 3.174 | 0.686 | 191.9 | 16.12 | 2.497 |
| 0.03 | 0.92 | 0.9971 | 567.0 | 3.141 | 0.687 | 191.7 | 16.13 | 2.506 |
| 0.04 | 1.23 | 0.9961 | 566.9 | 3.117 | 0.687 | 191.6 | 16.14 | 2.514 |
| 0.05 | 1.54 | 0.9951 | 566.8 | 3.099 | 0.688 | 191.5 | 16.15 | 2.523 |
| 0.1 | 3.10 | 0.9902 | 566.2 | 3.042 | 0.691 | 190.8 | 16.21 | 2.567 |
| 0.2 | 6.27 | 0.9802 | 565.1 | 2.984 | 0.697 | 189.4 | 16.33 | 2.657 |
| 0.3 | 9.50 | 0.9701 | 563.9 | 2.949 | 0.703 | 188.0 | 16.46 | 2.753 |
| 0.4 | 12.80 | 0.9598 | 562.8 | 2.924 | 0.710 | 186.5 | 16.59 | 2.854 |
| 0.5 | 16.17 | 0.9493 | 561.6 | 2.904 | 0.717 | 185.0 | 16.72 | 2.961 |
| 0.6 | 19.63 | 0.9386 | 560.4 | 2.887 | 0.724 | 183.5 | 16.87 | 3.075 |
| 0.7 | 23.17 | 0.9278 | 559.1 | 2.872 | 0.732 | 181.9 | 17.02 | 3.196 |
| 0.8 | 26.80 | 0.9167 | 557.9 | 2.859 | 0.740 | 180.3 | 17.17 | 3.326 |
| 0.9 | 30.52 | 0.9054 | 556.6 | 2.847 | 0.749 | 178.7 | 17.34 | 3.464 |
| 1.0 | 34.35 | 0.8938 | 555.3 | 2.836 | 0.759 | 177.0 | 17.51 | 3.612 |
| 1.5 | 55.37 | 0.8318 | 548.2 | 2.791 | 0.817 | 167.9 | 18.53 | 4.560 |
| 2.0 | 80.80 | 0.7600 | 539.9 | 2.752 | 0.910 | 157.1 | 19.94 | 4.119 |
| 2.5 | 1044.25 | 0.0736 | 388.4 | 2.365 | 1.252 | 381.7 | 0.66 | 4.632 |
| 3.0 | 1050.90 | 0.0877 | 388.0 | 2.363 | 1.229 | 392.2 | 0.60 | 4.399 |
| 3.5 | 1057.12 | 0.1017 | 387.6 | 2.361 | 1.210 | 402.2 | 0.54 | 4.198 |
| 4.0 | 1062.98 | 0.1155 | 387.3 | 2.359 | 1.192 | 411.8 | 0.49 | 4.023 |
| 4.5 | 1068.51 | 0.1293 | 387.0 | 2.357 | 1.177 | 420.9 | 0.44 | 3.868 |
| 5.0 | 1073.76 | 0.1430 | 386.8 | 2.355 | 1.163 | 429.7 | 0.40 | 3.730 |
| 5.5 | 1078.76 | 0.1566 | 386.6 | 2.354 | 1.151 | 438.2 | 0.37 | 3.606 |
| 6.0 | 1083.53 | 0.1700 | 386.4 | 2.352 | 1.139 | 446.5 | 0.33 | 3.494 |
| 6.5 | 1088.10 | 0.1834 | 386.2 | 2.350 | 1.129 | 454.4 | 0.30 | 3.392 |
| 7.0 | 1092.48 | 0.1967 | 386.0 | 2.349 | 1.119 | 462.1 | 0.27 | 3.298 |
| 7.5 | 1096.70 | 0.2100 | 385.9 | 2.347 | 1.111 | 469.7 | 0.24 | 3.212 |
| 8.0 | 1100.77 | 0.2232 | 385.8 | 2.346 | 1.103 | 477.0 | 0.22 | 3.133 |
| 8.5 | 1104.69 | 0.2363 | 385.6 | 2.345 | 1.095 | 484.1 | 0.19 | 3.060 |
| 9.0 | 1108.48 | 0.2493 | 385.5 | 2.343 | 1.088 | 491.1 | 0.17 | 2.991 |
| 9.5 | 1112.16 | 0.2623 | 385.5 | 2.342 | 1.082 | 497.9 | 0.15 | 2.928 |
| 10.0 | 1115.72 | 0.2752 | 385.4 | 2.341 | 1.075 | 504.5 | 0.13 | 2.868 |
| 11.0 | 1122.53 | 0.3009 | 385.3 | 2.338 | 1.064 | 517.4 | 0.09 | 2.759 |
| 12.0 | 1128.97 | 0.3264 | 385.2 | 2.336 | 1.054 | 529.8 | 0.06 | 2.663 |
| 13.0 | 1135.09 | 0.3517 | 385.1 | 2.333 | 1.045 | 541.8 | 0.03 | 2.576 |
| 14.0 | 1140.92 | 0.3788 | 385.1 | 2.331 | 1.037 | 553.3 | 0.01 | 2.498 |
| 15.0 | 1146.48 | 0.4017 | 385.1 | 2.329 | 1.029 | 564.5 | —0.02 | 2.427 |
| 16.0 | 1151.81 | 0.4265 | 385.1 | 2.327 | 1.023 | 575.3 | —0.04 | 2.362 |
| 17.0 | 1156.93 | 0.4512 | 385.2 | 2.325 | 1.016 | 585.9 | —0.06 | 2.302 |
| 18.0 | 1161.85 | 0.4757 | 385.3 | 2.323 | 1.011 | 596.1 | —0.08 | 2.247 |
| 19.0 | 1166.59 | 0.5001 | 385.4 | 2.321 | 1.005 | 606.1 | —0.10 | 2.196 |
| 20.0 | 1171.17 | 0.5244 | 385.5 | 2.319 | 1.000 | 615.8 | —0.11 | 2.149 |

**TABLE 23.** Thermodynamic Properties of Freon-21 in the Single-Phase Region (*Continued*)

$$T = 413.15 \text{ K}$$

| $p$ | $\rho$ | $z$ | $h$ | $s$ | $c_p$ | $w$ | $\mu$ | $\alpha \cdot 10^3$ |
|---|---|---|---|---|---|---|---|---|
| 0.01 | 0.30 | 0.9991 | 574.1 | 3.247 | 0.693 | 194.2 | 15.15 | 2.428 |
| 0.02 | 0.60 | 0.9982 | 574.0 | 3.191 | 0.694 | 194.1 | 15.16 | 2.436 |
| 0.03 | 0.90 | 0.9973 | 573.9 | 3.158 | 0.694 | 194.0 | 15.17 | 2.443 |
| 0.04 | 1.20 | 0.9964 | 537.8 | 3.134 | 0.695 | 193.9 | 15.18 | 2.451 |
| 0.05 | 1.50 | 0.9955 | 573.7 | 3.116 | 0.695 | 193.7 | 15.19 | 2.459 |
| 0.1 | 3.02 | 0.9910 | 573.1 | 3.059 | 0.698 | 193.1 | 15.24 | 2.498 |
| 0.2 | 6.10 | 0.9819 | 572.1 | 3.001 | 0.704 | 191.8 | 15.34 | 2.580 |
| 0.3 | 9.24 | 0.9726 | 571.0 | 2.967 | 0.709 | 190.5 | 15.45 | 2.666 |
| 0.4 | 12.44 | 0.9632 | 569.9 | 2.941 | 0.715 | 189.2 | 15.56 | 2.756 |
| 0.5 | 15.71 | 0.9536 | 568.8 | 2.921 | 0.722 | 187.8 | 15.67 | 2.850 |
| 0.6 | 19.05 | 0.9439 | 567.6 | 2.905 | 0.728 | 186.4 | 15.79 | 2.951 |
| 0.7 | 22.45 | 0.9341 | 566.5 | 2.890 | 0.735 | 185.0 | 15.91 | 3.056 |
| 0.8 | 25.94 | 0.9240 | 565.3 | 2.877 | 0.743 | 183.5 | 16.04 | 3.168 |
| 0.9 | 29.51 | 0.9138 | 564.1 | 2.866 | 0.750 | 182.1 | 16.18 | 3.288 |
| 1.0 | 33.17 | 0.9034 | 562.9 | 2.855 | 0.759 | 180.5 | 16.32 | 3.414 |
| 1.5 | 53.00 | 0.8480 | 556.3 | 2.810 | 0.808 | 172.4 | 17.13 | 4.197 |
| 2.0 | 76.30 | 0.7854 | 548.9 | 2.773 | 0.882 | 163.0 | 18.19 | 5.382 |
| 2.5 | 105.28 | 0.7115 | 540.0 | 2.738 | 1.007 | 151.7 | 19.68 | 7.469 |
| 3.0 | 1000.05 | 0.0899 | 400.8 | 2.394 | 1.353 | 339.6 | 0.99 | 5.654 |
| 3.5 | 1008.83 | 0.1040 | 400.2 | 2.392 | 1.314 | 351.9 | 0.88 | 5.253 |
| 4.0 | 1016.86 | 0.1179 | 399.6 | 2.389 | 1.283 | 363.4 | 0.79 | 4.925 |
| 4.5 | 1024.26 | 0.1316 | 399.2 | 2.387 | 1.257 | 374.2 | 0.72 | 4.652 |
| 5.0 | 1031.14 | 0.1453 | 398.7 | 2.385 | 1.234 | 384.5 | 0.65 | 4.419 |
| 5.5 | 1037.58 | 0.1588 | 398.4 | 2.382 | 1.214 | 394.3 | 0.59 | 4.217 |
| 6.0 | 1043.64 | 0.1723 | 398.0 | 2.380 | 1.197 | 493.7 | 0.54 | 4.041 |
| 6.5 | 1049.36 | 0.1856 | 397.7 | 2.379 | 1.182 | 412.7 | 0.49 | 3.885 |
| 7.0 | 1054.80 | 0.1988 | 397.4 | 2.377 | 1.168 | 421.4 | 0.45 | 3.746 |
| 7.5 | 1059.97 | 0.2120 | 397.2 | 2.375 | 1.156 | 429.8 | 0.41 | 3.621 |
| 8.0 | 1064.91 | 0.2251 | 397.0 | 2.373 | 1.145 | 437.9 | 0.37 | 3.508 |
| 8.5 | 1069.63 | 0.2381 | 396.8 | 2.372 | 1.134 | 445.7 | 0.34 | 3.405 |
| 9.0 | 1074.17 | 0.2510 | 396.6 | 2.370 | 1.125 | 453.4 | 0.30 | 3.311 |
| 9.5 | 1078.53 | 0.2639 | 396.4 | 2.369 | 1.116 | 460.8 | 0.28 | 3.224 |
| 10.0 | 1082.73 | 0.2767 | 396.3 | 2.367 | 1.108 | 468.0 | 0.25 | 3.144 |
| 11.0 | 1090.70 | 0.3022 | 396.0 | 2.364 | 1.094 | 482.0 | 0.20 | 3.001 |
| 12.0 | 1098.17 | 0.3274 | 395.8 | 2.362 | 1.081 | 495.3 | 0.16 | 2.876 |
| 13.0 | 1105.20 | 0.3524 | 395.7 | 2.359 | 1.070 | 508.0 | 0.12 | 2.766 |
| 14.0 | 1111.84 | 0.3773 | 395.6 | 2.357 | 1.060 | 520.3 | 0.09 | 2.668 |
| 15.0 | 1118.15 | 0.4019 | 395.5 | 2.354 | 1.051 | 532.1 | 0.06 | 2.581 |
| 16.0 | 1124.16 | 0.4265 | 395.5 | 2.352 | 1.043 | 543.5 | 0.03 | 2.502 |
| 17.0 | 1129.89 | 0.4508 | 395.5 | 2.350 | 1.035 | 554.6 | 0.00 | 2.430 |
| 18.0 | 1135.38 | 0.4750 | 395.5 | 2.348 | 1.029 | 565.3 | −0.02 | 2.364 |
| 19.0 | 1140.65 | 0.4991 | 395.5 | 2.346 | 1.022 | 575.7 | −0.04 | 2.304 |
| 20.0 | 1145.71 | 0.5230 | 395.5 | 2.344 | 1.017 | 585.9 | −0.06 | 2.248 |

**TABLE 23.** Thermodynamic Properties of Freon-21 in the Single-Phase Region (*Continued*)

$$T = 423.15 \ K$$

| $p$ | $\rho$ | $z$ | $h$ | $s$ | $c_p$ | $w$ | $\mu$ | $\alpha \cdot 10^3$ |
|---|---|---|---|---|---|---|---|---|
| 0.01 | 0.29 | 0.9992 | 581.1 | 3.263 | 0.701 | 196.5 | 14.28 | 2.370 |
| 0.02 | 0.59 | 0.9984 | 581.0 | 3.207 | 0.701 | 196.3 | 14.28 | 2.377 |
| 0.03 | 0.88 | 0.9975 | 580.9 | 3.174 | 0.702 | 196.2 | 14.29 | 2.384 |
| 0.04 | 1.17 | 0.9967 | 580.8 | 3.151 | 0.702 | 196.1 | 14.30 | 2.391 |
| 0.05 | 1.47 | 0.9959 | 580.7 | 3.133 | 0.703 | 196.0 | 14.31 | 2.398 |
| 0.1 | 2.95 | 0.9917 | 580.2 | 3.076 | 0.705 | 195.4 | 14.35 | 2.434 |
| 0.2 | 5.95 | 0.9834 | 579.1 | 3.018 | 0.710 | 194.2 | 14.44 | 2.507 |
| 0.3 | 9.00 | 0.9749 | 578.1 | 2.984 | 0.715 | 193.0 | 14.53 | 2.584 |
| 0.4 | 12.11 | 0.9663 | 577.1 | 2.959 | 0.721 | 191.8 | 14.62 | 2.665 |
| 0.5 | 15.28 | 0.9575 | 576.0 | 2.939 | 0.727 | 190.5 | 14.72 | 2.749 |
| 0.6 | 18.50 | 0.9487 | 574.9 | 2.922 | 0.733 | 189.2 | 14.82 | 2.838 |
| 0.7 | 21.79 | 0.9397 | 573.8 | 2.908 | 0.739 | 187.9 | 14.92 | 2.931 |
| 0.8 | 25.15 | 0.9306 | 572.7 | 2.895 | 0.745 | 186.6 | 15.03 | 3.029 |
| 0.9 | 28.58 | 0.9213 | 571.6 | 2.884 | 0.752 | 185.3 | 15.14 | 3.132 |
| 1.0 | 32.08 | 0.9118 | 570.4 | 2.873 | 0.760 | 183.9 | 15.26 | 3.242 |
| 1.5 | 50.90 | 0.8621 | 564.4 | 2.830 | 0.802 | 176.5 | 15.91 | 3.899 |
| 2.0 | 72.51 | 0.8069 | 557.6 | 2.794 | 0.862 | 168.3 | 16.72 | 4.833 |
| 2.5 | 98.31 | 0.7439 | 549.8 | 2.762 | 0.954 | 158.7 | 17.78 | 6.303 |
| 3.0 | 131.34 | 0.6682 | 540.3 | 2.729 | 1.123 | 147.1 | 19.27 | 9.080 |
| 3.5 | 948.42 | 0.1080 | 414.2 | 2.425 | 1.515 | 295.0 | 1.48 | 7.404 |
| 4.0 | 960.88 | 0.1218 | 413.2 | 2.421 | 1.441 | 309.9 | 1.30 | 6.603 |
| 4.5 | 971.79 | 0.1355 | 412.3 | 2.418 | 1.386 | 323.5 | 1.14 | 6.007 |
| 5.0 | 981.55 | 0.1490 | 411.6 | 2.415 | 1.343 | 336.1 | 1.02 | 5.544 |
| 5.5 | 990.40 | 0.1625 | 410.9 | 2.412 | 1.308 | 347.8 | 0.92 | 5.170 |
| 6.0 | 998.51 | 0.1758 | 410.4 | 2.410 | 1.278 | 358.9 | 0.83 | 4.861 |
| 6.5 | 1006.02 | 0.1890 | 409.9 | 2.408 | 1.254 | 369.3 | 0.75 | 4.601 |
| 7.0 | 1013.01 | 0.2022 | 409.4 | 2.405 | 1.232 | 379.3 | 0.68 | 4.377 |
| 7.5 | 1019.57 | 0.2152 | 409.0 | 2.403 | 1.214 | 388.8 | 0.62 | 4.183 |
| 8.0 | 1025.75 | 0.2282 | 408.7 | 2.401 | 1.197 | 397.9 | 0.57 | 4.012 |
| 8.5 | 1031.59 | 0.2410 | 408.3 | 2.399 | 1.183 | 406.6 | 0.52 | 3.860 |
| 9.0 | 1037.14 | 0.2539 | 408.1 | 2.397 | 1.169 | 415.1 | 0.48 | 3.725 |
| 9.5 | 1042.42 | 0.2666 | 407.8 | 2.396 | 1.157 | 423.3 | 0.44 | 3.602 |
| 10.0 | 1047.47 | 0.2793 | 407.6 | 2.394 | 1.147 | 431.2 | 0.40 | 3.491 |
| 11.0 | 1056.95 | 0.3045 | 407.1 | 2.391 | 1.127 | 446.3 | 0.33 | 3.297 |
| 12.0 | 1065.71 | 0.3294 | 406.8 | 2.388 | 1.111 | 460.7 | 0.28 | 3.132 |
| 13.0 | 1073.88 | 0.3541 | 406.5 | 2.385 | 1.097 | 474.4 | 0.22 | 2.990 |
| 14.0 | 1081.52 | 0.3787 | 406.3 | 2.382 | 1.085 | 487.4 | 0.18 | 2.867 |
| 15.0 | 1088.73 | 0.4031 | 406.1 | 2.380 | 1.074 | 500.0 | 0.14 | 2.757 |
| 16.0 | 1095.53 | 0.4273 | 406.0 | 2.377 | 1.064 | 512.1 | 0.11 | 2.660 |
| 17.0 | 1101.99 | 0.4513 | 405.9 | 2.375 | 1.055 | 523.7 | 0.08 | 2.573 |
| 18.0 | 1108.15 | 0.4752 | 405.8 | 2.372 | 1.047 | 535.0 | 0.05 | 2.494 |
| 19.0 | 1114.02 | 0.4989 | 405.8 | 2.370 | 1.040 | 545.9 | 0.02 | 2.422 |
| 20.0 | 1119.64 | 0.5226 | 405.8 | 2.368 | 1.033 | 556.4 | −0.00 | 2.357 |

**TABLE 23.** Thermodynamic Properties of Freon-21 in the Single-Phase Region (*Continued*)

$$T = 433.15 \text{ K}$$

| $p$ | $\rho$ | $z$ | $h$ | $s$ | $c_p$ | $w$ | $\mu$ | $\alpha \cdot 10^3$ |
|---|---|---|---|---|---|---|---|---|
| 0.01 | 0.29 | 0.9992 | 588.1 | 3.280 | 0.708 | 198.6 | 13.48 | 2.315 |
| 0.02 | 0.57 | 0.9985 | 588.0 | 3.224 | 0.708 | 198.5 | 13.49 | 2.321 |
| 0.03 | 0.86 | 0.9977 | 587.9 | 3.191 | 0.709 | 198.4 | 13.49 | 2.328 |
| 0.04 | 1.15 | 0.9970 | 587.8 | 3.167 | 0.709 | 198.3 | 13.50 | 2.334 |
| 0.05 | 1.43 | 0.9962 | 587.7 | 3.149 | 0.710 | 198.2 | 13.51 | 2.340 |
| 0.1 | 2.88 | 0.9924 | 587.2 | 3.092 | 0.712 | 197.7 | 13.54 | 2.373 |
| 0.2 | 5.88 | 0.9847 | 586.3 | 3.035 | 0.717 | 196.6 | 13.62 | 2.439 |
| 0.3 | 8.78 | 0.9769 | 585.3 | 3.000 | 0.721 | 195.4 | 13.70 | 2.509 |
| 0.4 | 11.80 | 0.9690 | 584.3 | 2.975 | 0.726 | 194.3 | 13.77 | 2.581 |
| 0.5 | 14.87 | 0.9611 | 583.3 | 2.956 | 0.732 | 193.2 | 13.86 | 2.657 |
| 0.6 | 17.99 | 0.9530 | 582.3 | 2.939 | 0.737 | 192.0 | 13.94 | 2.735 |
| 0.7 | 21.18 | 0.9448 | 581.2 | 2.925 | 0.743 | 190.8 | 14.03 | 2.818 |
| 0.8 | 24.42 | 0.9364 | 580.2 | 2.912 | 0.749 | 189.6 | 14.12 | 2.904 |
| 0.9 | 27.72 | 0.9280 | 579.1 | 2.901 | 0.755 | 188.3 | 14.21 | 2.995 |
| 1.0 | 31.08 | 0.9194 | 578.0 | 2.891 | 0.761 | 187.1 | 14.30 | 3.090 |
| 1.5 | 49.02 | 0.8745 | 572.4 | 2.848 | 0.799 | 180.5 | 14.83 | 3.649 |
| 2.0 | 69.24 | 0.8255 | 566.1 | 2.814 | 0.848 | 173.1 | 15.46 | 4.406 |
| 2.5 | 92.70 | 0.7708 | 559.2 | 2.783 | 0.919 | 164.9 | 16.25 | 5.506 |
| 3.0 | 121.12 | 0.7079 | 551.0 | 2.754 | 1.033 | 155.3 | 17.26 | 7.292 |
| 3.5 | 159.49 | 0.6311 | 540.9 | 2.722 | 1.256 | 143.6 | 18.64 | 10.875 |
| 4.0 | 883.39 | 0.1294 | 429.2 | 2.459 | 1.848 | 246.0 | 2.36 | 11.193 |
| 4.5 | 903.51 | 0.1423 | 427.3 | 2.453 | 1.661 | 265.6 | 1.95 | 9.070 |
| 5.0 | 919.68 | 0.1554 | 425.9 | 2.448 | 1.546 | 282.4 | 1.67 | 7.779 |
| 5.5 | 933.34 | 0.1684 | 424.7 | 2.445 | 1.466 | 297.4 | 1.45 | 6.895 |
| 6.0 | 945.22 | 0.1814 | 423.7 | 2.441 | 1.407 | 311.0 | 1.28 | 6.244 |
| 6.5 | 955.80 | 0.1944 | 422.9 | 2.438 | 1.361 | 323.6 | 1.14 | 5.741 |
| 7.0 | 965.34 | 0.2072 | 422.2 | 2.435 | 1.324 | 335.3 | 1.03 | 5.338 |
| 7.5 | 974.07 | 0.2201 | 421.5 | 2.432 | 1.293 | 346.4 | 0.93 | 5.006 |
| 8.0 | 982.11 | 0.2328 | 421.0 | 2.430 | 1.267 | 356.8 | 0.84 | 4.727 |
| 8.5 | 989.59 | 0.2455 | 420.5 | 2.428 | 1.245 | 366.7 | 0.77 | 4.489 |
| 9.0 | 996.59 | 0.2581 | 420.0 | 2.425 | 1.225 | 376.2 | 0.70 | 4.282 |
| 9.5 | 1003.16 | 0.2706 | 419.6 | 2.423 | 1.208 | 385.3 | 0.64 | 4.101 |
| 10.0 | 1009.37 | 0.2831 | 419.2 | 2.421 | 1.193 | 394.1 | 0.59 | 3.941 |
| 11.0 | 1020.85 | 0.3080 | 418.6 | 2.417 | 1.167 | 410.7 | 0.50 | 3.669 |
| 12.0 | 1031.29 | 0.3325 | 418.1 | 2.414 | 1.146 | 426.3 | 0.42 | 3.446 |
| 13.0 | 1040.88 | 0.3569 | 417.6 | 2.411 | 1.127 | 441.0 | 0.35 | 3.259 |
| 14.0 | 1049.77 | 0.3811 | 417.3 | 2.408 | 1.112 | 455.0 | 0.29 | 3.100 |
| 15.0 | 1058.06 | 0.4052 | 417.0 | 2.405 | 1.099 | 468.3 | 0.24 | 2.962 |
| 16.0 | 1065.83 | 0.4290 | 416.7 | 2.402 | 1.087 | 481.1 | 0.20 | 2.841 |
| 17.0 | 1073.16 | 0.4527 | 416.5 | 2.400 | 1.076 | 493.4 | 0.16 | 2.734 |
| 18.0 | 1080.08 | 0.4763 | 416.4 | 2.397 | 1.067 | 505.2 | 0.12 | 2.639 |
| 19.0 | 1086.66 | 0.4997 | 416.3 | 2.395 | 1.058 | 516.6 | 0.09 | 2.553 |
| 20.0 | 1092.92 | 0.5230 | 416.2 | 2.392 | 1.051 | 527.7 | 0.06 | 2.475 |

**TABLE 23.** Thermodynamic Properties of Freon-21 in the Single-Phase Region (*Continued*)

$$T = 443.15 \text{ K}$$

| $p$ | $\rho$ | $z$ | $h$ | $s$ | $c_p$ | $w$ | $\mu$ | $\alpha \cdot 10^3$ |
|---|---|---|---|---|---|---|---|---|
| 0.01 | 0.28 | 0.9993 | 595.2 | 3.296 | 0.715 | 200.8 | 12.75 | 2.262 |
| 0.02 | 0.56 | 0.9986 | 595.1 | 3.240 | 0.715 | 200.7 | 12.76 | 2.268 |
| 0.03 | 0.84 | 0.9979 | 595.0 | 3.207 | 0.716 | 200.6 | 12.76 | 2.274 |
| 0.04 | 1.12 | 0.9972 | 594.9 | 3.183 | 0.716 | 200.5 | 12.77 | 2.280 |
| 0.05 | 1.40 | 0.9965 | 594.8 | 3.165 | 0.716 | 200.4 | 12.77 | 2.286 |
| 0.1 | 2.81 | 0.9930 | 594.4 | 3.109 | 0.718 | 199.9 | 12.81 | 2.315 |
| 0.2 | 5.67 | 0.9859 | 593.5 | 3.051 | 0.723 | 198.9 | 12.87 | 2.376 |
| 0.3 | 8.56 | 0.9788 | 592.5 | 3.017 | 0.727 | 197.9 | 12.93 | 2.438 |
| 0.4 | 11.50 | 0.9716 | 591.6 | 2.992 | 0.732 | 196.8 | 13.00 | 2.504 |
| 0.5 | 14.49 | 0.9642 | 590.6 | 2.972 | 0.737 | 195.7 | 13.07 | 2.571 |
| 0.6 | 17.52 | 0.9568 | 589.6 | 2.956 | 0.742 | 194.7 | 13.14 | 2.642 |
| 0.7 | 20.60 | 0.9493 | 588.7 | 2.942 | 0.747 | 193.6 | 13.21 | 2.715 |
| 0.8 | 23.73 | 0.9417 | 587.7 | 2.929 | 0.752 | 192.5 | 13.29 | 2.792 |
| 0.9 | 26.92 | 0.9340 | 586.7 | 2.918 | 0.758 | 191.3 | 13.36 | 2.872 |
| 1.0 | 30.16 | 0.9263 | 585.7 | 2.908 | 0.763 | 190.2 | 13.44 | 2.955 |
| 1.5 | 47.31 | 0.8856 | 580.3 | 2.866 | 0.797 | 184.2 | 13.87 | 3.436 |
| 2.0 | 66.38 | 0.8417 | 574.6 | 2.833 | 0.839 | 177.6 | 14.97 | 4.063 |
| 2.5 | 88.00 | 0.7936 | 568.2 | 2.804 | 0.896 | 170.4 | 14.97 | 4.920 |
| 3.0 | 113.27 | 0.7399 | 561.1 | 2.777 | 0.979 | 162.3 | 15.69 | 6.182 |
| 3.5 | 144.23 | 0.6779 | 552.7 | 2.749 | 1.114 | 152.9 | 16.58 | 8.272 |
| 4.0 | 185.69 | 0.6017 | 542.1 | 2.718 | 1.392 | 141.7 | 17.73 | 12.597 |
| 4.5 | 782.35 | 0.1607 | 448.4 | 2.501 | 3.183 | 186.9 | 4.47 | 27.380 |
| 5.0 | 828.16 | 0.1687 | 443.7 | 2.489 | 2.159 | 217.6 | 3.16 | 15.014 |
| 5.5 | 856.19 | 0.1794 | 440.9 | 2.481 | 1.831 | 239.9 | 2.51 | 11.139 |
| 6.0 | 877.13 | 0.1911 | 438.9 | 2.476 | 1.658 | 258.3 | 2.09 | 9.127 |
| 6.5 | 894.07 | 0.2051 | 437.3 | 2.471 | 1.548 | 274.4 | 1.80 | 7.863 |
| 7.0 | 908.42 | 0.2153 | 436.1 | 2.467 | 1.471 | 288.9 | 1.57 | 6.983 |
| 7.5 | 920.93 | 0.2275 | 435.0 | 2.463 | 1.413 | 302.2 | 1.39 | 6.328 |
| 8.0 | 932.06 | 0.2398 | 434.1 | 2.460 | 1.367 | 314.4 | 1.24 | 5.819 |
| 8.5 | 942.11 | 0.2520 | 433.3 | 2.457 | 1.330 | 325.9 | 1.12 | 5.409 |
| 9.0 | 951.28 | 0.2643 | 432.6 | 2.454 | 1.299 | 336.7 | 1.01 | 5.071 |
| 9.5 | 959.74 | 0.2765 | 432.0 | 2.451 | 1.273 | 347.0 | 0.92 | 4.787 |
| 10.0 | 967.59 | 0.2887 | 431.4 | 2.449 | 1.251 | 356.8 | 0.84 | 4.543 |
| 11.0 | 981.83 | 0.3130 | 430.5 | 2.445 | 1.214 | 375.2 | 0.70 | 4.147 |
| 12.0 | 994.49 | 0.3371 | 429.7 | 2.441 | 1.185 | 392.2 | 0.59 | 3.836 |
| 13.0 | 1005.93 | 0.3610 | 429.1 | 2.437 | 1.162 | 408.1 | 0.50 | 3.584 |
| 14.0 | 1016.37 | 0.3848 | 428.5 | 2.433 | 1.142 | 423.1 | 0.43 | 3.376 |
| 15.0 | 1025.99 | 0.4084 | 428.1 | 2.430 | 1.125 | 437.3 | 0.36 | 3.200 |
| 16.0 | 1034.93 | 0.4319 | 427.7 | 2.427 | 1.111 | 450.8 | 0.31 | 3.048 |
| 17.0 | 1043.28 | 0.4552 | 427.4 | 2.424 | 1.098 | 463.8 | 0.26 | 2.917 |
| 18.0 | 1051.12 | 0.4784 | 427.2 | 2.422 | 1.087 | 476.2 | 0.21 | 2.801 |
| 19.0 | 1058.51 | 0.5014 | 427.0 | 2.419 | 1.077 | 488.1 | 0.17 | 2.698 |
| 20.0 | 1065.52 | 0.5243 | 426.8 | 2.416 | 1.068 | 499.7 | 0.14 | 2.606 |

**TABLE 23.** Thermodynamic Properties of Freon-21 in the Single-Phase Region (*Continued*)

$$T = 453.15 \ \text{K}$$

| $p$ | $\rho$ | $z$ | $h$ | $s$ | $c_p$ | $w$ | $\mu$ | $\alpha \cdot 10^3$ |
|---|---|---|---|---|---|---|---|---|
| 0.01 | 0.27 | 0.9994 | 602.4 | 3.312 | 0.721 | 202.9 | 12.08 | 2.212 |
| 0.02 | 0.55 | 0.9987 | 602.3 | 3.256 | 0.722 | 202.9 | 12.08 | 2.217 |
| 0.03 | 0.82 | 0.9981 | 602.2 | 3.223 | 0.722 | 202.8 | 12.09 | 2.223 |
| 0.04 | 1.10 | 0.9974 | 602.1 | 3.199 | 0.723 | 202.7 | 12.10 | 2.228 |
| 0.05 | 1.37 | 0.9968 | 602.0 | 3.181 | 0.723 | 202.6 | 12.10 | 2.233 |
| 0.1 | 2.75 | 0.9936 | 601.6 | 3.125 | 0.725 | 202.1 | 12.13 | 2.260 |
| 0.2 | 5.54 | 0.9871 | 600.7 | 3.067 | 0.729 | 201.2 | 12.18 | 2.315 |
| 0.3 | 8.36 | 0.9805 | 599.8 | 3.033 | 0.733 | 200.2 | 12.24 | 2.372 |
| 0.4 | 11.22 | 0.9738 | 598.9 | 3.008 | 0.737 | 199.3 | 12.29 | 2.431 |
| 0.5 | 14.12 | 0.9671 | 598.0 | 2.989 | 0.742 | 198.3 | 12.35 | 2.492 |
| 0.6 | 17.07 | 0.9603 | 597.1 | 2.973 | 0.746 | 197.3 | 12.41 | 2.556 |
| 0.7 | 20.06 | 0.9535 | 596.2 | 2.959 | 0.751 | 196.3 | 12.47 | 2.621 |
| 0.8 | 23.09 | 0.9465 | 595.2 | 2.946 | 0.756 | 195.3 | 12.53 | 2.689 |
| 0.9 | 26.17 | 0.9395 | 594.3 | 2.935 | 0.761 | 194.2 | 12.60 | 2.760 |
| 1.0 | 29.30 | 0.9324 | 593.3 | 2.925 | 0.766 | 193.2 | 12.66 | 2.834 |
| 1.5 | 45.76 | 0.8955 | 588.3 | 2.884 | 0.796 | 187.7 | 13.01 | 3.252 |
| 2.0 | 63.83 | 0.8560 | 582.9 | 2.852 | 0.832 | 181.8 | 13.41 | 3.779 |
| 2.5 | 83.97 | 0.8133 | 577.1 | 2.824 | 0.880 | 175.4 | 13.87 | 4.468 |
| 3.0 | 106.91 | 0.7666 | 570.7 | 2.798 | 0.944 | 168.4 | 14.41 | 5.415 |
| 3.5 | 133.83 | 0.7144 | 563.4 | 2.773 | 1.038 | 160.6 | 15.03 | 6.812 |
| 4.0 | 166.97 | 0.6545 | 554.9 | 2.747 | 1.192 | 151.8 | 15.75 | 9.122 |
| 4.5 | 211.44 | 0.5814 | 544.2 | 2.717 | 1.504 | 141.6 | 16.55 | 13.817 |
| 5.0 | 285.49 | 0.4784 | 528.1 | 2.677 | 2.597 | 129.4 | 17.11 | 30.199 |
| 5.5 | 699.98 | 0.2146 | 466.4 | 2.538 | 4.568 | 162.4 | 6.21 | 45.992 |
| 6.0 | 771.65 | 0.2124 | 458.5 | 2.519 | 2.467 | 196.3 | 4.03 | 19.117 |
| 6.5 | 808.71 | 0.2196 | 454.6 | 2.509 | 1.987 | 219.8 | 3.10 | 13.183 |
| 7.0 | 834.77 | 0.2291 | 452.0 | 2.502 | 1.758 | 239.0 | 2.54 | 10.413 |
| 7.5 | 855.17 | 0.2396 | 450.1 | 2.497 | 1.619 | 255.7 | 2.15 | 8.769 |
| 8.0 | 872.07 | 0.2506 | 448.5 | 2.492 | 1.526 | 270.6 | 1.86 | 7.666 |
| 8.5 | 886.56 | 0.2619 | 447.2 | 2.488 | 1.457 | 284.1 | 1.64 | 6.867 |
| 9.0 | 899.29 | 0.2734 | 446.1 | 2.484 | 1.404 | 296.7 | 1.45 | 6.257 |
| 9.5 | 910.67 | 0.2850 | 445.1 | 2.481 | 1.362 | 308.4 | 1.30 | 5.775 |
| 10.0 | 920.97 | 0.2966 | 444.3 | 2.478 | 1.327 | 319.5 | 1.18 | 5.382 |
| 11.0 | 939.12 | 0.3200 | 442.9 | 2.472 | 1.273 | 339.9 | 0.97 | 4.777 |
| 12.0 | 954.80 | 0.3433 | 441.8 | 2.467 | 1.233 | 358.5 | 0.82 | 4.330 |
| 13.0 | 968.64 | 0.3666 | 440.9 | 2.463 | 1.201 | 375.7 | 0.69 | 3.984 |
| 14.0 | 981.05 | 0.3898 | 440.1 | 2.459 | 1.176 | 391.8 | 0.59 | 3.707 |
| 15.0 | 992.34 | 0.4129 | 439.5 | 2.456 | 1.155 | 407.0 | 0.50 | 3.479 |
| 16.0 | 1002.69 | 0.4359 | 439.0 | 2.452 | 1.137 | 421.3 | 0.43 | 3.287 |
| 17.0 | 1012.26 | 0.4588 | 438.5 | 2.449 | 1.121 | 435.0 | 0.37 | 3.124 |
| 18.0 | 1021.18 | 0.4815 | 438.1 | 2.446 | 1.108 | 448.0 | 0.31 | 2.982 |
| 19.0 | 1029.53 | 0.5042 | 437.8 | 2.443 | 1.096 | 460.5 | 0.26 | 2.858 |
| 20.0 | 1037.38 | 0.5267 | 437.6 | 2.440 | 1.086 | 472.6 | 0.22 | 2.749 |

**TABLE 23.** Thermodynamic Properties of Freon-21 in the Single-Phase Region (*Continued*)

$$T = 463.15 \ \text{K}$$

| $p$ | $\rho$ | $z$ | $h$ | $s$ | $c_p$ | $w$ | $\mu$ | $\alpha \cdot 10^3$ |
|---|---|---|---|---|---|---|---|---|
| 0.01 | 0.27 | 0.9994 | 609.6 | 3.328 | 0.728 | 205.1 | 11.46 | 2.164 |
| 0.02 | 0.54 | 0.9988 | 609.5 | 3.271 | 0.728 | 205.0 | 11.47 | 2.169 |
| 0.03 | 0.80 | 0.9982 | 609.5 | 3.239 | 0.729 | 204.9 | 11.47 | 2.174 |
| 0.04 | 1.07 | 0.9976 | 609.4 | 3.215 | 0.729 | 204.8 | 11.48 | 2.178 |
| 0.05 | 1.34 | 0.9970 | 609.3 | 3.197 | 0.729 | 204.7 | 11.48 | 2.183 |
| 0.1 | 2.69 | 0.9941 | 608.9 | 3.140 | 0.731 | 204.3 | 11.50 | 2.208 |
| 0.2 | 5.41 | 0.9881 | 608.0 | 3.083 | 0.735 | 203.4 | 11.55 | 2.258 |
| 0.3 | 8.17 | 0.9820 | 607.2 | 3.049 | 0.739 | 202.5 | 11.60 | 2.310 |
| 0.4 | 10.95 | 0.9759 | 606.3 | 3.024 | 0.743 | 201.7 | 11.64 | 2.364 |
| 0.5 | 13.78 | 0.9698 | 605.4 | 3.005 | 0.747 | 200.8 | 11.69 | 2.419 |
| 0.6 | 16.64 | 0.9635 | 604.6 | 2.989 | 0.751 | 199.8 | 11.74 | 2.476 |
| 0.7 | 19.55 | 0.9573 | 603.7 | 2.975 | 0.755 | 198.9 | 11.79 | 2.535 |
| 0.8 | 22.49 | 0.9509 | 602.8 | 2.963 | 0.760 | 198.0 | 11.84 | 2.596 |
| 0.9 | 25.47 | 0.9445 | 601.9 | 2.952 | 0.764 | 197.0 | 11.90 | 2.659 |
| 1.0 | 28.50 | 0.9380 | 601.0 | 2.942 | 0.769 | 196.1 | 11.95 | 2.725 |
| 1.5 | 44.33 | 0.9044 | 596.2 | 2.901 | 0.796 | 191.1 | 12.24 | 3.091 |
| 2.0 | 61.54 | 0.8687 | 591.2 | 2.870 | 0.828 | 185.8 | 12.56 | 3.541 |
| 2.5 | 80.45 | 0.8305 | 585.8 | 2.843 | 0.868 | 180.1 | 12.92 | 4.107 |
| 3.0 | 101.58 | 0.7894 | 580.0 | 2.818 | 0.920 | 173.9 | 13.32 | 4.846 |
| 3.5 | 125.65 | 0.7445 | 573.5 | 2.795 | 0.990 | 167.2 | 13.77 | 5.857 |
| 4.0 | 153.90 | 0.6947 | 566.2 | 2.771 | 1.092 | 160.0 | 14.26 | 7.334 |
| 4.5 | 188.53 | 0.6380 | 557.8 | 2.747 | 1.256 | 152.0 | 14.77 | 9.709 |
| 5.0 | 234.21 | 0.5706 | 547.3 | 2.719 | 1.565 | 143.3 | 15.20 | 14.190 |
| 5.5 | 303.87 | 0.4838 | 532.9 | 2.684 | 2.367 | 134.3 | 15.14 | 25.663 |
| 6.0 | 454.80 | 0.3526 | 507.1 | 2.625 | 5.901 | 130.9 | 12.27 | 73.260 |
| 6.5 | 648.12 | 0.2681 | 481.3 | 2.567 | 3.875 | 158.3 | 6.74 | 38.710 |
| 7.0 | 722.79 | 0.2689 | 472.7 | 2.547 | 2.527 | 185.4 | 4.63 | 20.431 |
| 7.5 | 765.24 | 0.2620 | 468.1 | 2.533 | 2.053 | 207.1 | 3.58 | 14.286 |
| 8.0 | 795.13 | 0.2689 | 465.0 | 2.528 | 1.811 | 225.6 | 2.92 | 11.234 |
| 8.5 | 818.34 | 0.2776 | 462.7 | 2.522 | 1.662 | 241.9 | 2.46 | 9.397 |
| 9.0 | 837.40 | 0.2873 | 460.9 | 2.516 | 1.561 | 256.5 | 2.13 | 8.162 |
| 9.5 | 853.62 | 0.2975 | 459.3 | 2.512 | 1.486 | 270.0 | 1.87 | 7.270 |
| 10.0 | 867.79 | 0.3080 | 458.1 | 2.508 | 1.429 | 282.5 | 1.66 | 6.593 |
| 11.0 | 891.72 | 0.3297 | 456.0 | 2.501 | 1.347 | 305.2 | 1.34 | 5.626 |
| 12.0 | 911.57 | 0.3519 | 454.4 | 2.495 | 1.289 | 325.6 | 1.10 | 4.964 |
| 13.0 | 928.59 | 0.3742 | 453.1 | 2.490 | 1.246 | 344.3 | 0.93 | 4.478 |
| 14.0 | 943.54 | 0.3966 | 452.1 | 2.485 | 1.213 | 361.5 | 0.79 | 4.104 |
| 15.0 | 956.88 | 0.4190 | 451.2 | 2.481 | 1.186 | 377.7 | 0.67 | 3.806 |
| 16.0 | 968.96 | 0.4413 | 450.5 | 2.477 | 1.164 | 392.9 | 0.58 | 3.562 |
| 17.0 | 980.00 | 0.4636 | 449.8 | 2.474 | 1.146 | 407.2 | 0.50 | 3.358 |
| 18.0 | 990.19 | 0.4859 | 449.3 | 2.470 | 1.130 | 420.9 | 0.42 | 3.185 |
| 19.0 | 999.65 | 0.5080 | 448.9 | 2.467 | 1.116 | 434.0 | 0.36 | 3.035 |
| 20.0 | 1008.48 | 0.5301 | 448.5 | 2.464 | 1.104 | 446.5 | 0.31 | 2.905 |

**TABLE 23.** Thermodynamic Properties of Freon-21 in the Single-Phase Region (*Continued*)

$$T = 473. \ 15 \ K$$

| $p$ | $\rho$ | $z$ | $h$ | $s$ | $c_p$ | $w$ | $\mu$ | $\alpha \cdot 10^3$ |
|------|--------|--------|-------|-------|-------|-------|-------|---------|
| 0.01 | 0.26 | 0.9995 | 616.9 | 3.343 | 0.734 | 207.2 | 10.89 | 2.118 |
| 0.02 | 0.52 | 0.9989 | 616.9 | 3.287 | 0.735 | 207.1 | 10.90 | 2.122 |
| 0.03 | 0.79 | 0.9984 | 616.8 | 3.254 | 0.735 | 207.0 | 10.90 | 2.127 |
| 0.04 | 1.05 | 0.9978 | 616.7 | 3.231 | 0.735 | 206.9 | 10.90 | 2.131 |
| 0.05 | 1.31 | 0.9973 | 616.6 | 3.213 | 0.736 | 206.8 | 10.91 | 2.136 |
| 0.1 | 2.63 | 0.9945 | 616.2 | 3.156 | 0.737 | 206.4 | 10.93 | 2.158 |
| 0.2 | 5.29 | 0.9890 | 615.4 | 3.099 | 0.741 | 205.6 | 10.97 | 2.204 |
| 0.3 | 7.98 | 0.9835 | 614.6 | 3.065 | 0.744 | 204.8 | 11.01 | 2.252 |
| 0.4 | 10.70 | 0.9778 | 613.8 | 3.040 | 0.748 | 204.0 | 11.05 | 2.300 |
| 0.5 | 13.46 | 0.9722 | 612.9 | 3.021 | 0.752 | 203.2 | 11.09 | 2.351 |
| 0.6 | 16.24 | 0.9665 | 612.1 | 3.005 | 0.756 | 202.3 | 11.13 | 2.402 |
| 0.7 | 19.06 | 0.9607 | 611.3 | 2.991 | 0.760 | 201.5 | 11.17 | 2.455 |
| 0.8 | 21.92 | 0.9549 | 610.4 | 2.979 | 0.764 | 200.6 | 11.22 | 2.510 |
| 0.9 | 24.81 | 0.9490 | 609.5 | 2.968 | 0.768 | 199.8 | 11.26 | 2.567 |
| 1.0 | 27.74 | 0.9430 | 608.7 | 2.958 | 0.772 | 198.9 | 11.30 | 2.625 |
| 1.5 | 43.01 | 0.9124 | 604.2 | 2.918 | 0.796 | 194.4 | 11.54 | 2.949 |
| 2.0 | 59.46 | 0.8801 | 599.5 | 2.887 | 0.825 | 189.5 | 11.80 | 3.336 |
| 2.5 | 77.33 | 0.8458 | 594.4 | 2.861 | 0.859 | 184.4 | 12.08 | 3.810 |
| 3.0 | 96.99 | 0.8092 | 589.1 | 2.838 | 0.902 | 179.0 | 12.39 | 4.405 |
| 3.5 | 118.93 | 0.7700 | 583.2 | 2.816 | 0.957 | 173.2 | 12.72 | 5.176 |
| 4.0 | 143.87 | 0.7274 | 576.8 | 2.794 | 1.032 | 167.0 | 13.07 | 6.214 |
| 4.5 | 172.98 | 0.6806 | 569.7 | 2.772 | 1.137 | 160.4 | 13.42 | 7.692 |
| 5.0 | 208.20 | 0.6283 | 561.4 | 2.749 | 1.298 | 153.5 | 13.71 | 9.948 |
| 5.5 | 253.13 | 0.5685 | 551.6 | 2.724 | 1.572 | 146.4 | 13.83 | 13.745 |
| 6.0 | 313.14 | 0.4981 | 539.3 | 2.694 | 2.096 | 140.2 | 13.45 | 20.892 |
| 6.5 | 408.25 | 0.4166 | 523.0 | 2.657 | 3.082 | 137.7 | 11.85 | 33.619 |
| 7.0 | 530.13 | 0.3455 | 505.1 | 2.616 | 3.583 | 145.9 | 8.72 | 37.731 |
| 7.5 | 626.78 | 0.3131 | 492.7 | 2.588 | 2.895 | 163.9 | 6.36 | 26.500 |
| 8.0 | 687.97 | 0.3042 | 485.5 | 2.572 | 2.330 | 183.4 | 4.81 | 18.421 |
| 8.5 | 729.61 | 0.3048 | 480.9 | 2.560 | 2.002 | 201.6 | 3.84 | 13.971 |
| 9.0 | 760.64 | 0.3096 | 477.6 | 2.552 | 1.800 | 218.0 | 3.18 | 11.318 |
| 9.5 | 785.27 | 0.3165 | 475.8 | 2.545 | 1.665 | 233.1 | 2.70 | 9.585 |
| 10.0 | 805.68 | 0.3247 | 473.0 | 2.540 | 1.569 | 247.0 | 2.34 | 8.370 |
| 11.0 | 838.34 | 0.3433 | 469.9 | 2.530 | 1.440 | 271.9 | 1.83 | 6.777 |
| 12.0 | 864.05 | 0.3634 | 467.6 | 2.523 | 1.357 | 294.1 | 1.48 | 5.777 |
| 13.0 | 885.32 | 0.3842 | 465.8 | 2.517 | 1.299 | 314.1 | 1.22 | 5.087 |
| 14.0 | 983.51 | 0.4054 | 464.4 | 2.512 | 1.255 | 332.5 | 1.03 | 4.579 |
| 15.0 | 919.43 | 0.4268 | 463.2 | 2.507 | 1.221 | 349.6 | 0.87 | 4.188 |
| 16.0 | 933.61 | 0.4484 | 462.2 | 2.502 | 1.194 | 365.6 | 0.75 | 3.877 |
| 17.0 | 946.41 | 0.4700 | 461.4 | 2.498 | 1.172 | 380.7 | 0.64 | 3.623 |
| 18.0 | 958.09 | 0.4915 | 460.7 | 2.495 | 1.153 | 395.0 | 0.56 | 3.410 |
| 19.0 | 968.83 | 0.5131 | 460.1 | 2.491 | 1.137 | 408.6 | 0.48 | 3.230 |
| 20.0 | 978.79 | 0.5346 | 459.6 | 2.488 | 1.122 | 421.6 | 0.41 | 3.074 |

**TABLE 24.** Transport Properties of Freon-21 on the Saturation Line

| $T$ | $p$ | $\eta' \cdot 10^6$ | $\eta'' \cdot 10^6$ | $\nu' \cdot 10^6$ | $\nu'' \cdot 10^6$ | $\lambda' \cdot 10^6$ | $\lambda'' \cdot 10^3$ | $a' \cdot 10^6$ | $a'' \cdot 10^6$ | $Pr'$ | $Pr''$ |
|---|---|---|---|---|---|---|---|---|---|---|---|
| 303.15 | 0.2150 | 294 | 11.41 | 0.218 | 1.225 | 97.6 | 9.44 | 0.0738 | 1.586 | 2.96 | 0.773 |
| 305.15 | 0.2296 | 288 | 11.50 | 0.214 | 1.160 | 96.9 | 9.56 | 0.0737 | 1.500 | 2.91 | 0.773 |
| 307.15 | 0.2448 | 283 | 11.58 | 0.211 | 1.100 | 96.2 | 9.68 | 0.0736 | 1.421 | 2.87 | 0.774 |
| 309.15 | 0.2607 | 277 | 11.66 | 0.208 | 1.043 | 95.5 | 9.80 | 0.0734 | 1.347 | 2.83 | 0.775 |
| 311.15 | 0.2774 | 272 | 11.74 | 0.204 | 0.991 | 94.8 | 9.92 | 0.0732 | 1.278 | 2.80 | 0.775 |
| 313.15 | 0.2948 | 267 | 11.82 | 0.201 | 0.942 | 94.1 | 10.04 | 0.0730 | 1.214 | 2.76 | 0.776 |
| 315.15 | 0.3130 | 262 | 11.91 | 0.1986 | 0.896 | 93.4 | 10.16 | 0.0728 | 1.154 | 2.73 | 0.777 |
| 317.15 | 0.3321 | 258 | 11.99 | 0.1958 | 0.853 | 92.7 | 10.29 | 0.0726 | 1.098 | 2.70 | 0.777 |
| 319.15 | 0.3520 | 253 | 12.07 | 0.1931 | 0.813 | 91.8 | 10.41 | 0.0723 | 1.045 | 2.68 | 0.778 |
| 321.15 | 0.3728 | 248 | 12.16 | 0.1902 | 0.775 | 91.0 | 10.54 | 0.0720 | 0.995 | 2.65 | 0.779 |
| 323.15 | 0.3945 | 244 | 12.24 | 0.1879 | 0.739 | 90.5 | 10.66 | 0.0716 | 0.947 | 2.63 | 0.781 |
| 325.15 | 0.4172 | 240 | 12.33 | 0.1854 | 0.706 | 89.9 | 10.79 | 0.0713 | 0.903 | 2.61 | 0.781 |
| 327.15 | 0.4408 | 236 | 12.42 | 0.1830 | 0.674 | 89.2 | 10.92 | 0.0710 | 0.861 | 2.59 | 0.783 |
| 329.15 | 0.4655 | 232 | 12.50 | 0.1807 | 0.644 | 88.6 | 11.05 | 0.0706 | 0.822 | 2.57 | 0.784 |
| 331.15 | 0.4912 | 228 | 12.59 | 0.1784 | 0.616 | 87.6 | 11.18 | 0.0699 | 0.784 | 2.56 | 0.786 |
| 333.15 | 0.5180 | 225 | 12.68 | 0.1762 | 0.589 | 86.9 | 11.32 | 0.0694 | 0.748 | 2.54 | 0.788 |
| 335.15 | 0.5459 | 221 | 12.77 | 0.1741 | 0.564 | 86.2 | 11.45 | 0.0690 | 0.715 | 2.53 | 0.790 |
| 337.15 | 0.5749 | 217 | 12.86 | 0.1719 | 0.540 | 85.6 | 11.59 | 0.0686 | 0.683 | 2.51 | 0.791 |
| 339.15 | 0.6051 | 214 | 12.95 | 0.1699 | 0.518 | 84.9 | 11.72 | 0.0682 | 0.653 | 2.49 | 0.793 |
| 341.15 | 0.6365 | 210 | 13.05 | 0.1678 | 0.496 | 84.2 | 11.86 | 0.0677 | 0.624 | 2.48 | 0.795 |
| 343.15 | 0.6691 | 207 | 13.14 | 0.1658 | 0.476 | 83.6 | 12.00 | 0.0672 | 0.597 | 2.46 | 0.797 |
| 345.15 | 0.7030 | 203 | 13.24 | 0.1639 | 0.457 | 82.9 | 12.14 | 0.0667 | 0.571 | 2.45 | 0.800 |
| 347.15 | 0.7381 | 200 | 13.34 | 0.1619 | 0.439 | 82.3 | 12.29 | 0.0662 | 0.546 | 2.44 | 0.803 |

**TABLE 24.** Transport Properties of Freon-21 on the Saturation Line (*Continued*)

| T | p | $\eta'\cdot10^6$ | $\eta''\cdot10^6$ | $\nu'\cdot10^6$ | $\nu''\cdot10^6$ | $\lambda'\cdot10^6$ | $\lambda''\cdot10^6$ | $a'\cdot10^6$ | $a''\cdot10^6$ | Pr' | Pr" |
|---|---|---|---|---|---|---|---|---|---|---|---|
| 349.15 | 0.7745 | 196.8 | 13.43 | 0.1600 | 0.421 | 81.6 | 12.43 | 0.0657 | 0.524 | 2.43 | 0.805 |
| 351.15 | 0.8123 | 193.6 | 13.53 | 0.1581 | 0.405 | 81.0 | 12.58 | 0.0652 | 0.501 | 2.42 | 0.808 |
| 353.15 | 0.8514 | 190.3 | 13.64 | 0.1563 | 0.389 | 80.4 | 12.73 | 0.0647 | 0.480 | 2.41 | 0.811 |
| 355.15 | 0.8919 | 187.1 | 13.74 | 0.1544 | 0.375 | 79.7 | 12.88 | 0.0642 | 0.460 | 2.40 | 0.815 |
| 357.15 | 0.9338 | 184.0 | 13.85 | 0.1525 | 0.360 | 78.8 | 13.03 | 0.0635 | 0.441 | 2.40 | 0.818 |
| 359.15 | 0.9772 | 180.8 | 13.95 | 0.1507 | 0.347 | 78.2 | 13.19 | 0.0629 | 0.422 | 2.39 | 0.822 |
| 361.15 | 1.0220 | 177.7 | 14.06 | 0.1488 | 0.334 | 77.6 | 13.35 | 0.0624 | 0.405 | 2.38 | 0.826 |
| 363.15 | 1.0684 | 174.6 | 14.18 | 0.1470 | 0.322 | 77.0 | 13.51 | 0.0619 | 0.388 | 2.37 | 0.830 |
| 365.15 | 1.1162 | 171.6 | 14.29 | 0.1452 | 0.310 | 76.3 | 13.67 | 0.0614 | 0.372 | 2.36 | 0.834 |
| 367.15 | 1.1657 | 167.5 | 14.41 | 0.1426 | 0.299 | 75.7 | 13.83 | 0.0608 | 0.357 | 2.34 | 0.839 |
| 369.15 | 1.2167 | 163.6 | 14.53 | 0.1400 | 0.289 | 75.1 | 14.00 | 0.0603 | 0.342 | 2.32 | 0.845 |
| 371.15 | 1.2693 | 159.9 | 14.65 | 0.1376 | 0.279 | 74.5 | 14.17 | 0.0597 | 0.328 | 2.30 | 0.850 |
| 373.15 | 1.3236 | 156.3 | 14.78 | 0.1353 | 0.269 | 74.1 | 14.35 | 0.0592 | 0.315 | 2.28 | 0.856 |
| 375.15 | 1.3795 | 152.8 | 14.91 | 0.1330 | 0.260 | 73.5 | 14.52 | 0.0585 | 0.302 | 2.27 | 0.852 |
| 377.15 | 1.4372 | 149.5 | 15.04 | 0.1309 | 0.251 | 72.9 | 14.70 | 0.0580 | 0.289 | 2.25 | 0.868 |
| 379.15 | 1.4966 | 146.3 | 15.18 | 0.1289 | 0.243 | 72.3 | 14.89 | 0.0575 | 0.277 | 2.24 | 0.876 |
| 381.15 | 1.5557 | 143.2 | 15.32 | 0.1269 | 0.235 | 71.7 | 15.07 | 0.0569 | 0.266 | 2.23 | 0.883 |
| 383.15 | 1.6207 | 140.2 | 15.46 | 0.1251 | 0.227 | 71.2 | 15.26 | 0.0563 | 0.255 | 2.22 | 0.891 |
| 385.15 | 1.6855 | 137.3 | 15.61 | 0.1233 | 0.220 | 70.3 | 15.46 | 0.0555 | 0.244 | 2.22 | 0.900 |
| 387.15 | 1.7522 | 134.5 | 15.76 | 0.1216 | 0.213 | 69.7 | 15.66 | 0.0549 | 0.234 | 2.21 | 0.909 |
| 389.15 | 1.8207 | 131.8 | 15.92 | 0.1199 | 0.206 | 69.2 | 15.86 | 0.0543 | 0.224 | 2.21 | 0.919 |
| 391.15 | 1.8912 | 129.1 | 16.09 | 0.1184 | 0.1998 | 68.6 | 16.07 | 0.0537 | 0.215 | 2.20 | 0.929 |
| 393.15 | 1.9637 | 126.6 | 16.26 | 0.1168 | 0.1936 | 68.1 | 16.28 | 0.0531 | 0.203 | 2.20 | 0.911 |
| 395.15 | 2.038 | 124.1 | 16.11 | 0.1154 | 0.1877 | 67.5 | 16.50 | 0.0525 | 0.1969 | 2.20 | 0.953 |
| 397.15 | 2.115 | 121.6 | 16.62 | 0.1140 | 0.1820 | 67.0 | 16.73 | 0.0519 | 0.18982 | 2.20 | 0.967 |

**TABLE 24.** Transport Properties of Freon-21 on the Saturation Line (*Continued*)

| $T$ | $p$ | $\eta' \cdot 10^6$ | $\eta'' \cdot 10^6$ | $\nu' \cdot 10^6$ | $\nu'' \cdot 10^6$ | $\lambda' \cdot 10^6$ | $\lambda'' \cdot 10^6$ | $a' \cdot 10^6$ | $a'' \cdot 10^6$ | $Pr'$ | $Pr''$ |
|---|---|---|---|---|---|---|---|---|---|---|---|
| 399.15 | 2.193 | 119.3 | 16.81 | 0.1126 | 0.1766 | 66.5 | 16.96 | 0.0512 | 0.1800 | 2.20 | 0.981 |
| 401.15 | 2.274 | 116.9 | 17.01 | 0.1113 | 0.1713 | 66.0 | 17.19 | 0.0505 | 0.1719 | 2.20 | 0.997 |
| 403.15 | 2.357 | 114.6 | 17.22 | 0.1100 | 0.1663 | 65.4 | 17.44 | 0.0498 | 0.1639 | 2.21 | 1.014 |
| 405.15 | 2.442 | 112.4 | 17.44 | 0.1087 | 0.1615 | 64.9 | 17.69 | 0.0491 | 0.1564 | 2.21 | 1.032 |
| 407.15 | 2.530 | 110.2 | 17.67 | 0.1075 | 0.1569 | 64.3 | 17.95 | 0.0483 | 0.1489 | 2.22 | 1.054 |
| 409.15 | 2.620 | 107.9 | 17.91 | 0.1063 | 0.1524 | 63.8 | 18.21 | 0.0476 | 0.1417 | 2.23 | 1.076 |
| 411.15 | 2.712 | 105.8 | 18.16 | 0.1051 | 0.1481 | 63.2 | 18.49 | 0.0467 | 0.1345 | 2.25 | 1.101 |
| 413.15 | 2.806 | 103.6 | 18.43 | 0.1039 | 0.1439 | 62.5 | 18.78 | 0.0458 | 0.1276 | 2.27 | 1.128 |
| 415.15 | 2.903 | 101.4 | 18.71 | 0.1028 | 0.1399 | 61.9 | 19.08 | 0.0449 | 0.1207 | 2.29 | 1.159 |
| 417.15 | 3.003 | 99.2 | 19.00 | 0.1016 | 0.1361 | 61.3 | 19.39 | 0.0439 | 0.1141 | 2.31 | 1.193 |
| 419.15 | 3.105 | 97.0 | 19.31 | 0.1005 | 0.1324 | 60.8 | 19.71 | 0.0429 | 0.1075 | 2.34 | 1.232 |
| 421.15 | 3.210 | 94.8 | 19.65 | 0.0993 | 0.1288 | 60.2 | 20.1 | 0.0419 | 0.1010 | 2.37 | 1.275 |
| 423.15 | 3.317 | 92.6 | 20.0 | 0.0981 | 0.1253 | 59.6 | 20.4 | 0.0408 | 0.0946 | 2.40 | 1.325 |
| 425.15 | 3.428 | 90.3 | 20.4 | 0.0969 | 0.1219 | 58.9 | 20.8 | 0.0396 | 0.0882 | 2.45 | 1.382 |
| 427.15 | 3.541 | 88.0 | 20.8 | 0.0957 | 0.1186 | 58.3 | 21.2 | 0.0382 | 0.0819 | 2.51 | 1.448 |
| 429.15 | 3.657 | 85.6 | 21.2 | 0.0945 | 0.1154 | 57.6 | 21.6 | 0.0368 | 0.0757 | 2.57 | 1.526 |
| 431.15 | 3.776 | 83.2 | 21.7 | 0.0932 | 0.1124 | 56.9 | 22.0 | 0.0353 | 0.0694 | 2.64 | 1.618 |
| 433.15 | 3.898 | 80.7 | 22.2 | 0.0919 | 0.1093 | 56.2 | 22.5 | 0.0336 | 0.0632 | 2.74 | 1.731 |

**TABLE 25.** Transport Properties of Freon-21 in the Single-Phase Region

| | $T=303.15$ K | | | | $T=313.15$ K | | | |
|---|---|---|---|---|---|---|---|---|
| $p$ | $\eta \cdot 10^6$ | $\lambda \cdot 10^6$ | $a \cdot 10^6$ | Pr | $\eta \cdot 10^6$ | $\lambda \cdot 10^6$ | $a \cdot 10^6$ | Pr |
| 0.1 | 11.28 | 9.20 | 3.59 | 0.755 | 11.64 | 9.72 | 3.85 | 0.745 |
| 0.5 | 295 | 98.0 | 0.0741 | 2.94 | 268 | 94.3 | 0.0733 | 2.76 |
| 1.0 | 296 | 98.3 | 0.0744 | 2.94 | 269 | 94.6 | 0.0736 | 2.75 |
| 2.0 | 299 | 98.8 | 0.0748 | 2.95 | 272 | 95.2 | 0.0741 | 2.75 |
| 3.0 | 302 | 99.4 | 0.0753 | 2.95 | 274 | 95.8 | 0.0746 | 2.75 |
| 4.0 | 305 | 99.9 | 0.0758 | 2.96 | 277 | 96.4 | 0.0753 | 2.75 |
| 5.0 | 308 | 100.5 | 0.0763 | 2.96 | 280 | 97.0 | 0.0758 | 2.76 |
| 6.0 | 310 | 101.1 | 0.0768 | 2.96 | 283 | 97.6 | 0.0764 | 2.76 |
| 8.0 | 316 | 102.2 | 0.0778 | 2.97 | 288 | 98.8 | 0.0775 | 2.76 |
| 10.0 | 322 | 103.1 | 0.0786 | 2.98 | 293 | 99.8 | 0.0783 | 2.76 |
| 12.0 | 327 | 103.9 | 0.0793 | 2.99 | 298 | 100.6 | 0.0791 | 2.77 |
| 14.0 | 333 | 104.5 | 0.0799 | 3.00 | 304 | 101.3 | 0.0798 | 2.79 |
| 16.0 | 338 | 105.0 | 0.0804 | 3.02 | 309 | 101.8 | 0.0803 | 2.81 |
| 18.0 | 344 | 105.4 | 0.0808 | 3.06 | 314 | 102.1 | 0.0806 | 2.84 |
| 20.0 | 349 | 105.6 | 0.0810 | 3.09 | 319 | 102.4 | 0.0808 | 2.87 |

**TABLE 25.** Transport Properties of Freon-21 in the Single-Phase Region (*Continued*)

| | $T=323.15$ K | | | | $T=333.15$ K | | | |
|---|---|---|---|---|---|---|---|---|
| $p$ | $\eta \cdot 10^6$ | $\lambda \cdot 10^6$ | $a \cdot 10^6$ | Pr | $\eta \cdot 10^6$ | $\lambda \cdot 10^6$ | $a \cdot 10^6$ | Pr |
| 0.1 | 12.00 | 10.23 | 4.16 | 0.738 | 12.35 | 10.75 | 4.45 | 0.733 |
| 0.5 | 245 | 90.7 | 0.0718 | 2.62 | 12.66 | 11.31 | 0.783 | 0.781 |
| 1.0 | 246 | 91.0 | 0.0721 | 2.62 | 226 | 87.4 | 0.0700 | 2.52 |
| 2.0 | 250 | 91.6 | 0.0726 | 2.63 | 228 | 88.1 | 0.0706 | 2.52 |
| 3.0 | 251 | 92.3 | 0.0733 | 2.63 | 231 | 88.8 | 0.0714 | 2.52 |
| 4.0 | 254 | 92.9 | 0.0738 | 2.62 | 233 | 89.5 | 0.0720 | 2.51 |
| 5.0 | 256 | 93.6 | 0.0745 | 2.62 | 236 | 90.2 | 0.0727 | 2.51 |
| 6.0 | 259 | 94.2 | 0.0751 | 2.62 | 238 | 90.9 | 0.0734 | 2.51 |
| 8.0 | 264 | 95.4 | 0.0763 | 2.62 | 243 | 92.1 | 0.0746 | 2.51 |
| 10.0 | 269 | 96.5 | 0.0770 | 2.62 | 248 | 93.3 | 0.0758 | 2.51 |
| 12.0 | 274 | 97.4 | 0.0782 | 2.62 | 252 | 94.2 | 0.0766 | 2.51 |
| 14.0 | 279 | 98.0 | 0.0787 | 2.64 | 257 | 94.9 | 0.0773 | 2.52 |
| 16.0 | 284 | 98.6 | 0.0796 | 2.65 | 262 | 95.5 | 0.0779 | 2.53 |
| 18.0 | 288 | 99.0 | 0.0797 | 2.67 | 266 | 95.9 | 0.0784 | 2.55 |
| 20.0 | 293 | 99.3 | 0.0801 | 2.70 | 271 | 96.3 | 0.0788 | 2.58 |

**TABLE 25.** Transport Properties of Freon-21 in the Single-Phase Region (*Continued*)

| | $T=343{,}15$ K | | | | $T=353{,}15$ K | | | |
|---|---|---|---|---|---|---|---|---|
| $p$ | $\eta\cdot10^6$ | $\lambda\cdot10^6$ | $a\cdot10^6$ | Pr | $\eta\cdot10^6$ | $\lambda\cdot10^6$ | $a\cdot10^6$ | Pr |
| 0.1 | 12.70 | 11.26 | 4.75 | 0.728 | 13.04 | 11.77 | 5.05 | 0.724 |
| 0.5 | 13.00 | 11.80 | 0.851 | 0.770 | 13.33 | 12.29 | 0.920 | 0.759 |
| 1.0 | 208 | 83.9 | 0.0676 | 2.46 | 19.07 | 80.4 | 0.0648 | 2.41 |
| 2.0 | 210 | 84.6 | 0.0683 | 2.45 | 193.3 | 81.2 | 0.0657 | 2.40 |
| 3.0 | 212 | 85.3 | 0.0690 | 2.44 | 195.8 | 82.0 | 0.0665 | 2.40 |
| 4.0 | 215 | 86.1 | 0.0698 | 2.44 | 198.2 | 82.8 | 0.0674 | 2.38 |
| 5.0 | 217 | 86.8 | 0.0706 | 2.43 | 201 | 83.6 | 0.0682 | 2.38 |
| 6.0 | 220 | 87.6 | 0.0713 | 2.43 | 203 | 84.4 | 0.0691 | 2.37 |
| 8.0 | 224 | 88.9 | 0.0726 | 2.42 | 207 | 85.8 | 0.0704 | 2.36 |
| 10.0 | 229 | 90.1 | 0.0738 | 2.42 | 212 | 87.1 | 0.0718 | 2.35 |
| 12.0 | 233 | 91.1 | 0.0748 | 2.42 | 217 | 88.2 | 0.0729 | 2.35 |
| 14.0 | 238 | 91.9 | 0.0757 | 2.43 | 221 | 89.0 | 0.0738 | 2.35 |
| 16.0 | 243 | 92.6 | 0.0763 | 2.44 | 225 | 89.7 | 0.0745 | 2.36 |
| 18.0 | 247 | 93.0 | 0.0768 | 2.45 | 230 | 90.2 | 0.0750 | 2.38 |
| 20.0 | 251 | 93.4 | 0.0772 | 2.47 | 234 | 90.6 | 0.0755 | 2.40 |

**TABLE 25.** Transport Properties of Freon-21 in the Single-Phase Region (*Continued*)

| | $T=363{.}15$ K | | | | $T=373{,}15$ K | | | |
|---|---|---|---|---|---|---|---|---|
| $p$ | $\eta\cdot10^6$ | $\lambda\cdot10^6$ | $a\cdot10^6$ | Pr | $\eta\cdot10^6$ | $\lambda\cdot10^6$ | $a\cdot10^6$ | Pr |
| 0.1 | 13.38 | 12.29 | 5.30 | 0.729 | 13.72 | 12.80 | 5.69 | 0.717 |
| 0.5 | 13.66 | 12.79 | 0.990 | 0.749 | 13.99 | 13.28 | 1.069 | 0.743 |
| 1.0 | 14.10 | 13.42 | 0.426 | 0.816 | 14.40 | 13.88 | 0.4668 | 0.797 |
| 2.0 | 177.1 | 77.8 | 0.0629 | 2.36 | 160.9 | 74.6 | 0.0600 | 2.31 |
| 3.0 | 179.7 | 78.7 | 0.0639 | 2.35 | 163.9 | 75.5 | 0.0611 | 2.30 |
| 4.0 | 182.3 | 79.6 | 0.0648 | 2.34 | 166.8 | 76.4 | 0.0621 | 2.29 |
| 5.0 | 184.8 | 80.4 | 0.0657 | 2.33 | 169.5 | 77.3 | 0.0632 | 2.28 |
| 6.0 | 187.3 | 81.2 | 0.0666 | 2.32 | 172.1 | 78.2 | 0.0642 | 2.26 |
| 8.0 | 192.0 | 82.8 | 0.0682 | 2.30 | 177.1 | 79.9 | 0.0660 | 2.24 |
| 10.0 | 196.7 | 84.2 | 0.0702 | 2.28 | 181.9 | 81.3 | 0.0675 | 2.24 |
| 12.0 | 201 | 85.3 | 0.0718 | 2.26 | 186.5 | 82.6 | 0.0689 | 2.24 |
| 14.0 | 206 | 86.3 | 0.0719 | 2.30 | 191.0 | 83.6 | 0.0692 | 2.25 |
| 16.0 | 210 | 87.0 | 0.0726 | 2.31 | 195.3 | 84.4 | 0.0707 | 2.25 |
| 18.0 | 214 | 87.6 | 0.0733 | 2.32 | 199.6 | 85.1 | 0.0715 | 2.26 |
| 20.0 | 218 | 88.0 | 0.0737 | 2.33 | 203 | 85.6 | 0.0720 | 2.26 |

**TABLE 25.** Transport Properties of Freon-21 in the Single-Phase Region (*Continued*)

| | $T=383.15$ K | | | | $T=393.15$ K | | | |
|---|---|---|---|---|---|---|---|---|
| $p$ | $\eta \cdot 10^6$ | $\lambda \cdot 10^6$ | $a \cdot 10^6$ | Pr | $\eta \cdot 10^6$ | $\lambda \cdot 10^6$ | $a \cdot 10^6$ | Pr |
| 0.1 | 14.06 | 13.32 | 6.03 | 0.713 | 14.39 | 13.83 | 6.36 | 0.711 |
| 0.5 | 14.31 | 13.78 | 1.132 | 0.735 | 14.64 | 14.28 | 1.20 | 0.731 |
| 1.0 | 14.70 | 14.35 | 0.507 | 0.782 | 15.00 | 14.82 | 0.547 | 0.769 |
| 2.0 | 144.6 | 71.4 | 0.0568 | 2.26 | 129.3 | 68.3 | 0.0534 | 2.23 |
| 3.0 | 148.1 | 72.4 | 0.0581 | 2.25 | 133.1 | 69.3 | 0.0554 | 2.21 |
| 4.0 | 151.3 | 73.4 | 0.0594 | 2.24 | 136.5 | 70.4 | 0.0564 | 2.19 |
| 5.0 | 154.3 | 74.3 | 0.0605 | 2.22 | 139.8 | 71.4 | 0.0578 | 2.18 |
| 6.0 | 157.2 | 75.3 | 0.0616 | 2.21 | 142.9 | 72.4 | 0.0590 | 2.16 |
| 8.0 | 162.7 | 77.1 | 0.0637 | 2.19 | 148.8 | 74.3 | 0.0613 | 2.14 |
| 10.0 | 167.7 | 78.6 | 0.0654 | 2.18 | 154.2 | 76.0 | 0.0633 | 2.12 |
| 12.0 | 172.6 | 80.0 | 0.0669 | 2.18 | 159.2 | 77.4 | 0.0649 | 2.12 |
| 14.0 | 177.2 | 81.1 | 0.0681 | 2.18 | 163.9 | 78.7 | 0.0663 | 2.12 |
| 16.0 | 181.6 | 82.0 | 0.0690 | 2.19 | 168.6 | 79.6 | 0.0673 | 2.13 |
| 18.0 | 185.9 | 82.7 | 0.0698 | 2.20 | 173.0 | 80.5 | 0.0683 | 2.13 |
| 20.0 | 190.1 | 83.3 | 0.0704 | 2.21 | 177.2 | 81.1 | 0.0689 | 2.14 |

**TABLE 25.** Transport Properties of Freon-21 in the Single-Phase Region (*Continued*)

| | $T=403.15$ K | | | | $T=413.15$ K | | | |
|---|---|---|---|---|---|---|---|---|
| $p$ | $\eta \cdot 10^6$ | $\lambda \cdot 10^6$ | $a \cdot 10^6$ | Pr | $\eta \cdot 10^6$ | $\lambda \cdot 10^6$ | $a \cdot 10^6$ | Pr |
| 0.1 | 14.72 | 14.34 | 6.69 | 0.709 | 15.04 | 14.86 | 7.05 | 0.706 |
| 0.5 | 14.96 | 14.78 | 1.275 | 0.726 | 15.27 | 15.28 | 1.347 | 0.722 |
| 1.0 | 15.30 | 15.30 | 0.587 | 0.758 | 15.60 | 15.78 | 0.627 | 0.750 |
| 2.0 | 16.49 | 16.71 | 0.227 | 0.898 | 16.68 | 17.09 | 0.254 | 0.861 |
| 3.0 | 120.8 | 66.3 | 0.0514 | 2.24 | 106.5 | 63.3 | 0.0468 | 2.28 |
| 4.0 | 124.2 | 67.4 | 0.0532 | 2.20 | 111.0 | 64.5 | 0.0494 | 2.22 |
| 5.0 | 127.4 | 68.5 | 0.0548 | 2.16 | 115.1 | 65.6 | 0.0516 | 2.16 |
| 6.0 | 130.5 | 69.6 | 0.0564 | 2.14 | 118.7 | 66.8 | 0.0535 | 2.12 |
| 8.0 | 136.3 | 71.6 | 0.0590 | 2.10 | 124.9 | 68.9 | 0.0565 | 2.07 |
| 10.0 | 141.7 | 73.4 | 0.0612 | 2.08 | 130.5 | 70.9 | 0.0591 | 2.04 |
| 12.0 | 146.8 | 75.0 | 0.0630 | 2.06 | 135.6 | 72.5 | 0.0611 | 2.02 |
| 14.0 | 151.7 | 76.3 | 0.0645 | 2.06 | 140.5 | 74.0 | 0.0628 | 2.01 |
| 16.0 | 156.3 | 77.4 | 0.0657 | 2.07 | 145.0 | 75.2 | 0.0641 | 2.01 |
| 18.0 | 160.8 | 78.3 | 0.0667 | 2.08 | 149.2 | 76.1 | 0.0651 | 2.02 |
| 20.0 | 165.1 | 79.0 | 0.0674 | 2.09 | 153.3 | 76.9 | 0.0660 | 2.03 |

**TABLE 25.** Transport Properties of Freon-21 in the Single-Phase Region (*Continued*)

| | $T=423.15$ K | | | | $T=433.15$ K | | | |
|---|---|---|---|---|---|---|---|---|
| $p$ | $\eta \cdot 10^6$ | $\lambda \cdot 10^6$ | $a \cdot 10^6$ | Pr | $\eta \cdot 10^6$ | $\lambda \cdot 10^6$ | $a \cdot 10^6$ | Pr |
| 0.1 | 15.36 | 15.37 | 7.39 | 0.704 | 15.68 | 15.89 | 7.75 | 0.702 |
| 0.5 | 15.59 | 15.79 | 1.422 | 0.718 | 15.90 | 16.29 | 1.497 | 0.714 |
| 1.0 | 15.90 | 16.26 | 0.667 | 0.743 | 16.20 | 16.75 | 0.708 | 0.736 |
| 2.0 | 16.89 | 17.48 | 0.280 | 0.833 | 17.12 | 17.89 | 0.305 | 0.811 |
| 3.0 | 18.87 | 19.43 | 0.1317 | 1.092 | 18.81 | 19.59 | 0.1562 | 0.992 |
| 4.0 | 96.1 | 61.3 | 0.0442 | 2.26 | 81.6 | 56.1 | 0.0344 | 2.68 |
| 5.0 | 100.9 | 62.6 | 0.0470 | 2.18 | 88.3 | 58.6 | 0.0412 | 2.33 |
| 6.0 | 105.1 | 63.8 | 0.0488 | 2.16 | 93.0 | 60.5 | 0.0455 | 2.16 |
| 8.0 | 112.5 | 66.0 | 0.0528 | 2.08 | 101.0 | 63.3 | 0.0509 | 2.02 |
| 10.0 | 118.9 | 68.1 | 0.0567 | 2.00 | 107.8 | 65.5 | 0.0544 | 1.961 |
| 12.0 | 124.7 | 69.9 | 0.0590 | 1.983 | 113.8 | 67.2 | 0.0569 | 1.938 |
| 14.0 | 129.8 | 71.4 | 0.0608 | 1.974 | 119.1 | 68.9 | 0.0590 | 1.922 |
| 16.0 | 134.5 | 72.7 | 0.0624 | 1.966 | 123.9 | 70.4 | 0.0608 | 1.911 |
| 18.0 | 138.7 | 73.8 | 0.0636 | 1.967 | 128.2 | 71.6 | 0.0621 | 1.910 |
| 20.0 | 142.4 | 74.7 | 0.0646 | 1.967 | 132.4 | 73.0 | 0.0636 | 1.904 |

**TABLE 25.** Transport Properties of Freon-21 in the Single-Phase Region (*Continued*)

| | $T=443.15$ K | | | | $T=453.15$ K | | | |
|---|---|---|---|---|---|---|---|---|
| $p$ | $\eta \cdot 10^6$ | $\lambda \cdot 10^6$ | $a \cdot 10^6$ | Pr | $\eta \cdot 10^6$ | $\lambda \cdot 10^6$ | $a \cdot 10^6$ | Pr |
| 0.1 | 16.0 | 16.40 | 8.13 | 0.700 | 16.31 | 16.91 | 8.48 | 0.699 |
| 0.5 | 16.21 | 16.79 | 1.572 | 0.712 | 16.52 | 17.30 | 1.651 | 0.708 |
| 1.0 | 16.49 | 17.23 | 0.749 | 0.730 | 16.79 | 17.72 | 0.789 | 0.726 |
| 2.0 | 17.36 | 18.32 | 0.329 | 0.795 | 17.60 | 18.75 | 0.353 | 0.782 |
| 3.0 | 18.84 | 19.84 | 0.1791 | 0.930 | 18.93 | 20.1 | 0.1992 | 0.889 |
| 4.0 | 21.7 | 22.4 | 0.0867 | 1.35 | 21.2 | 22.2 | 0.1115 | 1.141 |
| 5.0 | 72.6 | 53.0 | 0.0296 | 2.96 | 26.7 | 26.8 | 0.0361 | 2.59 |
| 6.0 | 80.8 | 56.2 | 0.0386 | 2.39 | 64.5 | 50.1 | 0.0263 | 3.17 |
| 8.0 | 90.2 | 60.0 | 0.0471 | 2.05 | 80.2 | 56.4 | 0.0423 | 2.17 |
| 10.0 | 97.6 | 62.7 | 0.0518 | 1.946 | 88.6 | 59.8 | 0.0489 | 1.967 |
| 12.0 | 103.8 | 64.5 | 0.0547 | 1.907 | 95.1 | 61.8 | 0.0525 | 1.897 |
| 14.0 | 109.5 | 66.4 | 0.0572 | 1.883 | 100.7 | 64.0 | 0.0555 | 1.856 |
| 16.0 | 114.5 | 68.1 | 0.0592 | 1.868 | 106.0 | 65.9 | 0.0578 | 1.837 |
| 18.0 | 119.1 | 69.5 | 0.0608 | 1.863 | 110.4 | 67.4 | 0.0596 | 1.814 |
| 20.0 | 123.3 | 70.9 | 0.0623 | 1.857 | 114.9 | 68.9 | 0.0612 | 1.810 |

**TABLE 25.** Transport Properties of Freon-21 in the Single-Phase Region (*Continued*)

| | $T = 463.15$ K | | | | $T = 473.15$ K | | | |
|---|---|---|---|---|---|---|---|---|
| $p$ | $\eta \cdot 10^6$ | $\lambda \cdot 10^6$ | $a \cdot 10^6$ | Pr | $\eta \cdot 10^6$ | $\lambda \cdot 10^5$ | $a \cdot 10^6$ | Pr |
| 0.1 | 16.62 | 17.43 | 8.86 | 0.697 | 16.93 | 17.94 | 9.25 | 0.696 |
| 0.5 | 16.82 | 17.80 | 1.73 | 0.706 | 17.13 | 18.31 | 1.809 | 0.702 |
| 1.0 | 17.09 | 18.21 | 0.831 | 0.722 | 17.38 | 18.71 | 0.874 | 0.716 |
| 2.0 | 17.85 | 19.19 | 0.377 | 0.770 | 18.10 | 19.64 | 0.400 | 0.760 |
| 3.0 | 19.06 | 20.5 | 0.219 | 0.855 | 19.22 | 20.8 | 0.238 | 0.832 |
| 4.0 | 21.0 | 22.3 | 0.1326 | 1.030 | 20.9 | 22.4 | 0.1509 | 0.964 |
| 5.0 | 24.6 | 25.3 | 0.0690 | 1.520 | 23.7 | 24.8 | 0.0918 | 1.238 |
| 6.0 | 35.8 | 34.6 | 0.01289 | 6.11 | 28.8 | 29.0 | 0.0439 | 2.08 |
| 8.0 | 68.1 | 52.0 | 0.0361 | 2.37 | 54.6 | 46.5 | 0.0290 | 2.74 |
| 10.0 | 79.8 | 56.6 | 0.0456 | 2.01 | 70.0 | 53.2 | 0.0423 | 2.05 |
| 12.0 | 86.9 | 59.0 | 0.0502 | 1.898 | 78.3 | 56.0 | 0.0477 | 1.899 |
| 14.0 | 92.9 | 61.6 | 0.0538 | 1.830 | 85.2 | 59.1 | 0.0480 | 1.804 |
| 16.0 | 98.7 | 63.7 | 0.0565 | 1.803 | 91.4 | 61.5 | 0.0552 | 1.772 |
| 18.0 | 103.6 | 65.3 | 0.0584 | 1.791 | 96.8 | 63.3 | 0.0573 | 1.763 |
| 20.0 | 107.9 | 67.0 | 0.0602 | 1.777 | 101.5 | 65.1 | 0.0593 | 1.748 |

# THREE

## THERMOPHYSICAL PROPERTIES OF FREON-22

Freon-22 (difluorochloromethane) is the most widely investigated freon of the methane series. Up to the present time, 70 experimental works have been published containing very broad experimental information about the thermophysical properties of gaseous and liquid Freon-22 in wide intervals of temperature and pressure. In particular, $pvT$ relationships for Freon-22 were investigated in the range $T = 203-473$ K at pressure $p = 0.1-35$ MPa, and about 950 experimental points were obtained. A total of 250 experimental values of heat capacity $c_p$ and $c_v$ in the range $T = 232-473$ K at $p = 0.1-3.5$ MPa are known. Viscosity ($N_{exp} \approx 530$) and thermal conductivity ($N_{exp} \approx 360$) were measured in the interval $T = 115-473$ K and $p = 0.1-59$ MPa.

Only a relatively modest portion of the available experimental results about the thermophysical properties of Freon-22 are considered in the published handbooks [0.7, 0.9, 0.10, 0.45, 2.22], in A. V. Køletsky's monograph [0.18], and in the reference books [0.8, 0.34]. As a result, the aforementioned references are not only incomplete but also insufficiently accurate.

In order to develop more complete and reliable reference data, the authors carried out a series of experimental and computational investigations. The results have been partially published [0.13, 0.36, 1.2–1.4, 3.1–3.3, 3.9–3.11, 3.32] and provided as a basis for compiling the recommended tables of thermophysical properties of gaseous and liquid Freon-22.

### 3.1. A REVIEW OF PUBLISHED DATA
### ABOUT THE THERMODYNAMIC PROPERTIES

The fundamental experimental investigations of the thermodynamic properties of Freon-22, which were carried out in 18 laboratories, are given in Tables 26

**TABLE 26.** Experimental Investigations of Thermodynamic Properties of Freon-22 in the Single-Phase Region

| Year | Authors | Measured property | Temperature, K | Pressure, MPa | Phase[a] | Number of experimental points | Reference |
|------|---------|-------------------|----------------|---------------|----------|------------------------------|-----------|
| 1940 | Benning and McHarness | $\varrho$ | 298—412 | 0.03—2.1 | G | 40 | [2.25] |
| 1940 | Benning et al. | $c_p$ | 322—408 | 0.1 | G | 6 | [2.27] |
| 1956 | Miyahara and Richardson | $w$ | 300—475 | 0.1 | G | 9 | [3.53] |
| 1957 | Michels | $\varrho$ | 292—423 | 0.7—13 | G, L | 181 | [3.50] |
| 1961 | Hwang | $c_v$ | 323—473 | 3.5 | G | 58 | [3.54] |
| 1970 | Kletsky | $\varrho$ | 293—465 | 0.8—5.8 | G, L | 63 | [0.18] |
| 1966, 1967 | Lagutina | $\varrho$ | 273—393 | 0.2—6 | G | 135 | [3.22, 3.23] |
| 1967 | Novikov and Lagutina | $w$ | 273—393 | 0.2—6 | G | 145 | [3.23, 3.25] |
| 1968 | Zander | $\varrho$ | 287—473 | 0.3—35 | G, L | 270 | [3.66] |
| 1968, 1970 | Hajjar and MacWood | $B_1$ | 303—403 | 0.1 | G | 4 | [3.44, 3.45] |
| 1969 | Shumskaya and Gruzdev | $c_p$ | 300—460 | 0.1 | G | 9 | [2.20] |
| 1969 | Meyer | $w$ | 198—226 | 0.1 | L | 9 | [3.52] |
| 1970 | Ernst and Büsser | $c_p$ | 298—353 | 0.1—1.4 | G | 37 | [3.40] |
| 1971 | Suther and Cole | $B_1$ | 323—425 | 0.1 | G | 3 | [3.61] |
| 1971 | Haworth and Sutton | $B_1$ | 298—328 | 0.1 | G | 3 | [3.46] |
| 1971 | Kumagi and Iwasaki | $\varrho$ | 253—313 | 9.7—160 | L | 34 | [3.48] |
| 1971 | Date and Iwasaki | $\varrho$ | 298—398 | 0.1—6.5 | G | 48 | [0.51] |
| 1974 | Hirata et al. | $\varrho$ | 297—414 | 2.0—9.4 | G | 65 | [0.51] |
| 1974 | Bier et al. | $c_p$ | 333—353 | 0.2—3.4 | G | 16 | [3.36] |
| 1975 | Gruzdev and Shumskaya | $c_p$ | 301—452 | 0.2—2 | G | 92 | [2.9] |
| 1975 | Oguchi et al. | $\varrho$ | 253—363 | 0.2—8.9 | G, L | 63 | [3.56] |

[a] G, Gas; L, liquid.

and 29. As a rule, in each of these laboratories, the thermodynamic values were measured in both the single phase and on the saturation line. For comparative analysis, however, it is more convenient to group the experimental work according to the regions of the investigation. Therefore, the results of calorimetric and acoustic measurements, determination of thermodynamic functions in the ideal gas state, and the equations of state (used earlier to calculate the tables of thermodynamic properties of Freon-22) are discussed in separate subsections.

## Experimental Data in the Single-Phase Region

Table 26 and Fig. 15 show that the measurements of density (or coefficient of compressibility) cover a large region of the phase state. The maximum pressure and temperature were on the order of 35 MPa and 473 K. However, for the

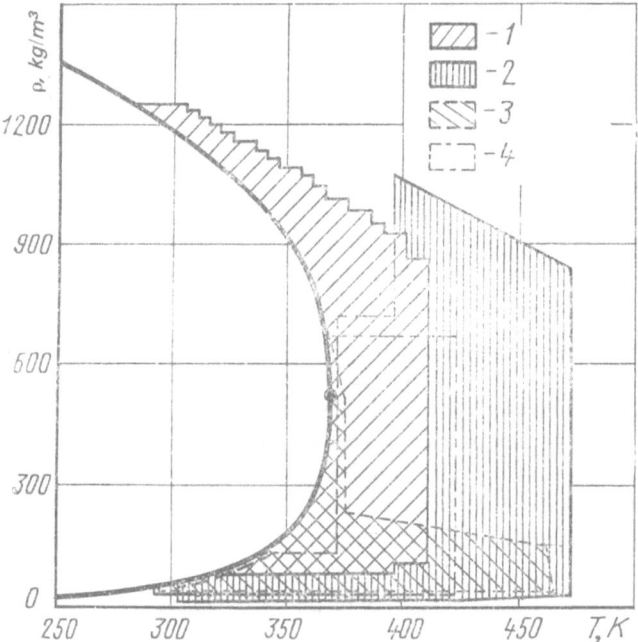

**FIG. 15.** The regions of experimental study of gas density of Freon-22 in references: 1, Zander (constant-volume measurements) [3.66]; 2, Zander (measurements by the Burnett method) [3.66]; 3, Kletsky [0.18]; 4, Michels [3.50].

gaseous state, tests were conducted at temperatures down to 253 K, and for the liquid state the minimum temperature was 287 K.

Three experimental studies were conducted by Japanese researchers in 1971–1974 and were not published. They became known recently from Ref. [0.51]. They include the work of Kumagi and Iwasaka [3.48], who measured the density of liquid Freon-22 in the interval $T = 253–313$ K at pressures up to 160 MPa.

It is pertinent to mention that the extent of coverage of experimental points in the investigated region of state is not even, and most of the experimental values of density are related to the gas phase at pressures up to 6 MPa. The statistical analysis of experimental $pvT$ data published up to 1972, which we conducted [3.1], showed that the actual accuracy of these results are on occasions much lower than that assessed by the authors of the experimental studies themselves. Thus, in the works of L. M. Lagutina [3.22, 3.23], the maximum error in the measured values of density of gaseous Freon-22 is estimated to be 0.2–0.3%. In reality, the standard deviation of these data from the methodically independent and concordant measurements made by Zander [3.66] and A. V. Kletsky [0.18] constitute $\sigma_z = 0.85\%$ and $\sigma_p = 1.27\%$. Even if we exclude from the data in [3.22, 3.23] the 20 points or so which strongly fall out (by more than 0.5–1.0%), the value of $\sigma_{exp}$ for the remaining 115 points is 0.38%. The deviation of experimental data of Benning and McHarness [2.25] is even more: $\sigma_z \approx 1.8\%$. Unfortunately, some of the data in later experimental works have an error exceeding the mean level for standard measurements. This concerns mainly Ref. [3.66] and partly Refs. [0.18, 3.56]. From the aggregate of overlapping measurements [0.18, 3.50, 3.56, 3.66], it is possible to make a representative choice of concordant $pvT$ data and, based on their statistical analysis, build an accurate equation of state.

While constructing a weight function, the need arises to operate with relative errors of single measurements. It is imperative in this case to make preliminary assessment of fundamental groups of experimental results.

In the 1960s, in an Amsterdam laboratory under the supervision of A. Michels, $pvT$ behavior of Freons-12, -13, and -22 was investigated in the interval $T$ = 273–423 K at pressures up to 40 MPa. The study was carried out on a well-known experimental apparatus which allowed the use of the pressure-relieved variable-volume piezometer. The samples contained no less than 99.95% pure freon. The error in experimental data of $pv$ values was usually evaluated to be ±0.05%. The experimental results for Freons-12 and -13 obtained at this laboratory were published in 1966 and used to construct the tables of thermodynamic properties in Ref. [0.28]. However, the results of direct measurements for Freon-22 were not published and were known from the computational work of Martin [3.50].

At the Leningrad Technological Institute for the Refrigeration Industry $pvT$ behavior of Freon-22 was investigated in an apparatus allowing the use of the constant-volume piezometer with ballasted clearance [0.18, 3.17]. The piezometer, without pressure relief ($v \approx 456$ cm³), and the glass mercury differential manometer were spatially separated and connected with a steel capillary tube. The piezometer and the differential manometer had automatic thermostatic control, and the connecting capillary had an external heater. To measure the pressure of the investigated substance, a dead-weight manometer with class 0.02 accuracy was employed. The temperature of the main thermostat was measured with a resistance platinum thermometer, and the resistance of the thermometer was measured by a potentiometer PMC-18 with class A accuracy. The sample mass was determined by a detachable ampule-metering device which could be weighed on an analytical balance. The initial sample, as shown by chromatography, contained impurities not more than 0.15% by volume and was subjected to further purification. The probable error in the experimental data of A. V. Kletsky [0.18] was about 0.15%. Comparison with the data of Michels and Zander confirmed the indicated assessment.

Zander [3.66] carried out a very detailed investigation of $pvT$ behavior of liquid and gaseous Freon-22 on two experimental apparatuses. Compressibility of gaseous Freon-22 was determined by the Burnett method and produced 107 experimental values $z(p, T)$ on seven isotherms in the interval $T$ = 303–473 K at pressure of 0.3–34 MPa. According to Zander's assertions, the maximum error in the results does not exceed ±0.25%. The second apparatus was a constant-volume piezometer adapted for isocharic measurements. The volume of the piezometer at room conditions was 63 cm$^3$ determined by calibration using mercury. The piezometer was assembled with a mercury seal in a single thermostat. The level of mercury in the seal was fixed by an electrical contact, and the possible uncontrollable volume variation $\Delta v/v$ was less than $2 \cdot 10^{-5}$. The results of direct measurements were corrected for piezometer deformation as a function of temperature and pressure and were given on 35 isochores in the range $\rho = 80-1250$ kg/m$^3$ at $p = 1.9-21$ MPa and $T = 287-413$ K ($N_{exp} = 163$). The error in this group of measurements, by our assessment, was equal to 0.3–0.4% (for $z_{exp}$) on the average. However, since the second group of experimental data belongs mainly to the liquid phase, the corresponding error can be considered as entirely acceptable. It is important to note Zander's statement that based on the calorimetric measurement data on the melting point, the purity of the investigated Freon-22 was about 99.97 mol %.

Oguchi et al. [3.56] carried out experimental investigations of $pvT$ behavior of Freon-22 in the range $T = 253-363$ K and $p = 0.2-8.9$ MPa on an apparatus they described earlier [3.62]. A spherical piezometer without pressure relief employing a zero-indicator membrane housed in a single thermostat were used. The internal volume of the piezometer part was determined in calibration tests using distilled water at $p < 12$ MPa. To calculate the corrections for temperature deformation of the piezometer, data on coefficients of linear expansion were obtained in the laboratory. The metal used was the same steel from which the piezometer was made. Considering all these corrections, the specific volume (defined as $v = V/m$) can be found, according to the authors, with an error on the order of 0.01%. The back pressure in the zero indicator, caused by gaseous nitrogen, was held constant with a special servosystem and measured (at $p_{exp} \leqslant 0.4$ MPa) by a mercury or dead-weight manometer. The dead-weight manometer was calibrated by vapor pressure of pure $CO_2$ at $T = 273.15$ K. The accuracy of the zero indicator [3.62] was not less than 0.2 kPa, and the error in determining $p_{exp}$ was estimated as 0.01%. Reference [3.56] indicated the following errors in measuring the pressure: 0.014% at 10 MPa, 0.018% at 5 MPa, and 0.05% at 1 MPa. The temperature was measured with a probable error of 0.01 K at $T > 293$ K and 0.015 K at $T < 293$ K. The overall error in determining the specific volume of Freon-22 (based on cumulative errors), according to the authors of [3.56], was equal to 0.03% in the gas phase, 0.04% in the liquid phase, and increased up to 0.2% at $v = 100$ cm$^3$/g. The investigated sample contained, as shown by chromatography, 99.19% pure Freon-22 by mass and only 0.81% by mass of Freons-23, -32, and air.

In Ref. [3.56] $pvT$ measurements were made on four isochores in the gas phase ($v \approx 10, 40, 70,$ and $100$ cm$^3$/g) and one isochore in the liquid phase (0.86 cm$^3$/g). These measurements can undoubtedly be called the most accurate for Freon-22.

From the experimental investigations enumerated in Table 26, only in Ref. [3.56] are the experimental data presented in accordance with IPTS-68. In much

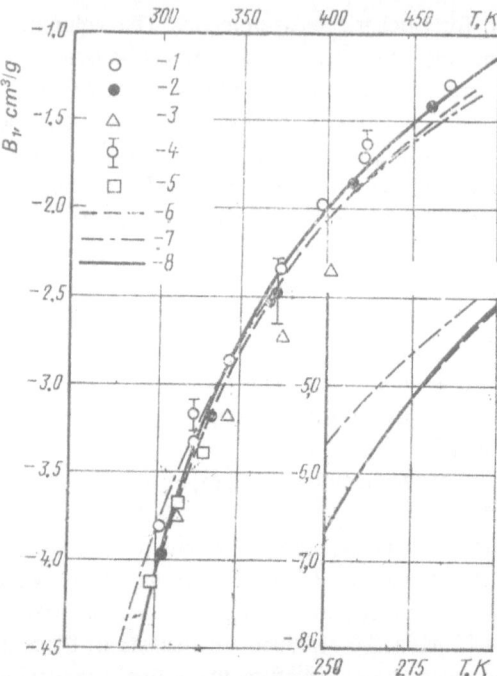

**FIG. 16.** The second virial coefficient of Freon-22. Experimental data: 1, Zander [3.66]; 2, Kletsky [0.18]; 3, Hajjar and MacWood [3.45]; 4, Suther and Cole [3.61]; 5, Haworth and Sutton [3.46]. Calculations from equation of state: 6, Kletsky [0.18]; 7, Zander [3.66]; 8, Altunin and Gadetsky [3.1].

earlier investigations IPTS-48 was used. This state of affairs must be taken into consideration when comparing and analyzing experimental data.*

The values of the second virial coefficient $B_1$, which were obtained in Refs. [3.44–3.46, 3.61], are presented in Fig. 16. Also given there are the values of $B_1$, found from $pvT$ data [0.18, 3.66], the calculated values from three different equations of state (recommended in Refs. [0.18, 3.1, 3.66] for compressed Freon-22. From Fig. 16, it follows that the best results in the investigated interval ($T = 293–473$ K) are given by the equation of V. V. Altunin and O. G. Gadetsky [3.1].

Heat capacity at constant pressure of Freon-22 was measured in the range $T = 298–460$ K at $p = 0.1–3.4$ MPa (see Table 26). Analysis performed earlier [3.1] showed that the experimental data on heat capacity $c_p$ at atmospheric pressure, which were published before 1970, have low accuracy, and consequently the works [2.20, 2.27] are not discussed here. In much later work [2.9, 3.36, 3.40], not only Freon-22 was investigated, but also polyatomic gases.

* Within the temperature interval under consideration, the difference between IPTS-68 and IPTS-48 amounts to 0.026 K at 230 K, to 0.0 K at 273.15 and 373.15 K, and to 0.011 K at 320 K.

At the Institute of Thermodynamics, University of Karlsruhe, two series of $c_p$ measurements with the following adiabatic calorimeter method were carried out [3.36, 3.40]. The total error in the measured values of $c_p$, according to the researchers, is about 0.1%. Their experimental data agree very well with each other, and the discrepancy in $c_p$ values on the isotherms 333 and 353 K at $p = 0.1 - 1.4$ MPa does not exceed 0.2 - 0.3%.

V. A. Gruzdev and A. I. Shumskaya [2.9] measured $c_p$ for Freon-22 in an original flowing capillary calorimeter with an error, according to the authors, of not more than 0.4 - 0.7%. Comparison of the experimental data in [2.9] and [3.36, 3.40] indicates that the discrepancy is within the limits of measurement errors at relatively low pressures. As the pressure increases, the discrepancy increases so that at 2 MPa it exceeds 3.5%. Note that the data in [2.9] are somewhat low; see Fig. 17.

Inasmuch as the calorimetric data [3.36, 3.40] at low pressures have high accuracy, they can be used to select the acceptable values of thermodynamic

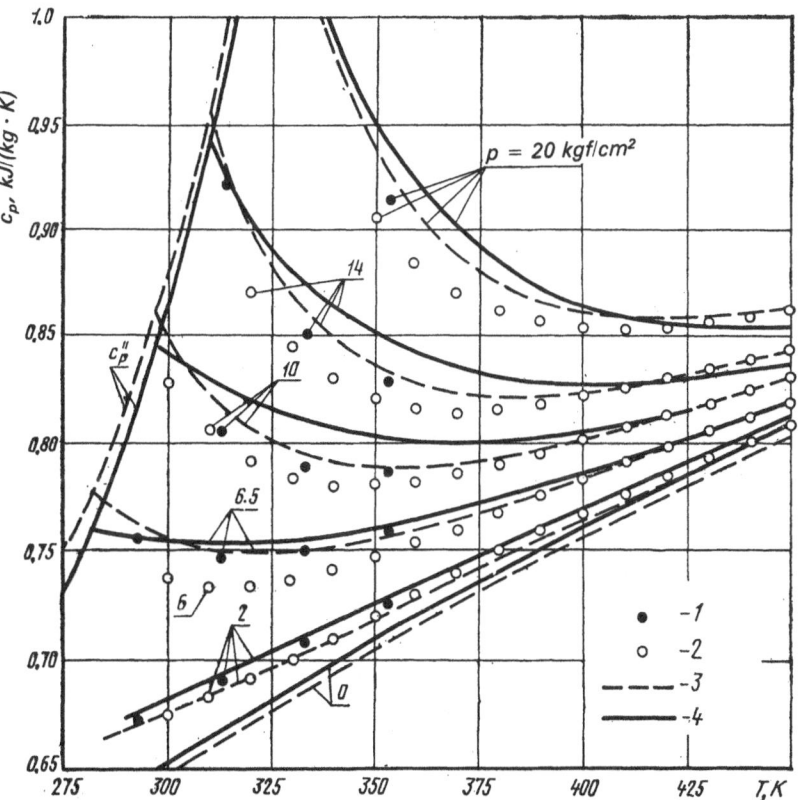

**FIG. 17.** Constant-pressure heat capacity of gaseous Freon-22 at elevated pressures. Experimental data: 1, Ernst et al. [3.36, 3.40]; 2, Gruzdev and Shumskaya [2.9]. Tabular data: 3, [0.51]; 4, Altunin and Gadetsky [3.1].

functions for gaseous Freon-22 in the ideal gas state. These ideal gas functions were tabulated earlier in several publications [0.37, 1.7, 1.39, 3.41, 3.63]. In most of them, calculations were carried out in the approximated rigid rotator-harmonic oscillator model (RRHO). The fundamental vibration frequencies in each cited reference for Freon-22 are practically the same, and the structural parameters differ only slightly (see Table 27). However, in Barho's work [0.37], the anharmonic oscillations are taken into account by semiempirical formula with the assumption that the anharmonic constant $\chi$ for Freon-22 and that for Freon-12 are identical.

In Table 28 the values of heat capacity $C_p^0$, enthalpy ($H_T^0 - H_0^0$), and entropy $S_T^0$ are compared. The differences between data in [1.7, 3.41, 3.63] are small; for enthalpy they do not exceed 4 cal/mol, and for entropy they do not exceed 0.035 cal/(mol · K). The results of Barho [0.37] for $C_p^0$ are, on the average, higher by 0.5% than the calculated values in [3.41, 3.63]. As Fig. 18 shows, they agree well with the calorimetric data [3.40] at $T = 293-353$ K.

**TABLE 27.** Molecular Constants for Freon-22

| Symbol, dimension | Numerical values listed in the references | | | | |
|---|---|---|---|---|---|
| | [3.41] | [3.63] | [1.7] | [0.37] | [1.39] |
| $\nu_1$, cm$^{-1}$ | 3023 | 3025 | 3023 | 3025 | 3024 |
| $\nu_2$, cm$^{-1}$ | 1311 | 1312 | 1311 | 1312 | 1312 |
| $\nu_3$, cm$^{-1}$ | 1178 | 1178 | 1178 | 1178 | 836 |
| $\nu_4$, cm$^{-1}$ | 809 | 808 | 809 | 808 | 812 |
| $\nu_5$, cm$^{-1}$ | 595 | 596 | 595 | 596 | 598 |
| $\nu_6$, cm$^{-1}$ | 422 | 421 | 422 | 421 | 417 |
| $\nu_7$, cm$^{-1}$ | 1347 | 1345 | 1347 | 1345 | 1350 |
| $\nu_8$, cm$^{-1}$ | 1116 | 1116 | 1116 | 1116 | 1108 |
| $\nu_9$, cm$^{-1}$ | 365 | 360 | 365 | 360 | 400 |
| $r_{C-F}$, Å | 1.34 | 1.36 | 1.35±0.03 | — | 1.35 |
| $r_{C-C}$, Å | 1.75 | 1.73 | 1.77±0.03 | — | 1.74 |
| $r_{C-H}$, Å | 1.10 | 1.093 | 1.10±0.03 | — | 1.09 |
| $I_x I_y I_z \cdot 10^{117}$, (g · cm$^2$)$^3$ | 3440 | — | 3585.3 | — | 3434.33 |
| $\sigma$ | 1 | 1 | 1 | 1 | — |
| $\chi_e$ | — | — | — | 0.035 | — |

**TABLE 28.** Values of Thermodynamic Functions of Freon-22 in the Ideal Gas State from the Data of Different Researchers

| $T$, K | $C_P^0$, cal/(mol·K) | | | | $H_T^0 - H_0^0$, cal/mol | | | | $S_T^0$, cal/(mol·K) | | | |
|---|---|---|---|---|---|---|---|---|---|---|---|---|
| | [0.37] | [3.41] | [3.53] | [1.39] | [1.7] | [3.41] | [3.63] | [1.39] | [1.7] | [3.41] | [3.63] | [1.39] |
| 100 | — | 8.44 | 8.43 | 8.368 | — | 804 | 803 | 802 | — | 55.75 | 55.77 | 55.720 |
| 150 | — | 9.60 | — | — | — | 1253 | — | — | — | 59.38 | — | — |
| 200 | 10.92 | 10.89 | 10.88 | 10.930 | — | 1765 | 1764 | 1758 | — | 62.32 | 62.33 | 62.245 |
| 250 | 12.22 | 12.16 | — | — | — | 2342 | — | — | — | 64.88 | — | — |
| 298.16 | — | 13.35 | 13.34 | 13.653 | 2956 | 2956 | 2952 | 2967 | 67.165 | 67.13 | 67.13 | 67.126 |
| 300 | 13.47 | 13.40 | 13.39 | 13.701 | — | 2981 | 2979 | 2992 | — | 67.21 | 67.21 | 67.211 |
| 400 | 15.74 | 15.63 | 15.62 | 16.039 | 4435 | 4435 | 4432 | 4484 | 71.416 | 71.38 | 71.38 | 71.485 |
| 500 | 17.58 | 17.49 | 17.45 | 17.849 | 6093 | — | 6090 | 6182 | 75.108 | 75.26 | 75.17 | 75.268 |
| 600 | 19.02 | 18.87 | 18.87 | 19.221 | 7912 | 7912 | 7908 | 8039 | 78.421 | 78.38 | 78.38 | 78.649 |
| 700 | 20.15 | 19.97 | 19.97 | 20.271 | 9856 | 9856 | 9849 | 10016 | 81.415 | 81.39 | 81.38 | 81.694 |
| 800 | 21.04 | 20.84 | 20.88 | 21.093 | 11900 | 11900 | 11904 | 12085 | 84.143 | 84.11 | 84.12 | 84.457 |
| 1000 | 22.35 | 22.11 | 22.12 | 22.286 | 16202 | 16200 | 16220 | 16431 | 88.938 | 88.90 | 89.94 | 89.331 |
| 1300 | — | 23.29 | — | — | 23027 | 23030 | — | — | 94.900 | 94.86 | — | — |

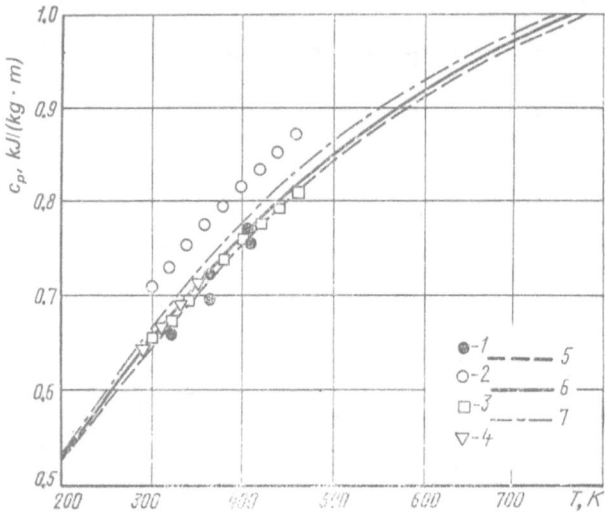

**FIG. 18.** Temperature dependence of isobaric heat capacity of Freon-22. Experimental data: 1, Benning et al. [2.27]; 2, Shumskaya and Gruzdev [2.20]; 3, Gruzdev and Shumskaya [2.9]; 4, Ernst and Büsser [3.40]. Calculated data: 5, Weissman et al. [3.63]; 6, Barho [0.37]. Tabular data: 7, [1.39].

Data in [1.39] indicate that the temperature dependence of heat capacity, enthalpy, and entropy are somewhat steeper. The values of $C_p^0$, $(H_T^0 - H_0^0)$ and $S_T^0$ are 1.5–2% higher than the data in [0.37, 3.40] at $T = 300$–500 K and 2–3% higher than the data in [2.9, 3.41, 3.63].

Because of these reasons, when calculating thermodynamic tables for compressed Freon-22, preference should be given to the function $C_p^0(T)$ found in Barho's work [0.37]. It is important that the reference data in [0.28, 3.1] were constructed using exactly these values of thermodynamic functions of Freon-22 in the ideal gas state. Contained in Ref. [0.21] are the coefficients of Eqs. (1.12)–(1.14) for the calculation of $C_p^0$, $(H_T^0 - H_0^0)$ and $S_T^0$ from data by Barho at $T \leqslant 500$ K. The values of these coefficients for Freon-22 when $C_p^0$ is in kJ/(kmol · K) are equal to

$$a_0 \cdot 10^{-2} = 0.2589346; \qquad a_3 \cdot 10^6 = -0.04123227;$$
$$a_1 \cdot 10^0 = 0.04288737; \qquad a_4 \cdot 10^{-2} = 1.049551.$$
$$a_2 \cdot 10^3 = 0.03381799;$$

A review by Morsy [3.54] presents the experimental data by Huang about constant-volume heat capacity of gaseous Freon-22 in the temperature interval 323–473 K at pressures up to 3.4 MPa. Most probably, the investigation was carried out on the experimental apparatus built by Nevers and Martin and was used by them to determine $c_v$ of Freon-C318 and propylene at $T = 330$–420 K and $\rho \leqslant 160$ kg/m³. This work also indicates that the maximum error in $c_v$ could reach 3.2% at low density, and decreases to 1.1% at the highest possible density for this investigation. In our opinion, more reliable information about the function $c_v(v,T)$ for Freon-22 can be obtained by analyzing the results of constant-pressure heat capacity measurements (see above) and the data of L. M. Lagutina and I. I. Novikov [3.23, 3.25]. In these studies, detailed measurements of speed of sound in superheated and saturated vapors of Freon-22 were carried out using Pierce's acoustic interferometer. The measurements were made at frequencies of 500 and 150 kHz with the measuring cell arranged as a constant-volume piezometer. Along with the acoustic measurements, $pvT$ behavior was also investigated. This allowed determination with the experimental points of the adiabatic index

$$K_A = \frac{v}{p}\left(\frac{\partial p}{\partial v}\right)_s = \frac{w^2}{pv}$$

and to control, by the $K_A(p, T)$ graph, the extent of agreement of two groups of measurements: $w(p, T)$ and $pv(p, T)$. According to the authors' assessment, the error in the measured values of $w$ amounts to 0.2%. We should note that this evaluation was undoubtedly based only on instrument errors and the actual error will be noticeably higher. It was established in Ref. [3.1] that the values of $w$ computed from $pvT$ data [0.18, 3.66] agree with the measurements of L. M. Lagutina with deviation in the region 0.3–1.0%. But in the vicinity of the critical point, the discrepancy increased appreciably.

Meyer [3.52] employed the method of Pierce's acoustic interferometer for measuring the speed of sound in liquid freons. For Freon-22 the experimental data were approximated with an error of $\pm 0.1\%$ by an equation for $w$ liquid as

$$w = 662.3 - 4.615t - 15.80 \cdot 10^{-4}t^2 - 3.830 \cdot 10^{-5}t^3 \qquad (3.1)$$

where $w$ is in m/s and $t$ is in °C. From Eq. (3.1) at $T_{NBP}$, $w'_{0\text{-}f} \approx 850$ m/s ($w'_{0\text{-}f}$ means speed of sound in liquid single-phase region near the saturation line).

## Experimental Data on the Saturation Line

Measured on the phase equilibrium line were the saturated vapor pressure $p_s$, orthobaric density of the liquid $\rho'$, heat capacity of the saturated liquid $C'_s$, and speed of sound in vapor $w''_{0\text{-}f}$ in a wide interval of temperatures; heat of vaporization $\gamma$ measured only at $T_{NBP}$ (Table 29). The temperature dependence of saturated vapor pressure of Freon-22 was thoroughly investigated in the interval $T$ from 203 K to the critical point (369.30 K) [3.17, 3.55, 3.56, 3.66]. The measurement results in [3.17, 3.66] were presented according to IPTS-48 and in [3.56] according to IPTS-68. It is important to note that in these works, function $p_s(T)$ was investigated with highly pure samples.

Figure 19 shows that the agreement of the indicated experimental data,

**TABLE 29.** Experimental Investigations of Thermodynamic Properties of Freon-22 on the Saturation Line

| Year | Authors | Measured properties | Temperature, K | Phase[a] | Number of measured points | Reference |
|---|---|---|---|---|---|---|
| 1935 | Booth and Swinehart | $p_s$ | 295—358 | V | 18 | [3.37] |
| 1940 | Benning and McHarness | $p_s$ | 212—366 | V | 7 | [2.24] |
| 1940 | Benning et al. | $\varrho$ | 204—360 | L | 8 | [2.26] |
| 1940 | | $c_s$ | 256—328 | L | 4 | [2.27] |
| 1957 | Neilson and White | $c_s$ | 115.7—232.5 | L | 13 | [3.55] |
| | | $r$ | 232.5 | L–V | 6 | |
| 1960 | Yakobson | $p_s$ | 238—333 | V | 6 | [3.65] |
| 1960 | Steinle | $\sigma$ | 248—323 | L–V | Graphic | [3.60] |
| 1964 | Kletsky | $\varrho$ | 230—366 | V | 14 | [3.16, 3.17] |
| 1970 | | $\varrho$ | 359—366 | L | 5 | [0.18] |
| 1967 | Lagutina | $p_s$ | 287—369 | V | 22 | [3.23] |
| | | $w$ | 287—369 | V | 22 | |
| 1968 | Zander | $p_s$ | 203—369 | V | 23 | [3.66] |
| | | $\varrho$ | 279—365 | L | 31 | |
| 1969 | Dorokhov et al. | $\sigma$ | 232—323 | L–V | 22 | [1.8, 3.13] |
| 1973 | Oguchi et al. | $p_s$ | 233—363 | V | 22 | [3.56, 3.62] |
| 1975 | | $\varrho$ | 305 | L | 1 | |
| 1975 | Hirata et al. | $p_s$ | 298—368 | V | 35 | [0.51] |
| 1976 | Zhelezny | $\sigma$ | 150—330 | L–V | 19 | [3.14] |

[a] L, Liquid; V, vapor.

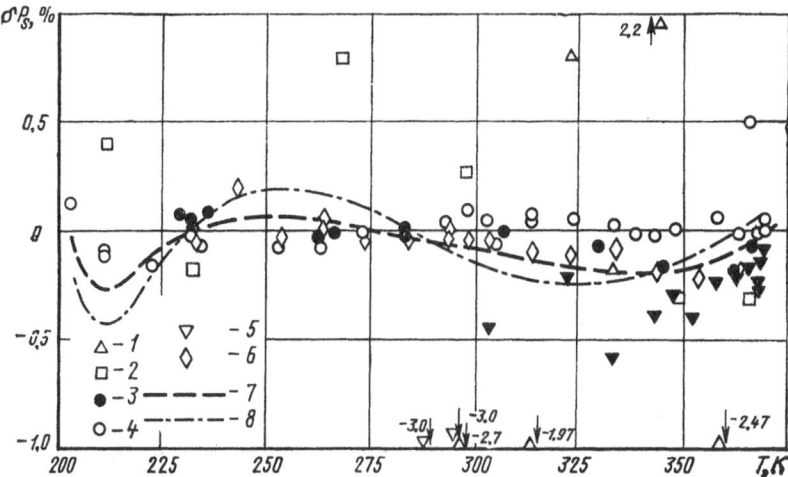

**FIG. 19.** Deviation of the values of the saturated vapor pressure of Freon-22 from the adopted values in the present work. Experimental data: 1, Booth and Swinehart [3.37]; 2, Benning and McHarness [2.24]; 3, Kletsky [3.17]; 4, Zander [3.66]; 5, Lagutina [3.23]; 6, Oguchi et al. [3.56]. Tabular data: 7, Kletsky [0.18]; 8, Perelshtein [0.28].

which are represented according to IPTS-48, is very good and discrepancy does not exceed 0.15%.

There are several known versions of interpolation equations $p_s(T)$ of types (0.10), (0.11), and (0.21), which are based on experimental data by A. V. Kletsky and Zander [0.18, 0.28, 3.16–3.19, 3.66]. According to information presented in Ref. [1.15], however, it is possible to use the generalized equation (1.2) to calculate $p_s$ for Freon-22 according to IPTS-48 in the range $T = 116$–369 K. The characteristic parameters of this equation in the case reviewed here are equal to $T_{cr} = 369.28$ K, $p_{cr} = 4.990$ MPa, Ri = 6.7964; $p_\alpha = -0.1644$.

For the calculations according to IPTS-68 in the range $T = 194$–369.30 K the following equation can apply

$$\ln \pi_s = [-7.0340913(1-\tau) + 1.4030736(1-\tau)^{1.5} - 4.9605880(1-\tau)^3 + 8.8828089(1-\tau)^4 - 10.600638(1-\tau)^5] \quad (3.2)$$

where $p_{cr} = 4.988$ MPa, $T_{cr} = 369.30$ K. This equation is recommended in Ref. [0.51], and it approximates the experimental data of A. V. Kletsky, Zander, Oguchi, and Hirata (about 100 experimental values for $p_s$) with a standard deviation of 0.07% and maximum deviation of 0.19%.

It must be noted that tables of $p_s$, which are given in Refs. [0.7, 0.45, 2.21] and in the first edition of the handbook by N. B. Vargaftik (0.10), differ considerably from the calculated values in [0.18, 0.28, 3.1] and in the present work; the discrepancy reaching as high as 2–4%. This is explained by the fact that the first group of reference data reproduces the tables of Graham and McHarness

(1946). These tables were computed using inaccurate equations derived as early as 1940.

The direct measurements of density on the saturation curve of Freon-22 were carried out only for the liquid phase (see Table 29). In Refs. [0.18, 3.66] the density of the saturated liquid was measured using the pycnometer method with an error of the order 0.1–0.2%. In the investigations of Benning and McHarness [2.26], the measurement of $\rho'$ was made using the hydrostatic weighing method and the error was, apparently, of the order 0.5–1%. Reference [3.1] indicates that in the $\rho$, $T$ diagram, the data of [2.26] are situated higher than the data in [3.66] and the discrepancy reaches 1–1.5%.

For Freon-22 the following interpolation equation was derived in Ref. [0.28]:

$$\rho' = 496.3 + 1{,}935780\,\theta + 173{,}97500\,\theta^{1/3} - 2 \cdot 10^{-3}\theta^2$$
$$- 17.8467 \cdot \theta^{1/2} \tag{3.3}$$

where $\rho'$ is in kg/m$^3$ and $\theta = (369.28 - T)$ K. The values of $\rho'$ calculated from Eq. (3.3) agree well with the recommended values in the present work, with the exception of a narrow region adjacent to the critical point.

In Ref. [3.21] an interpolation equation of the following form is derived from the same initial data to solve for $T$ saturation ($T_s$):

$$T_s = T_{cr} + \sum_{i=2}^{6} A_i (\rho - \rho_{cr})^i \tag{3.4}$$

where at $T_{cr} = 369.28$ K and $\rho_{cr} = (1/1.95)$ g/cm$^3$.

$$A_2 = -6.05; \qquad A_5 = -47.06;$$
$$A_3 = -357.30; \qquad A_6 = 1.68.$$
$$A_4 = 237.65;$$

Equations (3.3) and (3.4) are suitable for calculations using IPTS-48. For calculations using IPTS-68 in the range $T = 194$–369.30 K, Eq. (0.13) with $m = 4$ can be used with the coefficients

$$d_1 = 1.8877394 \cdot 10^0; \qquad d_3 = -7.1134041 \cdot 10^{-2};$$
$$d_2 = 5.9858531 \cdot 10^{-1}; \qquad d_4 = 4.0327650 \cdot 10^{-1}.$$

The coefficients were taken from Ref. [0.51] and are valid for $T_{cr} = 369.30$ K and $\rho_{cr} = 513$ kg/m$^3$.

Presented in Table 30 are experimentally determined parameters of key points on the saturation line and the recommended parameters in up-to-date handbooks. Some chosen values were given in Table 3, and they correspond with the adopted values in Refs. [3.1, 3.56] within the limits of tolerance.

**TABLE 30.** The Parameters of Key Points on the Saturation Line of Freon-22[a]

| Year | Authors | Critical parameters | | | $T_{NBP}$, K | $T_0$, K | Reference |
|---|---|---|---|---|---|---|---|
| | | $T_{cr}$, K | $p_{cr}$, MPa | $\rho_{cr}$, kg/m³ | | | |
| 1935 | Booth and Swinehart | 369.55 | 4.912 | — | — | — | [3.37] |
| 1940 | Benning and McHarness | 369.15 | 4.934 | 525.0 | 232.36 | — | [2.26] |
| 1957 | Neilson and White | — | — | — | 232.48±0.07 | 115.73±0.01 | [3.55] |
| 1964 | Kletsky | 369.28 | 4.985 | 515.5 | 232.30 | — | [3.18] |
| 1968 | Kudchadker et al. | 369.15 | 4.977 | 525.0 | — | — | [1.40] |
| 1968 | Zander | 369.33 | 4.990 | 513.0 | — | — | [3.66] |
| 1970 | Phillips and Murphy | — | — | — | 232.35 | — | [1.50] |
| 1973 | Altunin | 369.30 | 4.988 | 514 | 232.35 | 115.74 | [3.1] |
| 1971 | Perelshtein | 369.28 | 4.989 | 496.3 | — | — | [0.28] |
| 1971 | Küper and Löffler | 369.33 | 4.99 | 513.1 | — | — | [0.41] |
| 1975 | Hirata | 369.30 | 4.988 | — | — | — | [0.51] |
| 1975 | Oguchi et al. | 369.30 | 4.988 | 513.0 | — | — | [3.56] |
| 1976 | Perelshtein, Parushin | 369.28 | 4.99 | 537 | — | — | [1.15] |

[a] The values for the parameters used in this book are indicated in Table 3.

Heat capacity of liquid Freon-22 was measured in two studies [2.27, 3.55]. As noted in Ref. [3.1], both sets of experimental data agree with the values computed by the thermal equation of state within a discrepancy of ±3%.

Heat capacity of solid Freon-22 was measured in only one study [3.55] from 16 K to $T_{fusion}$ (Fig. 20). The investigated sample contained 99.98 mol % Freon-22. Therefore, the investigators did not link the effect of early fusion with the presence of impurities. At a temperature of 59 K, a λ-type anomaly is observed with a heat change of around 16 ± 2 cal/mol. The Debye temperature, according to Ref. [3.55], is $\theta_D = 70$ K. The calorimetric entropy of the gas at the normal boiling point, computed from data in [3.55], was found to be equal to 63.919 cal/(mol · K) and to agree well with the results of calculating $S^0_{T_{NBP}}$ from the spectroscopic data. Thus, for example, the corresponding entropy values from the tables in [1.7, 3.41, 3.63] are 64.09, 63.959, and 63.99 cal/mol · K, respectively.

Experimental data about the surface tension of liquid Freon-22 are presented in Refs. [0.45, 1.8, 3.14, 3.60]. Figure 21 indicates that the scatter of experimental data is extremely large. According to the assessment of some investigators, the errors of the measured values of surface tension for Freon-22 are large and, for example, amount to ±8% in Ref. [3.60]. The authors of Ref. [1.8] report in paper [3.13] that they measured the surface tension of Freon-22 by using samples of technical purity. In these studies it was considered expedient to compute the tabular values of σ from the generalized equation (1.9), in which the constant $n = 1.249$ and $B = -5.024 \cdot 10^{-3}$ N/m for Freon-22 [1.2].

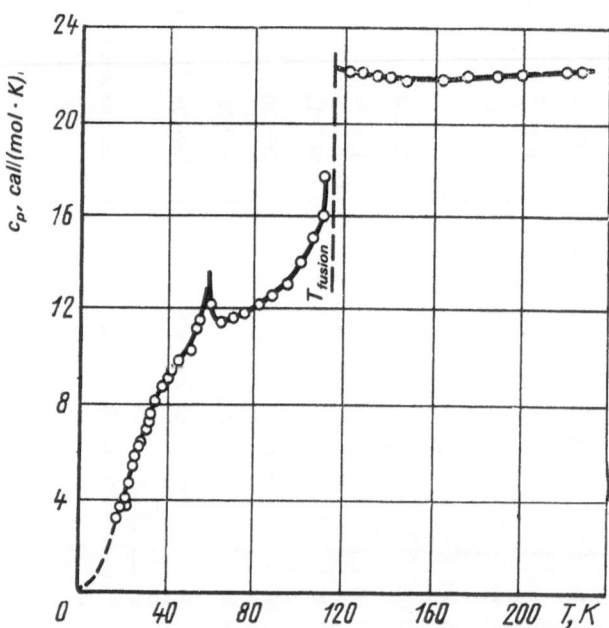

**FIG. 20.** Heat capacity of solid Freon-22 from data by Neilson and White [3.55].

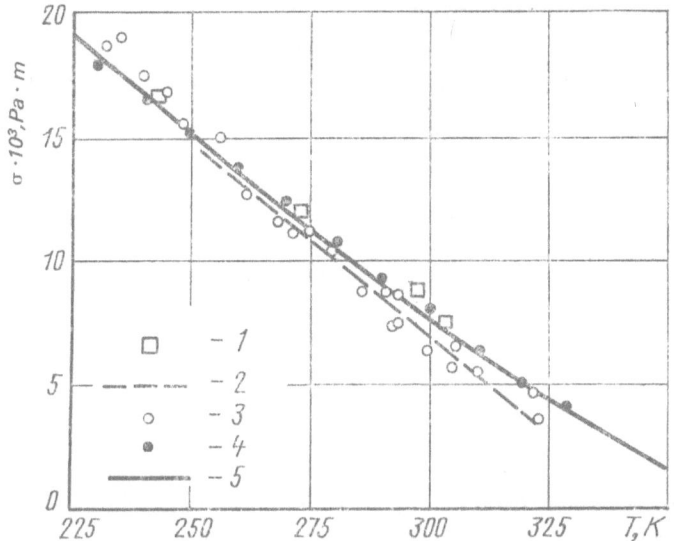

**FIG. 21.** Surface tension of Freon-22 as a function of temperature according to: 1, Plank [0.45]; 2, Steinle [3.60]; 3, Kiriyanenko and others [3.13]; 4, Zhelezny [3.14]; 5, calculated results from the generalized equation (1.9).

The error in the calculated values of $\sigma$ are evaluated to be not in excess of $\pm 3\%$. Recent published data [3.14] about surface tension of pure Freon-22 confirm this evaluation.

## Equations of State and Tables

The first tables of thermodynamic properties of Freon-22 were compiled by Graham and McHarness in 1945 based on the analysis of data by Benning and coworkers [2.23–2.26]. These tables were published practically unaltered in a number of handbooks [0.7, 0.45, 2.22, and others]. They include the values of $v$, $h$, and $s$ on the saturation line at $T = 170-325$ K and in the region of super-heated vapor at $T \leqslant 430$ K and $p \leqslant 1.8$ MPa.

A. V. Kletsky [0.18] compiled more comprehensive and accurate tables of thermodynamic properties which include data about $v$, $h$, and $s$ (at $T = 168-523$ K and $p = 0.002-6.5$ MPa) and heat capacity $c_p$ and $c_v$ (at $T = 233-513$ K and $p = 0.1-4$ MPa) for gaseous Freon-22. Computation of the tables for gaseous phase was done by the following equation of state

$$pv/RT_{cr} = \alpha_0 + \alpha_1 \tau + \beta/\tau^3 \qquad (3.5)$$

where $$\alpha_0 = \sum_{i=1}^{7} a_i \omega^i; \quad \alpha_1 = 1 + \sum_{i=1}^{7} b_i \omega^i; \quad \beta = \sum_{i=1}^{7} c_i \omega^i$$

The 21 coefficients* of the equation were obtained "manually" from experimental $pvT$ data ($T = 293-465$ K and $p = 0.8-5.8$ MPa).

Given also in Ref. [0.18] are tables of the values $v'$, $h'$, $s'$, $c'_p$, $c'_s$, and $\sigma$ for $T = 173-369$ K. These values were not obtained from the equation of state but from the computational and graphical analysis of the results that were available then. As a whole, the thermodynamic tables in Ref. [0.18] which were later included in handbooks [0.8-0.10] can be considered highly reliable, at least in the region of application of the equation of state.

One of two versions of the following equation of state

$$z = 1 + \rho \sum_{i=0}^{7} \left[ \sum_{j=0}^{s_i} c_{ij} \left( \frac{T_{cr}}{T} - 1 \right)^j \right] (\rho - \rho_{cr})^i \qquad (3.6)$$

was derived by Zander [3.66] (at $s_i = 2$, $\Sigma s_i = 24$) and constructed using the computer method of Stein. As reported by the author, this equation approximates his $pvT$ data, which were obtained by the Burnett method with a mean error $\delta\rho = 0.18\%$ and $\delta p = 0.34\%$. Data obtained by the piezometric method have a mean error $\delta\rho = 0.69\%$ and $\delta p = 0.48\%$.

The second version of Eq. (3.6) was obtained by V. V. Altunin and O. G. Gadetsky [3.1] also using a digital computer. The computer method in this case was, however, more universal and capable of handling more representative initial experimental data. In this equation $\Sigma s_i = 30$, and the values of $c_{ij}$ are determined on the basis of simultaneous statistical analysis of experimental $pvT$ data [0.18, 3.66] (that is, about 330 points in the region $T = 283-473$ K and $p = 0.3-35$ MPa), experimental data about the virial coefficient $B_1(T)$, and the smoothed-out data about $\rho'(T)$ and $c'_p(T_s)$ in the range $T = 203-360$ K.

Reference [3.1] shows that the equation of state obtained at MEI corresponds with the experimental data existing up to 1972 about thermodynamic properties of Freon-22 far better and in a wider region of state. From this equation, detailed tables of thermodynamic properties of liquid and gaseous Freon-22 were computed at $T = 273-553$ K and $p = 0.1-35$ MPa. These tables include the values of $\rho$, $h$, $s$, $\theta$, $f$, $c_p$, $c_v$, $w$, $\delta_T$, $\mu$, $p_t$, $\alpha$, and $\gamma$ in the single phase and on the saturation line.

In the work of Küper and Löffler [0.41] a system of equations is derived, which include the interpolation formulas $p_s(T)$ and $\rho'(T)$ and thermal equation of state of type (0.4) with 11 coefficients. These coefficients were obtained from $pvT$ data in [2.25, 3.22, 3.50, 3.66] and from the unpublished data of Braash (1969) at low temperatures. In all, only 510 points were used in the region $T = 173-472$ K and $\rho = 0.001-0.82$ g/cm³ ($p_{max} = 160$ bar). The standard deviation from experimental data was 0.54%. This system of equations was used by

---

* In Ref. [3.17] a similar equation of state was derived with a smaller number of coefficients. This equation was used to calculate tables and $h$, lg $p$ diagrams [3.18, 3.19].

the researchers in [0.41] for the calculation of $h$–$\lg p$ diagrams for gaseous Freon-22 and tables of the values $v$, $h$, and $s$ on the liquid saturation line. Values of $h'$ and $s'$ were calculated from the Clausius–Clapeyron equation with the help of auxiliary equations $\rho'(T)$ and $p_s(T)$ of the types (0.13) and (0.12).

In the work of I. I. Perelshtein [0.28] the tables of thermodynamic properties of saturated and superheated vapor of Freon-22 were calculated from the following equation:

$$z = 1 + \sum_{i=1}^{3} \sum_{j=0}^{3} \overline{b}_{ij} \ \rho^i/\tau^j + \sum_{i=4,68} \left[ \sum_{j=0}^{1} b_{ij} \ 1/\tau^j \right] \rho^i \quad (3.7)$$

The region of applicability of Eq. (3.7) was not specified, but the initial experimental data were reportedly taken from Refs. [0.18, 3.66]. The tables of the values were calculated in the gaseous phase only at pressures up to 6 MPa.*

The coefficients of the equation of state were determined by computer method. As a result of special analysis of $pvT$ data at low density values (see book [0.5]), the first two temperature functions are identified with the virial coefficients $B_1(\tau)$ and $B_2(\tau)$.

Given in paper [3.27] in graphical form are values [calculated by Eq. (3.7)] of the speed of sound and the adiabatic index (isentropic) $k_A$ for superheated vapors of Freon-22 at a pressure up to 6 MPa. Later, in the work of U. P. Aleshin and coworkers [0.1], diagrams of $w$–$T$ and $k_A$–$T$ for the same region of state were published. The calculations were carried out from an equation of state of type (0.7), where $r = 9$, $s_i = 3$, and $\Sigma s_i = 24$. The equation was not tested, and the region of its applicability and composition of the initial experimental data were not specified.

In September 1975 at the Fourteenth International Congress on Refrigeration, Oguchi presented a new equation of state for Freon-22 in a form of polynomial expansion (0.7), where $r = 11$, $s_i^{max} = 6$, $\Sigma s_i = 28$. The coefficients of this equation were obtained using a digital computer on the basis of analyzing heterogeneous data. In contrast to work [3.1], experimental data about $c_p$ in the single-phase region are included in the minimized quadratic functional relationship.

The authors of the equation report [3.56] that 539 $z(\rho, T)$ points were used as initial data including data from [0.18, 3.50] (242 points out of 243), part of data [3.66] (181 points out of 270), and part of data [3.56] (36 points out of 63). Additionally, 27 points of $\rho'$ measurements by Zander and 57 points of $c_p$ in the gaseous state from data by Ernst and coworkers [3.36, 3.40] were used.

The quality of approximation of the initial $pvT$ data is high and the standard deviation $\delta z$ ranges from 0.09 to 0.2%. For data [3.66] obtained by the piezometric method, the deviation increased to 0.57%. The calculated and experimental values of $c_p$ and $w$ in the gaseous phase agree, with a deviation of 1%.

---

* Tabular data [0.28] for liquid freon on the saturation line were obtained with the help of the independent system of Eqs. (0.9), (0.10), and (3.3).

But near the critical point and along the isotherm 373.15 K the discrepancy of $w_{calc}$ and $w_{exp}$ is, as expected, large. Thus, this equation belongs to the accurate equations of state of Freon-22. However, the region of its applicability was not fully clear, since about 35% of $z(\rho, T)$ points from Refs. [3.56, 3.66] were excluded from the analysis. The data for $\rho'$ and $c'_p$ at $T < 273$ K are not discussed in Ref. [3.56].

Much later, it became evident from Ref. [0.51] that Oguchi's equation is applicable in the temperature interval 287–473 K in the gaseous phase at a pressure up to 15 MPa, and in the liquid phase up to 20 MPa and $v > 0.8$ cm³/g ($w \leqslant 2.4$). However, the computed tabular values of $v$, $h$, and $s$ from this equation are given at a pressure of 15 MPa. Heat capacity $c_p$ is given at 10 MPa for superheated vapor only in the range $T = 233$–473 K. In addition, tables of $w$, $c_v$, $\gamma = c_p/c_v$ are also presented for superheated vapor in the range $T = 243$–363 K at pressures up to 4.5 MPa, and an $h$–lg $p$ diagram.

Figure 17 shows the presence of a systematic discrepancy in a certain temperature region. This discrepancy is not only between the experimental data in [2.9] and in [3.36, 3.40], but also between the tabular values of $c_p$ which are calculated in Refs. [0.51] and [3.1]. In tables [0.51] the values of $c_p^0$ used are lower (by about 1%); consequently, the deviation from the experimental data in [3.36, 3.40] at elevated pressures is decreased to 1–1.3%, whereas tabular values in [3.1] are higher by 1.5–2.8%. The compared tabular data on $c_p$ in this region of state are higher than the experimental values [2.9] by 2.5–3.2% and 2–4.2% respectively. Because of this, the probable error in the reference data about heat capacity $c_p$ of the compressed freon in the gaseous state is assessed as equal to 2–3%.

A. V. Kletsky and T. N. Tsuranova [3.21] derived the following equation of state for liquid Freon-22

$$p = p_s + b\,(T - T_s) + c(T - T_s)^2 \qquad (3.8)$$

where
$$b = 0.628 + 0.211\,\rho + 2.842\,\rho^4$$
$$c = -0.02629 + 0.09457\,\rho - 0.09879\,\rho^2 + 0.03191\,\rho^3$$

The coefficients are obtained by analyzing part of Zander's data in the interval $\rho = 500$–1250 kg/m³ ($\omega = 1.02$–2.43, $p_{max} = 20$ MPa, and $N_{exp} = 119$). The description of the initial data about density is not very good (standard deviation, 0.15%, maximum, 0.75%). Nonetheless, tables for the values of $\rho$ at $T = 263$–423 K and $p = 0.5$–30 MPa are compiled in Ref. [3.21].

It should be pointed out that earlier, in Ref. [2.3], attempts were made to construct equation of isotherms of the liquid in the form $(\rho - \rho_s) = f_1(p - p_s)$ and $p - p_s = f_2(\rho - \rho_s)$.

The abovementioned thermal equations of state are in fact approximations of the first derivative of Helmholtz's function $F = U - TS$ since $p = \rho^2(\partial F/$

$\partial \rho)_T$. However, some engineering problems are easier to solve in the $p$, $S$ variables instead of $\rho$, $T$ variables. In this case an equation of state of the form $H = f(p, S)$ should be sought. For Freon-22 such an equation was obtained in the work of Baehr and coworkers [3.34] and named the canonical equation:

$$y - y^0 = (a_0 + a_1 x + a_2 x^2 + a_3 x^3)\tilde{p} + (a_4 + a_5)x\tilde{p}^2 \qquad (3.9)$$

where $y = (H - H_0)/RT_0$; $x = (S - S_0)/R + \ln(p/p_0)$; $\tilde{p} = p/p_0$; $y^0 =$

$$\sum_{j=0}^{3} c_j x^j; \; p_0 = 1 \text{ bar}; \; T_0 = 500 \text{ K}.$$

The coefficients $c_j$ of the relation between nondimensional enthalpy and entropy in the ideal gas state are obtained from Barho's data. The coefficients $a_i$ are found by analyzing Zander's $pvT$ data in the gaseous phase at $T = 303-473$ K, $p \leqslant 2$ MPa.

Researchers in [3.34] report that the values of $v$, calculated from Eq. (3.9) using the thermodynamic relationship

$$v = [(\partial Y/\partial p)x + (1/\tilde{p})(\partial Y/\partial X)_{\tilde{p}}](RT_0/p_0) \qquad (3.10)$$

correspond well with the initial data ($\delta v \approx 0.08\%$).

When calculating the phase equilibria of solutions, the volatility of the components must be known. In the refrigerating industry, at low pressures, binary solutions with freons are employed, the components of which often having widely differing boiling temperatures. In such cases, the knowledge of the fugacity of the pure low-boiling component is sufficient.

The tables of the values $f$ for Freon-22 in the range $T = 273-553$ K and $p = 0.1-30$ MPa are presented in Ref. [3.1]. For the determination of fugacity of gaseous Freon-22, an interesting analytical method of direct calculation of $f$ from $pvT$ data is employed in Ref. [3.24]. Other thermodynamic properties of the abovementioned class of gaseous solutions can be easily evaluated from the known properties of the components [3.4].

Up to now, the discussion has been about thermodynamic properties of gaseous and liquid Freon-22. The thermodynamic functions of condensed-phase Freon-22, including the solid, are tabulated in Ref. [3.55] at $T = 15-232$ K (Table 31). This table also includes the values for entropy, enthalpy, and Gibb's function in the vapor phase at $T_{NBP}$. Specifically, $(H^0_T - H^0_0)_G = 9482.8$ cal/mol at $T_{NBP}$. On the other hand, data in [3.41, 3.63] show that $H^0_T - H^0_0 = 2135 \pm 2$ cal/mol at the normal boiling temperature of Freon-22. Consequently, heat of sublimation of the Freon-22 crystal at 0 K is $\Delta H^0_{s_0} = 7348$ cal/mol $= 355.5$ kJ/kg.

**TABLE 31.** Thermodynamic Properties in the Condensed Phase [3.55]

| $T$, K | $c_p$ | $s$ | $(h-h_0^0)$ | $-(\Phi-h_0^0)$ | |
|---|---|---|---|---|---|
| | kJ/(kg · K) | | kJ/kg | | Phase[a] |
| 15 | 0.1195 | 0.05512 | 0.5893 | 0.2375 | S |
| 20 | 0.1889 | 0.09865 | 1.350 | 0.6230 | S |
| 30 | 0.3372 | 0.2049 | 3.987 | 2.160 | S |
| 40 | 0.4362 | 0.3162 | 7.929 | 4.719 | S |
| 50 | 0.4952 | 0.4208 | 12.626 | 8.414 | S |
| 80 | 0.5845 | 0.6878 | 33.474 | 20.530 | S |
| 90 | 0.6215 | 0.7587 | 35.642 | 32.641 | S |
| 100 | 0.6602 | 0.8259 | 42.035 | 40.555 | S |
| 110 | 0.6987 | 0.8900 | 48.831 | 49.069 | S |
| 115.73 | 0.7209 | 0.9256 | 52.898 | 54.222 | S |
| 115.73 | 1.0713 | 1.3376 | 100.582 | 54.218 | L |
| 120 | 1.0693 | 1.3765 | 105.153 | 60.027 | L |
| 140 | 1.0598 | 1.5406 | 126.437 | 89.247 | L |
| 160 | 1.0575 | 1.6818 | 147.596 | 121.492 | L |
| 180 | 1.0634 | 1.8067 | 168.798 | 156.408 | L |
| 200 | 1.0695 | 1.9190 | 190.137 | 193.663 | L |
| 220 | 1.0733 | 2.0212 | 211.556 | 233.108 | L |
| 232.50 | 1.0758 | 2.0803 | 224.994 | 258.67 | L |
| 232.50 | — | 3.0859* | 458.807* | 258.67* | G |

[a] G, Gas; L, liquid; S, solid.

## 3.2. SYSTEM OF EQUATIONS FOR THE CALCULATION OF THERMODYNAMIC PROPERTIES

When deriving the equation of state for Freon-22, a computer method is used for the simultaneous analysis of heterogeneous experimental data covering the thermodynamic properties of the liquid and gases [0.5, 3.1]. As initial $pvT$ data, the experimental results of Martin [3.50], Kletsky [0.18], and Zander [3.66] are adopted. These results cover the region $T = 285$–473 K and $p = 0.3$–35 MPa in the gaseous and liquid phases, and more than 500 measured values of $z(\rho, T)$. The initial data also included smoothed-out experimental data about saturated vapor pressure $p_s$ in the interval $T_{NBP}$–$T_{cr}$, density $\rho'$, and heat capacity $c_p'$ of saturated liquid at low temperature. They also included experimental data about the temperature dependence of $B_1$ in the range $T = 275$–473 K.

The approximating expression was of the form (0.8), but the adopted version of the equation of state is transformed into the form (0.15) having the coefficients shown in Table 32.

Table 33 shows the coefficients of the interpolation equations (0.18)–(0.21) for the temperature dependencies of saturated vapor pressure and for three functions in the ideal gas state, enthalpy, heat capacity, and entropy. Since the state of crystal equilibrium at 0 K is taken as a beginning for enthalpy calculations, the numerical value of coefficient $\alpha_1^0$ in equation (0.18) is replaced by the value $\alpha_1 = \alpha_1^0 + \Delta H_{s_0}^0/RT_{cr}$ shown in Table 33.

**TABLE 32.** Coefficients of Equation of State (0.15) for Freon-22 (for $\rho$ in g/cm$^3$, $\tau = T/369.30$)

| $i$ | Values $b_{ij}$ at $j$ | | | | |
|---|---|---|---|---|---|
| | 0 | 1 | 2 | 3 | 4 |
| 0 | $0.154550033 \cdot 10^2$ | $-0.588169570 \cdot 10^2$ | $0.803989910 \cdot 10^2$ | $-0.501193823 \cdot 10^2$ | $0.106801017 \cdot 10^2$ |
| 1 | $-0.930456235 \cdot 10^2$ | $-0.327924916 \cdot 10^3$ | $-0.392420099 \cdot 10^3$ | $0.178147023 \cdot 10^3$ | $-0.195951367 \cdot 10^2$ |
| 2 | $0.137958558 \cdot 10^3$ | $-0.501523216 \cdot 10^3$ | $0.591126065 \cdot 10^3$ | $-0.220793269 \cdot 10^3$ | — |
| 3 | $0.158360768 \cdot 10^3$ | $-0.260340529 \cdot 10^3$ | $-0.468209639 \cdot 10^2$ | $0.127798687 \cdot 10^3$ | — |
| 4 | $-0.629643372 \cdot 10^3$ | $0.153334391 \cdot 10^4$ | $-0.863205629 \cdot 10^3$ | — | — |
| 5 | $0.663861736 \cdot 10^3$ | $-0.173618388 \cdot 10^4$ | $0.101955726 \cdot 10^4$ | — | — |
| 6 | $-0.284427636 \cdot 10^3$ | $0.859607943 \cdot 10^3$ | $-0.537060980 \cdot 10^3$ | — | — |
| 7 | $0.257036934 \cdot 10^2$ | $-0.132941408 \cdot 10^3$ | $0.964219190 \cdot 10^2$ | — | — |

Таблица 33

**TABLE 33.** Coefficients of Auxiliary Interpolation Equations (0.18)–(0.21) for Freon-22 (at $\tau = T/369.30$)

| $j$ | $A_{s_j}$ | $\alpha_j$ | $\beta_j$ | $\gamma_j$ |
|---|---|---|---|---|
| 0 | $0.249778028 \cdot 10^3$ | $0.120965268 \cdot 10^2$ | $0.141882115 \cdot 10^2$ | $0.700887000 \cdot 10^2$ |
| 1 | $-0.221085343 \cdot 10^4$ | $-0.1549J719 \cdot 10^1$ | $-0.856536102 \cdot 10^1$ | $-0.132753979 \cdot 10^3$ |
| 2 | $0.803322308 \cdot 10^4$ | $-0.325182809 \cdot 10^1$ | $0.18076197 \cdot 10^1$ | $0.273182973 \cdot 10^3$ |
| 3 | $-0.152712208 \cdot 10^5$ | $0.266630926 \cdot 10^2$ | $-0.134051841 \cdot 10^1$ | $-0.352929006 \cdot 10^3$ |
| 4 | $0.159603132 \cdot 10^5$ | $-0.309765069 \cdot 10^2$ | $0.29741914 \cdot 10^1$ | $0.269112619 \cdot 10^3$ |
| 5 | $-0.872481170 \cdot 10^4$ | $0.154076186 \cdot 10^2$ | $-0.1894974 \cdot 10^1$ | $-0.109500485 \cdot 10^3$ |
| 6 | $0.201342736 \cdot 10^4$ | $-0.286439441 \cdot 10^1$ | $0.35567870 \cdot 10^0$ | $0.181391902 \cdot 10^2$ |

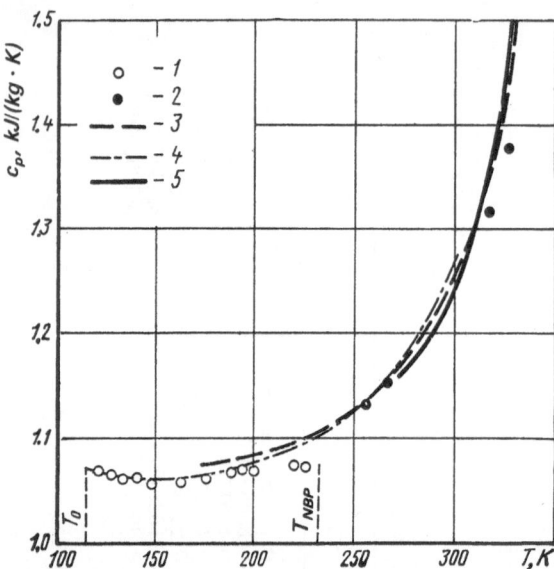

**FIG. 22.** Heat capacity of liquid Freon-22 on the saturation line. Experimental data: 1, Neilson and White [3.55]; 2, Benning et al. [2.27]. Calculated data: 3, Kletsky [0.18]; 4, in table of [0.51]; 5, present work.

Table 34 shows that the equation of state agrees well with the main groups of experimental $pvT$ data in the whole investigated region of temperature and pressure. The standard deviation of experimental data for speed of sound stays within the limit 1–2%. The agreement with the experimental values $p_s$, $c_p$, and $w$ on the saturation line are also satisfactory (see Figs. 19, 22, and 23). The discrepancy shown in Fig. 23 between the recommended values of $w'_{0\text{-}f}$ in this book and in Ref. [0.51] is considered large. This is, however, explainable. In [0.51] the dependence is calculated from a formula which artificially ties up the experimental data in the interval $T = 198$–226 K [3.52] with the evaluated values of $w'$ at 298 K in Ref. [3.47]. On the other hand, the deviation of data about heat capacity $c_p$ of gaseous Freon-22 at elevated pressures is difficult to explain. It is advisable, therefore, to ascribe to the values of $c_p$ calculated from the thermal equation of state a systematic error of 2–3%.

The surface tension of liquid Freon-22 is computed from the generalized equation [1.9]. Judging from Fig. 21, the probable error of the recommended values $\sigma$ apparently does not exceed 1.5–3%.

## 3.3. A REVIEW OF PUBLISHED DATA
## ON VISCOSITY
## AND THERMAL CONDUCTIVITY

The reference data [0.8, 0.10, 0.18, 0.34, 3.7] on viscosity of gas at low pressures and of liquid on the saturation line are practically the same. They are based

TABLE 34. Comparative Results of the Calculated Data from Eq. (0.15) and the Experimental Data for Thermodynamic Properties of Freon-22 in the Gaseous Region[a]

| Authors, reference | Measured property | $T_{exp}$, K | $p_{exp}$, MPa | Number of experimental points | $\sigma_w$ | $\sigma_p$ | $\sigma_\rho$ |
|---|---|---|---|---|---|---|---|
| | | | | | | % Error | |
| Benning and McHarness [2.25] | $z$ | 298–412 | 0.3–21 | 40 | — | 1.79 | 1.97 |
| Martin [3.50] | $z$ | 292–423 | 0.7–13 | 181 | — | 0.18 | — |
| Kletsky [0.18] | $z$ | 294–465 | 0.8–5,8 | 63(55) | — | 0.19(0.16) | 2.94(0.46) |
| Lagutina [3.23] | $z$ | 273–393 | 0.2–6 | 135(116) | — | 0.85(0.35) | 1.27 |
| Zander [3.66] | $z$ | 303–473 | 0.3–35 | 107 | — | 0.34 | 0.43 |
| | $z$ | 287–413 | 1.4–20,8 | 163 | — | 0.33(2.11*) | 0.81(0.54*) |
| Oguchi et al. [3.56] | $z$ | 253–363 | 0.2–8,9 | 63 | — | 0.26 | — |
| Novikov and Lagutina [3.25] | $w$ | 273–393 | 0.2–6 | 145(133) | 2.7(1.1) | — | — |

[a] The values in parentheses represent some of the experimental points and their corresponding standard deviation. The values of ($\sigma$) for the points in the liquid phase only are marked by an asterisk.

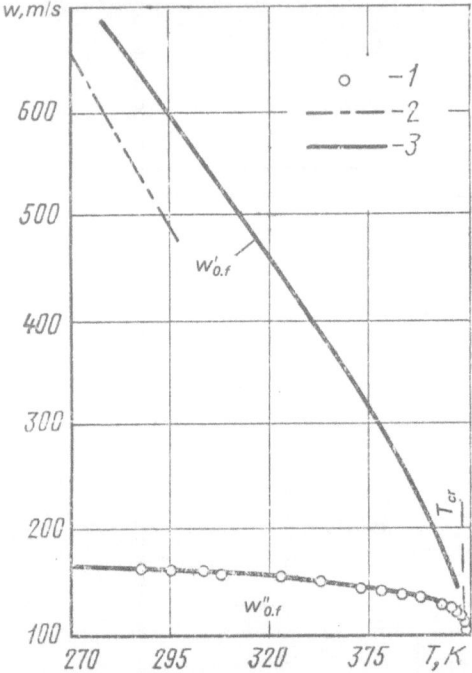

**FIG. 23.** Speed of sound on the saturation line of Freon-22 (from the side of the single-phase region) from data by: 1, Novikov and Lagutina [3.23, 3.25]; 2, tables in [0.51]; 3, present work.

on limited experimental information obtained by the rolling-ball method and published before 1962 (Table 35). So, when Kletsky [0.18] compiled tables, he used the experimental results of Makita [2.32] and Benning and Markwood [3.35] as initial data. These data, as was discovered later, are very inaccurate. Soon after this discovery, A. V. Kletsky and C. T. Butierskaya [3.20] and V. V. Altunin [3.2] compiled detailed tables of viscosity of liquid and gaseous Freon-22 in the range $T = 233-553$ K at pressures up to $30-50$ MPa. These tables, which take into consideration the measurements of C. T. Butierskaya [3.6] made using the capillary method, are more comprehensive and reliable than those published earlier. For example, according to the new measurements and tables [3.2, 3.20], at $T = 320$ K the viscosities for liquid Freon-22 on the saturation line and for gaseous freon at pressures $1-2$ MPa in comparison with the data in [0.10, 0.18, 0.34, 2.32, 3.35] are lower by $50-60\%$ and $10-15\%$ respectively. Presently, there are appreciably more experimental data, and the region of measurements is wider and experimentally verified tables of viscosity for Freon-22 are being compiled anew.

In Ref. [3.2], based on computer experiments with 43 experimental values for $\eta_T$ the following formula was obtained:

**TABLE 35.** Experimental Investigations of Viscosity of Freon-22

| Year | Authors | Tempera-ture, K | Pressure, MPa | Phase[a] | Number of experimental points | Method[b] | Reference |
|------|---------|-----------------|---------------|----------|------------------------------|-----------|-----------|
| 1939 | Benning and Markwood | 275—353 | 0.1—0.9 | G | 8 | RB | [3.35] |
|      |                      | 240—318 | $p_s$   | L | 5 | RB | [3.35] |
| 1954 | Makita | 298—473 | 0.1—2 | G | 23 | RB | [2.32] |
| 1954 | Coughlin | 239—329 | 0.1 | G | 5 | RB | [2.38] |
| 1956 | Kinser | 208—233 | $p_s$ | L | 4 | RB | [2.38] |
| 1959 | Kamien and Witzell | 303—363 | 0.1—2 | G | 13 | RB | [2.30] |
| 1959 | Tsui | 363—423 | 0.1 | G | 3 | RB | [2.38] |
| 1961 | Wilbers | 254—288 | 0.1 | G | 3 | RB | [2.38] |
| 1970 | Srichand et al. | 300 | 0.1 | G | 1 | OD | [3.59] |
| 1969 | Gordon et al. | 246—312 | $p_s$ | L | 8 | Ca | [0.40] |
| 1970 | Latto et al. | 233—365 | 0.1 | G | 14 | Ca | [3.49] |
| 1970 | Phillips and Murphy | 200—299 | $p_s$ | L | 8 | Ca | [1.50] |
| 1970, | Butierskaya | 298—473 | 0.1—49 | G, L | 200 | Ca | [3.5, |
| 1971 |           | 253—369 | $p_s$   | L | 10 |   | 3.6] |
| 1970 | Iwasaki and Takahashi | 273—398 | 0.1—6.0 | G | 48 | OD | [0.51] |
| 1973 | Geller et al. | 255—369 | 1—59 | L | 57 | Ca | [3.9] |
| 1977 | Tchaikovsky et al. | 115—313 | 1.5—59 | L | 95 | Ca | [3.32] |
| 1976 | Geller et al. | 192—228 | $p_s$ | L | 6 | Ca | [3.11, 3.12] |
| 1977 | Sagaedakova | 293—323 | 0.1—9.8 | L | 22 | Ca | [4.11] |

[a] G, Gas; L, liquid.
[b] Ca, Capillary; OD, oscillating disk; RB, rolling ball.

$$\eta_T = \left(23.6136 - 9.54140\,\frac{1}{\tau} + 1.67446\,\frac{1}{\tau^2}\right)\sqrt{\tau}, \qquad (3.11)$$

where $\eta$ is in MPa $\cdot$ s.

In order to unify the calculations in the present work, the temperature dependence of viscosity at atmospheric pressure is approximated by Eq. (1.18), in which

$$\eta_{T_{cr}} = 15.8 \text{ MPa} \cdot \text{s}; \qquad\qquad a_2 = -0.06696;$$
$$\qquad\qquad\qquad\qquad\qquad a_3 = 0.000654.$$
$$a_0 = 0.73796;$$
$$a_1 = 0.22430;$$

From Fig. 24 it follows that most of the experimental values of $\eta_T$ are in agreement, with a deviation of $\pm 1\%$. The experimental data in [3.49] are widely scattered and the data in [2.32] are systematically lower, with the discrepancy reaching 4%.

The calculated values of $\eta_T$ in the gas phase, using Eqs. (1.18) and (3.11), are practically the same both inside and outside the measurement region.

For the calculation of viscosity of gaseous and liquid Freon-22 at elevated

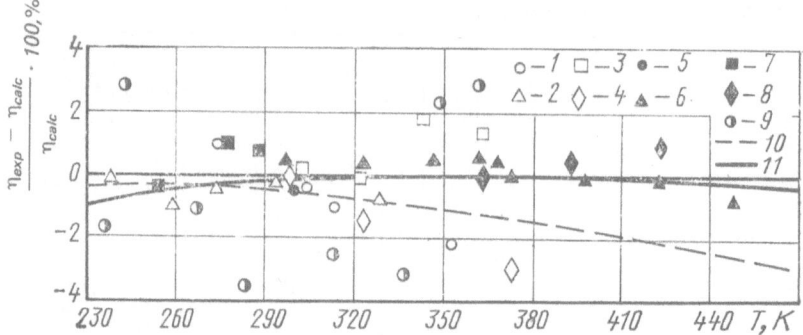

**FIG. 24.** Deviation of the values of viscosity of gaseous Freon-22 at atmospheric pressure from the values adopted in the present work. Experimental: 1, Benning and Markwood [3.35]; 2, Witzell and Jonson [2.38]; 3, Kamien and Witzell [2.30]; 4, Makita [2.32]; 5, Srichand et al. [3.59]; 6, Butierskaya [3.5, 3.6]; 7, Tsui [2.38]; 8, Wilbers [2.38]; 9, Latto et al. [3.49]. Tables: 10, Kletsky [0.18]; 11, Altunin [3.2].

pressures, two generalized equations are employed, (0.35) and (0.41). Their coefficients, presented in Table 4, are obtained by analyzing concordant experimental data about viscosity of three freons. The initial set of data included ~360 experimental values for viscosity of Freon-22 in the range $T = 115–473$ K and $p = 0.1–59$ MPa. Viscosity at the reference point ($\tau = 0.7$, $\pi = 0.7$) is $\eta_{0.7} = 254$ $\mu$Pa · s. Equation (0.35) is recommended for use at $\omega \leqslant 1.9–2.0$, whereas Eq. (0.41) is recommended for use at $\omega > 2.0$ and up to the solidification curve.

On the surface, the use of generalized equations to compute viscosity tables for Freon-22 might look unusual since it is the most widely investigated of the freons. Indeed, the experimental information about the viscosity of Freon-22 is far more extensive than that for other freons in the group under consideration. But often there were noticeable discrepancies in experimental data obtained with the same method. Here are some examples:

Viscosity of liquid Freon-22 on the saturation line was measured by Gordon and coworkers [0.40] utilizing the capillary-viscometer method with a "floating" level. The data in [0.40] lie 3–7% lower than the results of Phillips and Murphy [1.50], obtained using the same method. The discrepancy in the experimental results obtained using the rolling-ball method is even greater (Fig. 25).

C. T. Butierskaya [3.5, 3.6] measured the viscosity in a capillary viscometer designed by I. F. Golubev (version V). A similar method was used by V. Z. Geller and coworkers [3.9]. It appears that at moderate temperatures (both near the saturation line and in the single-phase region), the data in [3.6] are systematically higher (by 8–10%) than the results in [3.9]. Later, V. F. Tchaikovsky and coworkers [3.32] made new measurements, using essentially the same research method as in Refs. [3.6, 3.9]. In the overlapping region of state, the measurements in [3.9, 3.32] agree, with a deviation of around 1.1%, an

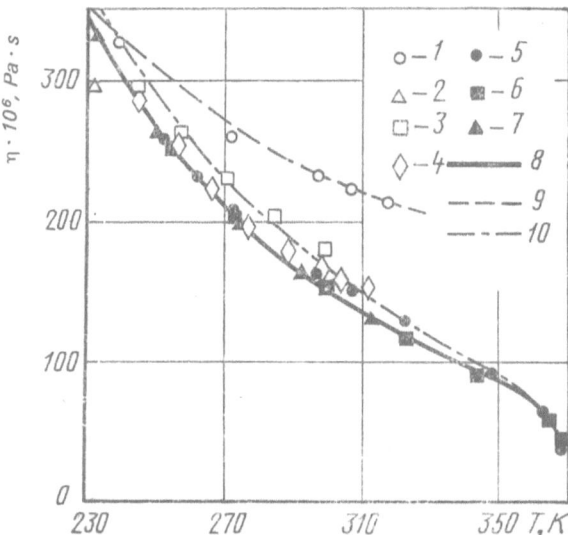

**FIG. 25.** Viscosity of liquid Freon-22 near the saturation line. Experimental data: 1, Benning and Markwood [3.35]; 2, Kinser [2.38]; 3, Phillips and Murphy [1.50]; 4, Gordon et al. [0.40]; 5, Butierskaya [3.6]; 6, Geller et al. [3.9]; 7, Tchaikovsky et al. [3.32]. Tabular data: 8, present work; 9, Kletsky [0.18]; 10, Altunin [3.2, 3.3].

indication of a systematic error in the data of Butierskaya. To confirm this, we point out one more fact. If the data in [3.6] about viscosity of Freons-21 and -22 in the liquid phase are used, a generalized equation of the type (0.41) cannot be derived. From data obtained at OTIFI [0.15, 3.9, 3.32] such an equation was derived and turned out to be quite accurate. Thus, the "generalized approach" to the analysis of measurements of viscosity for Freon-22 at high pressures is entirely justified, at least for the present time.

It is also useful to elaborate on the forms of the calculation equations. In several references (for example, [3.6, 3.20, 3.29]) for the calculation of viscosity of compressed Freon-22, equations of type (0.33) are recommmended. These equations were derived by using the experimental data of C. T. Butierskaya. In a later investigation, Eq. (0.33) at $m_1 = 4$ was used to calculate tables. However, at low temperatures the tabular data were obtained, not by analytical means, but on the basis of graphical analysis of experimental data.

The equation of type (0.37), recommended in Ref. [3.2], also includes four coefficients. However, this equation puts into effect a two-parameter function of excess viscosity $\Delta\eta(\rho, T)$ and is suitable for the analytical calculation of liquid and gaseous Freon-22 in the whole region of $\rho$, $T$ variables covered by the tables in [3.20].

In the introduction to this book it was indicated that equations of type (0.37) have several advantages in comparison with other forms of analytical functions $\Delta\eta(\rho, T)$. However, viscosity of Freon-22 is investigated in far wider ranges of

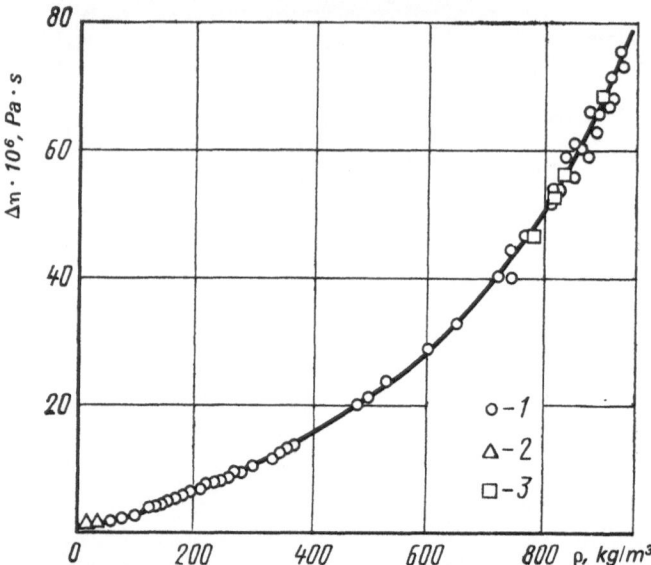

**FIG. 26.** Excess viscosity of Freon-22 as a function of density from: 1, Geller et al. [3.9]; 2, Kamien and Witzell [2.30]; 3, Tchaikovsky et al. [3.32].

temperature and pressure than is density. Therefore, the capabilities of Eq. (0.37) cannot be completely realized. In view of this, it was considered expedient to use simple functions $\Delta\eta(\rho)$ whenever such a simplification did not lead to noticeable errors (Fig. 26). At low temperatures in the liquid phase (for $\omega \geqslant$ 1.9) Eq. (0.41), in which viscosity is linked functionally to the directly measured pressure and temperature during the experiment, is used.

Figure 27 shows that the lines of constant viscosity of liquid Freon-22 on a $T$, $p$ diagram are straight lines in relatively wide temperature and pressure ranges, and they can be approximated with certainty by Eq. (0.41).*

The results of comparison of experimental viscosity values and analytical values computed from the generalized equations (0.35) and (0.41) are presented in Fig. 28.

The maximum deviation of calculated results from experimental data in [4.11] is 3.3%. The experimental data of Iwasaki and Takahashi were not published, and they became known from the reference book [0.51], where only the coefficients of the approximating polynomial are given

$$\eta = \sum_{i=0}^{4} \sum_{j=0}^{4} d_{ij}\, T^j\, p^i \qquad (3.12)$$

The equation contains 25 coefficients and describes a relatively narrow region of state in the gaseous phase at pressures up to 6 MPa.

---

* As the lines of constant viscosity get closer to the critical isotherm, they become distorted, and more complex equations (0.42)–(0.43) should be used.

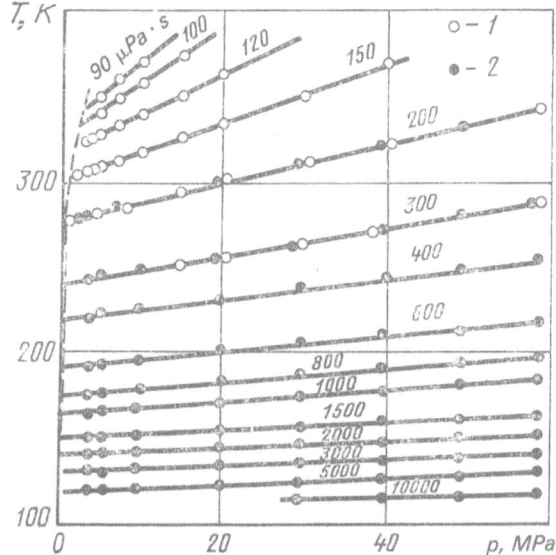

**FIG. 27.** The lines of constant viscosity of liquid Freon-22 in a diagram from: 1, Geller et al. [3.9]; 2, Tchaikovsky et al. [3.32].

Researchers in [0.20, 4.11] established that the constant-viscosity lines for a substance in a single-phase region can be represented in the $p,T$ behavior by parabolic power functions, and for the gaseous phase at $\eta \leq \eta_{cr}$ they suggested an equation of the following form

$$p = 1 + \sum_{i=1}^{2} \sum_{j=0}^{s_i} b_{ij} (\theta - T)^i (\eta - \eta_{cr})^j \qquad (3.13)$$

where $\theta = \sum_{k=0}^{2} c_k \eta^k$ = line of atmospheric pressure. Equation (3.13) contains roughly half the number of coefficients, for appreciably wider temperature and pressure ranges compared with Eq. (3.12).

In the present work, viscosity of liquid and gaseous Freon-22 are calculated from Eqs. (0.35) and (0.41), the coefficients of which were presented in Table 4. Viscosity of Freon-22 at the reference point ($\tau = 0.7$, $\pi = 0.7$) $\eta_{0.7}$ equals 254 $\mu$Pa $\cdot$ s.

The coefficients $\bar{b}_{\eta,i}$, $\bar{\alpha}_{\eta,i}$, and $\beta_{\eta,i}$ were obtained by analyzing concordant experimental data for three freons. In the general set of initial data were included ~370 experimental viscosity values for Freon-22 in the interval $T = 115-473$ K and $p = 0.1-59$ MPa, obtained in Refs. [1.50, 2.30, 3.6, 3.9, 3.32]. The error in the tabular values of viscosity is assessed to be 2–3% in the single-phase region and 4–5% on the saturation line.

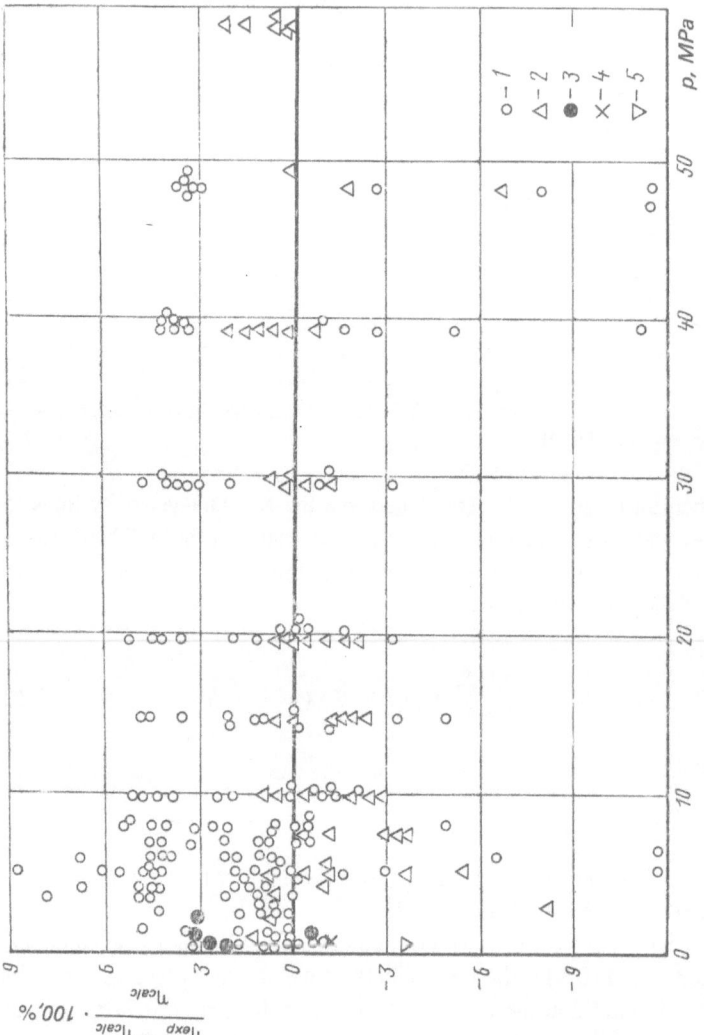

**FIG. 28.** Deviation of experimental values of viscosity for Freon-22 at elevated pressure from the values adopted in the present work: 1, Butierskaya [3.6]; 2, Geller et al. [3.9]; 3 and 4, Kamien and Witzell [2.30]; 5, Benning and Markwood [3.35].

**TABLE 36.** Experimental Investigations of Thermal Conductivity for Freon-22

| Year | Authors | Temperature, K | Pressure, MPa | Phase[a] | Number of experimental points | Method[b] | Reference |
|------|---------|-------------|----------|-------|----------------------|--------|-----------|
| 1943 | Markwood and Benning | 303; 363 | 0.1 | G | 2 | H | [2.33] |
|      |         | 273; 313 | $\sim p_s$ | L | 2 | CC | [2.33] |
| 1953 | Tchernieva | 243—353 | 0.16—1.1 | G | 18 | H | [3.33] |
|      |         | 198—293 | $\sim p_s$ | L | 11 | H | |
| 1959 | Powell and Challoner | 253—293 | $\sim p_s$ | L | 3 | F | [2.35] |
| 1962 | Widmer | 153—208 | $\sim p_s$ | L | 6 | NH | [3.64] |
| 1962 | Grassman et al. | 193—296 | $\sim p_s$ | L | 6 | NH | [3.42] |
| 1965 | Grassman and Jobst | 235—304 | $\sim p_s$ | L | 8 | NH | [3.43] |
| 1964 | Sale | 248—287 | $\sim p_s$ | L | 4 | NH | [3.58] |
| 1966 | Djalalian | 216—275 | $\sim p_s$ | L | 4 | H | [3.39] |
| 1964 | Masia et al. | 278—407 | 0.003—0.12 | G | 8 | H | [3.51] |
| 1965 | Tsvetkov | 194—343 | $\sim p_s$ | L | 9 | RR | [3.30] |
| 1966 | Powell et al. | 256 | $\sim p_s$ | L | 1 | TC | [2.36] |
| 1966 | Puranasamriddhi | 253—293 | $\sim p_s$ | L | — | H | [3.57] |
| 1967 | Tauscher | 148—303 | $\sim p_s$ | L | — | NH | [1.63] |
| 1969 | Gruzdev et al. | 315—450 | 0.15—1.6 | G | 16 | CC | [2.8] |
| 1971 | Sadiekov et al. | 126—228 | 1 | L | 14 | H | [3.28] |
| 1973, 1975 | Geller et al. | 193—433 | 0.1—59 | G, L | 158 | H | [3.9, 3.26] |
| 1975, 1977 | Geller et al. | 113—295 | 0.1—59 | L | 91 | H | [0.13, 3.10] |
| 1975 | Donaldson | 289—347 | 0.1 | G | 4 | H | [3.38] |

[a] G, Gas; L, liquid.
[b] CC, Coaxial cylinder; F, flat plate; H, heated filament; NH, nonstationary method of heated filament; RR, regular regime.

Detailed experimental investigations of thermal conductivity of compressed Freon-22 in the gaseous and liquid phases were carried out in the mid-1970s (Table 36). In part because they were done this recently, only the values of $\lambda_T$ and $\lambda_s$ are given in the reference publications [0.8–0.10, 0.18, 0.34]. For the region of elevated pressures the tables of thermal conductivity were compiled in Refs. [0.35, 3.3, 3.31].

Tables of thermal conductivity for gaseous Freon-22 at $T = 273$–473 K and $p = 0.1$–6 MPa are presented in the works of O. B. Tsvetkov [0.35, 3.31]. Calculations of the tables were carried out from the generalized equation (0.44), in which the constants $\eta$, $c_1$, and $c_2$ were considered equal for Freons-11, -12, -13, -14, and -22. But the numerical value of the individual constant $A$ for Freon-22 is not indicated. The experimental data of V. Z. Geller and coworkers [3.9] published in 1973 are not mentioned in either [0.35] or [3.31]. These references mention only that tabular values of $\lambda$ at $p = 1.6$ MPa are systematically lower (by $\sim$11%) than the data of V. A. Gruzdev and others [2.8].

Comparison of the recommended values of $\lambda$ in this book with tables in [3.31] showed that the latter are lower by 12–18%.

In the work of V. V. Altunin an experimentally based equation of type (0.40) is derived with the following coefficients:

$$a_{10} = -0.4040215; \qquad a_{20} = 0.1935286;$$
$$a_{11} = 1.451927; \qquad a_{21} = -0.3274052.$$

The initial experimental data included measured values of $\lambda$ [1.63, 2.8, 3.39] and experimental values of $\Delta\lambda$ [3.9, 3.33] at pressures up to 30 MPa. The temperature dependence of thermal conductivity at atmospheric pressure as determined from experimental data [2.8, 3.51] is represented by Eq. (0.31) with the coefficients

$$a_0 = -3.03677; \quad a_1 = 12.5878; \quad a_2 = 5.38463$$

where $\lambda_T \cdot 10^6$ is in kW/(m $\cdot$ K).

From Eqs. (0.31) and (0.40) tables of thermal conductivity of liquid and gaseous Freon-22 are calculated in the range $T = 273-553$ K and $p = 0.1-30$ MPa. In paper [3.3] tables of $\lambda'$ and $\lambda''$ are given in the range $T = 233-365$ K. It is noted there also that the experimental data in [2.8, 3.9, 3.33] do not agree at relatively low densities (Fig. 29).

In this handbook tables of thermal conductivity are computed from a system of equations derived by V. Z. Geller [1.4] on the basis of analyzing and generalizing all available experimental data (see Table 36).

Temperature dependence of thermal conductivity at atmospheric pressure is represented by Eq. (1.20) with the coefficients

$$a_0 = -8.269 \cdot 10^{-3}; \quad a_1 = 6.33 \cdot 10^{-5.}$$

where $\lambda_T$ is in W/(m $\cdot$ K). Figure 30 shows that the equation based on experimental data in [2.8, 3.51] approximates them well.

A brief analysis of the experimental investigations of Markwood and Benning [2.33], Powell et al. [2.35, 2.36], Masia et al. [3.51], Tauscher [1.63], and Gruzdev et al. [2.8] was presented in Chap. 2. Other investigations of thermal conductivity of compressed gas and liquid will be discussed next.

In the investigation of L. I. Tchernieva [3.33], in order to exclude the effect of convection, experiments were carried out at temperature gradients on two measuring cells (with clearances of 0.34 and 0.43 mm) with the thermostats being set at equal temperatures. The temperature drop in the experiments was small. The results of the measurements agree with the data of other researchers, and the discrepancy corresponds with the evaluated experimental error ($\pm 2.5\%$).

In a series of two measurements of thermal conductivity for liquid Freon-22 under the supervision of Grassman [3.42, 3.43], an experimental apparatus with two measuring cells was used. One of the cells was filled with a reference sub-

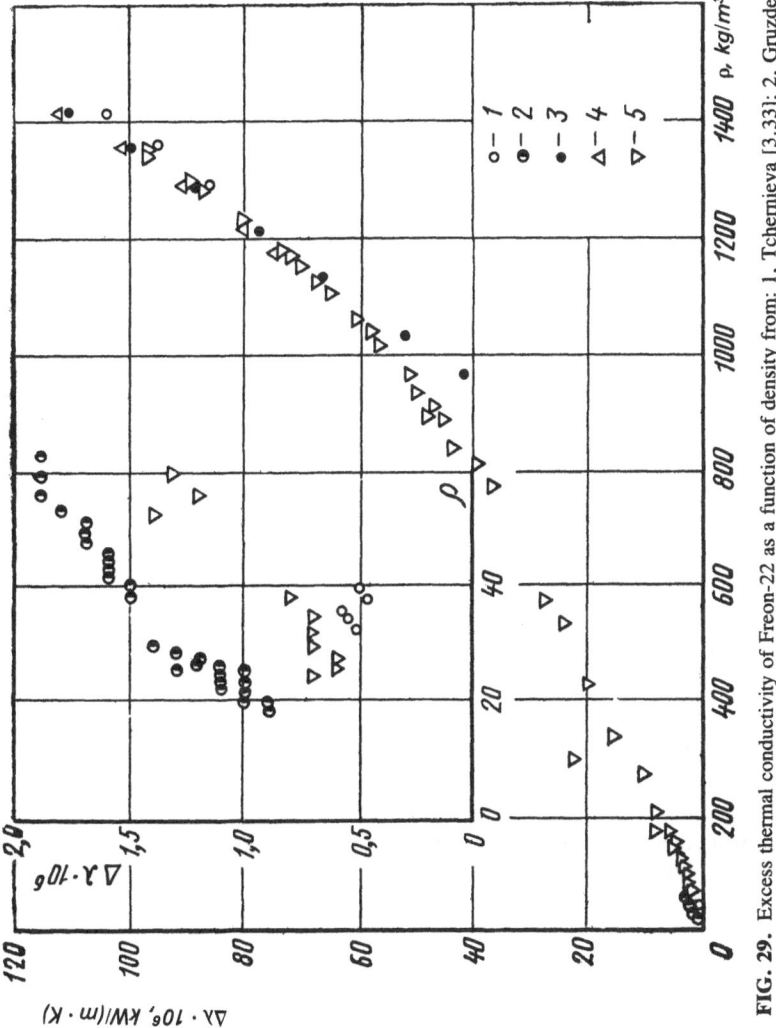

FIG. 29. Excess thermal conductivity of Freon-22 as a function of density from: 1, Tchernieva [3.33]; 2, Gruzdev et al. [2.8]; 3, Tsvetkov [3.30]; 4, Tauscher [1.63]; 5, Geller et al. [3.9].

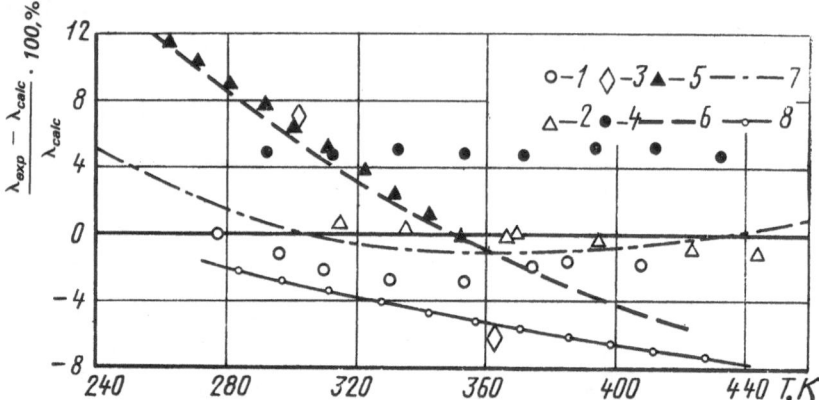

**FIG. 30.** Deviation of values of thermal conductivity of gaseous Freon-22 at atmospheric pressure from the adopted values in this book. Experimental: 1, Masia et al. [3.51]; 2, Gruzdev et al. [2.8]; 3, Markwood and Benning [2.33]; 4, Geller et al. [3.9]; 5, Tchernieva [3.33]; 6, Kletsky, tabular values, [0.18]; 7, Altunin [3.2]; 8, Tsvetkov and Polyakova [3.31].

stance while the other contained the investigated substance. As a reference, high-viscosity oil was used for which the thermal conductivity was determined at $T = 293$ K by the method relative to water. Although the authors assess the measurement error as $\pm 1\%$, in the dual relative method the error will be three times the error $\lambda_{H_2O}$, that is, $\sim 3\%$. Indeed, in the overlapping temperature interval, the experimental data in [3.64] are higher than the results in [3.42] by 4–2.5%. Grassman, Jobst, and Seich, in Refs. [3.43, 3.57], employed the absolute variant of the same method. The experimental results they obtained differ more from the measurements by other researchers, and the discrepancy amounts to 8–12%. Djalalian and Puranasamriddhi [3.39, 3.57] measured the thermal conductivity of liquid Freon-22 using the steady-heated-filament method, with an estimated error of $\pm 2\%$. Their experimental cell, however, was so arranged that the measured effective temperature difference could be too high [3.8, 3.15]. This, in turn, would lead to a decrease in the values of $\lambda$. At low temperature, the experimental data under discussion are lower than the results obtained at OTIFI by 4–5%. In the work of O. B. Tsvetkov [3.30] thermal conductivity of liquid freons is measured with an estimated error of $\pm 1.5\%$. However, the derived temperature dependence for liquid Freon-22 was unusually high in the interval $T = 313$–343 K. In our opinion, such a sharp increase in the derivative $(d\lambda_s/dT)$ a long way from the critical temperature could be caused by either the release of gases dissolved in the freon or by "falling" into the two-phase region. In Ref. [3.28] a classical version of the steady-state method of heated filament was used, and the results obtained are quite reliable. Also, it is essential to note that the measurements in [3.28] indicated a noticeable decrease of the derivative $(\partial\lambda/\partial T)_p$ near the solidification temperature of Freon-22. In the works of V. Z. Geller and others [0.13, 3.9, 3.10] two independent series of measurements using the absolute steady-state method of heated filament were made near the saturation line as well as in a wide temperature and pressure range in both the gaseous and liquid phases. As indicated by chromatography, the content of impurities in the freon did not exceed 0.1%. The error in experimental data is estimated at 1.2% and in the vicinity of the critical point it is estimated at 3–5%. The values obtained in the two-series experiments agree well with each other, and the deviation is within the stipulated measurement error.

The experimental data for thermal conductivity of liquid Freon-22 on the saturation line were analyzed according to the following scheme.

First, the temperature dependence of thermal conductivity was calculated on the basis of two independent series of experiments [0.13]. In order to achieve agreement with other measurements and bring out the most probable values of $\lambda_s$, all the experimental data were divided into three groups. The first group included the results having a probable error of 1–2%; the second group, 2–3%; and the third group, 3–5%. The first group included the results in [3.30] (except for the range $T = 310–343$ K) and all the results in [3.28]; the second group included the results in [1.63, 3.33]; and the third group included the results in [3.39, 3.42, 3.43, 3.57, 3.64]. The results in [2.33, 3.58] were excluded as they differed considerably from the rest of the results. Excluded also were data in [2.35], which, even when corrected on the basis of reliable values of $\lambda$ for toluene, were still higher by 6–7%. Next, the deviations from the function $\lambda_s(T)$ obtained beforehand were calculated. When averaging, the results of the first group of data were considered with a weight of 1; the second, 0.5; and the third, 0.25 (this operation was performed taking into account the signs of the deviations). The standard deviation obtained by this method was taken into account with a corresponding coefficient during the analytical representation of $\lambda_s$. The initial data were compiled from 113 experimental points in the range $T = 113–360$ K. The result had the form of Eq. (0.46). The optimum approximation is achieved when

$$\lambda_{cr} = 362 \cdot 10^{-4} \text{ W/(m} \cdot \text{K)};$$

$$
\begin{aligned}
c_1 &= 0.53089; & c_4 &= -8.36907; \\
c_2 &= -1.06620; & c_5 &= 6.92396; \\
c_3 &= 4.58756; & c_6 &= -1.98931.
\end{aligned}
$$

The results of comparing calculated values of $\lambda_s$ from the equation with the experimental data are shown in Fig. 31. The dependence of thermal conductivity of liquid Freon-22 on pressure is determined from the equation generalized for the group of freons under consideration (0.45), the coefficients of which were given in Table 5. The equation can be used in the region of liquid state for $\omega \geq 1.9$ and pressure to 60 MPa up to the solidification line.

The calculation of thermal conductivity of the gas and the liquid at $\omega < 1.9$ was carried out using Eq. (0.34) with individual coefficients ($\lambda \cdot 10^4$ in W/(m $\cdot$ K):

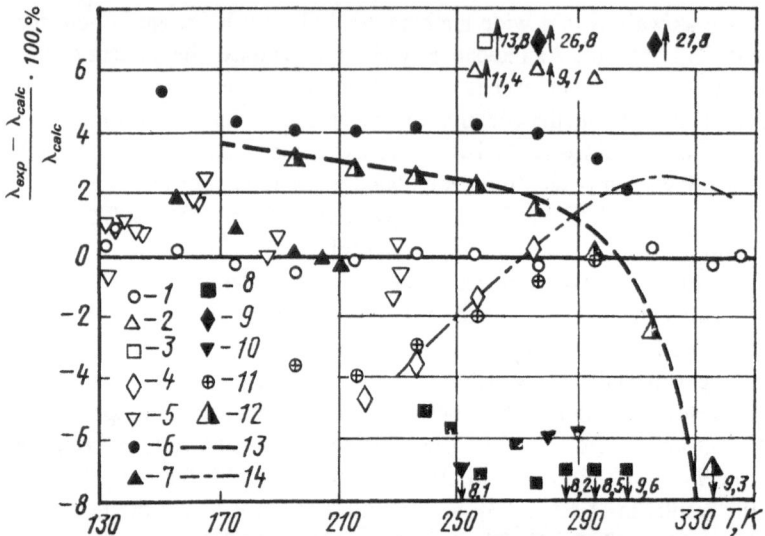

**FIG. 31.** Deviation of the values of thermal conductivity for liquid Freon-22 along the saturation line from the adopted values in the present work. Experimental: 1, Geller et al. [3.9]; 2, Powell and Challoner [2.35]; 3, Powell et al. [2.36]; 4, Djalalian [3.39]; 5, Sadiekov et al. [3.28]; 6, Tauscher [1.63]; 7, Widmer [3.64]; 8, Grassman and Jobst [3.43]; 9, Markwood and Benning [2.33]; 10, Sale [3.58]; 11, Tchernieva [3.33]; 12, Tsvetkov [3.30]. Tabular: 13, Kletsky [0.18]; 14, Altunin [3.3].

$$b_1 = 0.034072; \qquad b_3 = 0.028852;$$
$$b_2 = -0.000541; \qquad b_4 = -0.007593.$$

The coefficients of the equation are obtained by analyzing 118 experimental points [2.8] in the region $\rho = 10-900$ kg/m³.

The errors in the values of thermal conductivity of Freon-22, computed from the recommended system of equations, are estimated to be $\pm 2\%$ in the single-phase region and 3–4% on the saturation line. In the vicinity of the critical point the error could amount to 6–7%.

## 3.4. TABLES OF THERMOPHYSICAL PROPERTIES

The tables of thermodynamic properties of gaseous and liquid Freon-22 are computed from equations presented in Sec. 3.2 and cover the region $T = 233-473$ K and $p = 0.01-20$ MPa (Tables 37 and 38).

Since the thermodynamic properties of liquid Freon-22 at low temperatures were investigated only on the saturation line (see Sec. 3.1), the tables for boiling–condensation lines are compiled to the normal boiling temperature ($T_{NBP}$). In the single-phase region, experimental data were available at $T > 273$ K, and thermodynamic tables were compiled in the range 273–473 K.

The recommended tables of transport properties of Freon-22 in this book (Tables 39 and 40) cover the same region of state as the thermodynamic tables.

**TABLE 37.** Thermodynamic Properties of Freon-22 on the Saturation Line

| $T$ | $p$ | $\rho'$ | $\rho''$ | $h'$ | $h''$ | $s'$ | $s''$ | $c_p'$ | $c_p''$ | $r$ | $\sigma \cdot 10^3$ |
|---|---|---|---|---|---|---|---|---|---|---|---|
| 233.15 | 0.1054 | 1416.0 | 4.89 | 231.8 | 464.4 | 2.000 | 3.000 | 1.124 | 0.580 | 232.6 | 17.84 |
| 235.15 | 0.1155 | 1409.8 | 5.33 | 233.8 | 465.1 | 2.014 | 3.000 | 1.131 | 0.586 | 231.3 | 17.52 |
| 237.15 | 0.1264 | 1403.4 | 5.79 | 235.9 | 465.8 | 2.029 | 3.000 | 1.136 | 0.591 | 229.9 | 17.19 |
| 238.15 | 0.1321 | 1400.0 | 6.04 | 236.9 | 466.1 | 2.036 | 3.000 | 1.139 | 0.594 | 229.2 | 17.03 |
| 239.15 | 0.1380 | 1397.2 | 6.29 | 238.0 | 466.5 | 2.043 | 3.000 | 1.141 | 0.596 | 228.5 | 16.87 |
| 241.15 | 0.1505 | 1390.8 | 6.83 | 240.1 | 467.1 | 2.057 | 3.000 | 1.144 | 0.602 | 227.0 | 16.54 |
| 243.15 | 0.1639 | 1384.3 | 7.40 | 242.1 | 467.8 | 2.071 | 3.001 | 1.147 | 0.608 | 225.6 | 16.22 |
| 245.15 | 0.1782 | 1377.7 | 8.00 | 244.2 | 468.4 | 2.085 | 3.001 | 1.149 | 0.615 | 224.2 | 15.93 |
| 247.15 | 0.1934 | 1371.3 | 8.64 | 246.3 | 469.0 | 2.098 | 3.001 | 1.151 | 0.621 | 222.7 | 15.58 |
| 248.15 | 0.2014 | 1368.0 | 8.98 | 247.4 | 469.3 | 2.105 | 3.001 | 1.152 | 0.625 | 222.0 | 15.42 |
| 249.15 | 0.2097 | 1364.7 | 9.33 | 248.4 | 469.7 | 2.112 | 3.001 | 1.152 | 0.628 | 221.2 | 15.26 |
| 251.15 | 0.2269 | 1358.1 | 10.05 | 250.5 | 470.3 | 2.125 | 3.001 | 1.153 | 0.635 | 219.7 | 14.95 |
| 253.15 | 0.2453 | 1351.4 | 10.82 | 252.7 | 470.9 | 2.138 | 3.002 | 1.153 | 0.642 | 218.2 | 14.63 |
| 255.15 | 0.2647 | 1344.7 | 11.63 | 254.8 | 471.5 | 2.150 | 3.002 | 1.154 | 0.649 | 216.7 | 14.32 |
| 257.15 | 0.2854 | 1338.0 | 12.49 | 256.9 | 472.1 | 2.163 | 3.001 | 1.154 | 0.657 | 215.2 | 14.00 |
| 258.15 | 0.2961 | 1334.6 | 12.93 | 258.0 | 472.4 | 2.169 | 3.001 | 1.154 | 0.661 | 214.4 | 13.85 |
| 259.15 | 0.3072 | 1331.2 | 13.39 | 259.1 | 472.7 | 2.175 | 3.001 | 1.154 | 0.665 | 214.4 | 13.69 |
| 261.15 | 0.3303 | 1324.5 | 14.35 | 261.2 | 473.3 | 2.187 | 3.001 | 1.155 | 0.673 | 212.1 | 13.38 |
| 263.15 | 0.3548 | 1317.7 | 15.36 | 263.4 | 473.9 | 2.199 | 3.000 | 1.155 | 0.681 | 210.6 | 13.08 |
| 265.15 | 0.3805 | 1310.8 | 16.43 | 265.5 | 474.5 | 2.210 | 3.000 | 1.156 | 0.689 | 209.0 | 12.77 |

**TABLE 37.** Thermodynamic Properties of Freon-22 on the Saturation Line (*Continued*)

| $T$ | $p$ | $\rho'$ | $\rho''$ | $h'$ | $h''$ | $s'$ | $s''$ | $c_p'$ | $c_p''$ | $r$ | $\sigma \cdot 10^3$ |
|---|---|---|---|---|---|---|---|---|---|---|---|
| 267.15 | 0.4077 | 1304.0 | 17.55 | 267.7 | 475.1 | 2.221 | 2.999 | 1.157 | 0.698 | 207.4 | 12.46 |
| 268.15 | 0.4218 | 1300.5 | 18.13 | 268.8 | 475.4 | 2.227 | 2.999 | 1.158 | 0.702 | 206.6 | 12.31 |
| 269.15 | 0.4363 | 1297.1 | 18.73 | 269.9 | 475.7 | 2.232 | 2.998 | 1.159 | 0.707 | 205.8 | 12.16 |
| 271.15 | 0.4664 | 1290.2 | 19.97 | 272.1 | 476.3 | 2.243 | 2.998 | 1.161 | 0.716 | 204.2 | 11.86 |
| 273.15 | 0.4981 | 1283.2 | 21.28 | 274.3 | 476.9 | 2.254 | 2.997 | 1.163 | 0.725 | 202.6 | 11.56 |
| 275.15 | 0.5313 | 1276.2 | 22.65 | 276.5 | 477.5 | 2.264 | 2.996 | 1.166 | 0.735 | 201.0 | 11.28 |
| 277.15 | 0.5662 | 1269.1 | 24.10 | 278.7 | 478.0 | 2.274 | 2.994 | 1.169 | 0.744 | 199.3 | 10.98 |
| 278.15 | 0.5842 | 1265.6 | 24.84 | 279.8 | 478.3 | 2.279 | 2.994 | 1.171 | 0.749 | 198.5 | 10.83 |
| 279.15 | 0.6027 | 1262.0 | 25.61 | 281.0 | 478.6 | 2.284 | 2.993 | 1.173 | 0.754 | 197.6 | 10.68 |
| 281.15 | 0.6410 | 1254.9 | 27.20 | 283.2 | 479.2 | 2.294 | 2.992 | 1.177 | 0.764 | 196.0 | 10.39 |
| 283.15 | 0.6811 | 1247.7 | 28.87 | 285.5 | 479.8 | 2.303 | 2.991 | 1.182 | 0.775 | 194.2 | 10.10 |
| 285.15 | 0.7231 | 1240.4 | 30.62 | 287.8 | 480.3 | 2.313 | 2.989 | 1.188 | 0.785 | 192.5 | 9.83 |
| 287.15 | 0.7669 | 1233.0 | 32.45 | 290.2 | 480.9 | 2.322 | 2.988 | 1.193 | 0.796 | 190.7 | 9.51 |
| 288.15 | 0.7895 | 1229.3 | 33.40 | 291.3 | 481.1 | 2.327 | 2.987 | 1.197 | 0.802 | 189.9 | 9.37 |
| 289.15 | 0.8127 | 1225.6 | 34.38 | 292.4 | 481.4 | 2.331 | 2.986 | 1.200 | 0.808 | 189.0 | 9.23 |
| 291.15 | 0.8604 | 1218.1 | 36.39 | 294.8 | 481.9 | 2.341 | 2.984 | 1.207 | 0.819 | 187.2 | 8.94 |
| 293.15 | 0.9103 | 1210.6 | 38.50 | 297.2 | 482.5 | 2.350 | 2.983 | 1.215 | 0.831 | 185.3 | 8.66 |
| 295.15 | 0.9622 | 1202.8 | 40.70 | 299.5 | 483.0 | 2.358 | 2.981 | 1.223 | 0.843 | 183.5 | 8.37 |
| 297.15 | 1.0163 | 1195.1 | 43.03 | 301.9 | 483.5 | 2.367 | 2.979 | 1.231 | 0.856 | 181.6 | 8.09 |
| 298.15 | 1.0441 | 1191.1 | 44.23 | 303.2 | 483.8 | 2.372 | 2.978 | 1.236 | 0.863 | 180.6 | 7.95 |
| 299.15 | 1.0726 | 1187.2 | 45.45 | 304.4 | 484.0 | 2.376 | 2.977 | 1.241 | 0.869 | 179.6 | 7.81 |
| 301.15 | 1.1311 | 1179.2 | 48.00 | 306.8 | 484.5 | 2.385 | 2.975 | 1.251 | 0.883 | 177.7 | 7.54 |
| 303.15 | 1.1920 | 1171.1 | 50.66 | 309.3 | 484.9 | 2.393 | 2.973 | 1.261 | 0.897 | 175.7 | 7.26 |
| 305.15 | 1.2553 | 1162.9 | 53.45 | 311.8 | 485.4 | 2.402 | 2.971 | 1.273 | 0.911 | 173.6 | 6.99 |
| 307.15 | 1.3210 | 1154.5 | 56.38 | 314.3 | 485.8 | 2.410 | 2.969 | 1.285 | 0.927 | 171.5 | 6.72 |

**TABLE 37.** Thermodynamic Properties of Freon-22 on the Saturation Line (*Continued*)

| T | p | ρ' | ρ'' | h' | h'' | s' | s'' | $c_p'$ | $c_p''$ | r | $\sigma \cdot 10^3$ |
|---|---|----|-----|----|-----|----|-----|--------|---------|---|---------------------|
| 308.15 | 1.3548 | 1150.3 | 57.90 | 315.6 | 486.0 | 2.415 | 2.968 | 1.291 | 0.934 | 170.5 | 6.58 |
| 309.15 | 1.3892 | 1146.0 | 59.45 | 316.8 | 486.2 | 2.419 | 2.967 | 1.297 | 0.942 | 169.4 | 6.45 |
| 311.15 | 1.4600 | 1137.4 | 62.66 | 319.4 | 486.6 | 2.427 | 2.965 | 1.311 | 0.959 | 167.2 | 6.19 |
| 313.15 | 1.5335 | 1128.6 | 65.04 | 322.0 | 487.0 | 2.436 | 2.963 | 1.325 | 0.971 | 165.0 | 5.92 |
| 315.15 | 1.6096 | 1119.6 | 69.59 | 324.6 | 487.3 | 2.444 | 2.960 | 1.339 | 0.995 | 162.7 | 5.66 |
| 317.15 | 1.6885 | 1110.5 | 73.32 | 327.3 | 487.6 | 2.452 | 2.958 | 1.355 | 1.015 | 160.4 | 5.40 |
| 318.15 | 1.7290 | 1105.9 | 75.25 | 328.6 | 487.8 | 2.456 | 2.957 | 1.363 | 1.025 | 159.2 | 5.27 |
| 319.15 | 1.7702 | 1101.2 | 77.23 | 330.0 | 487.9 | 2.461 | 2.956 | 1.372 | 1.036 | 157.9 | 5.14 |
| 321.15 | 1.8548 | 1091.2 | 81.36 | 332.7 | 488.2 | 2.469 | 2.953 | 1.389 | 1.058 | 155.5 | 4.89 |
| 323.15 | 1.9424 | 1082.1 | 85.70 | 335.4 | 488.4 | 2.477 | 2.951 | 1.408 | 1.082 | 152.9 | 4.64 |
| 325.15 | 2.033 | 1072.2 | 90.27 | 338.2 | 488.6 | 2.486 | 2.948 | 1.428 | 1.108 | 150.4 | 4.39 |
| 327.15 | 2.127 | 1062.0 | 95.10 | 341.0 | 488.7 | 2.494 | 2.946 | 1.449 | 1.136 | 147.7 | 4.14 |
| 328.15 | 2.175 | 1056.9 | 97.61 | 342.5 | 488.7 | 2.498 | 2.944 | 1.461 | 1.151 | 146.3 | 4.02 |
| 329.15 | 2.224 | 1051.6 | 100.19 | 343.9 | 488.8 | 2.503 | 2.943 | 1.472 | 1.166 | 144.9 | 3.90 |
| 331.15 | 2.324 | 1041.0 | 105.59 | 346.8 | 488.8 | 2.511 | 2.940 | 1.497 | 1.199 | 142.0 | 3.66 |
| 333.15 | 2.427 | 1030.1 | 111.32 | 349.7 | 488.8 | 2.519 | 2.937 | 1.524 | 1.237 | 139.1 | 3.42 |
| 335.15 | 2.534 | 1018.8 | 117.40 | 352.6 | 488.7 | 2.528 | 2.934 | 1.554 | 1.278 | 136.0 | 3.19 |
| 337.15 | 2.645 | 1007.2 | 123.85 | 355.7 | 488.5 | 2.536 | 2.930 | 1.587 | 1.324 | 132.9 | 2.96 |
| 338.15 | 2.701 | 1001.3 | 127.24 | 357.2 | 488.4 | 2.541 | 2.929 | 1.605 | 1.350 | 131.2 | 2.84 |
| 339.15 | 2.759 | 995.3 | 130.74 | 358.7 | 488.3 | 2.545 | 2.927 | 1.623 | 1.376 | 129.6 | 2.73 |
| 341.15 | 2.877 | 982.9 | 138.10 | 361.8 | 487.9 | 2.554 | 2.923 | 1.665 | 1.436 | 126.1 | 2.51 |
| 343.15 | 2.998 | 970.0 | 146.00 | 365.0 | 487.5 | 2.563 | 2.919 | 1.712 | 1.505 | 122.5 | 2.29 |
| 345.15 | 3.124 | 956.7 | 154.49 | 368.2 | 487.0 | 2.571 | 2.915 | 1.766 | 1.585 | 118.7 | 2.07 |
| 347.15 | 3.253 | 942.7 | 163.73 | 371.5 | 486.3 | 2.580 | 2.911 | 1.830 | 1.680 | 114.8 | 1.86 |
| 348.15 | 3.319 | 935.4 | 168.56 | 373.2 | 485.9 | 2.585 | 2.907 | 1.866 | 1.734 | 112.7 | 1.76 |

**TABLE 37.** Thermodynamic Properties of Freon-22 on the Saturation Line (*Continued*)

| $T$ | $p$ | $\rho'$ | $\rho''$ | $h'$ | $h''$ | $s'$ | $s''$ | $c_p'$ | $c_p''$ | $r$ | $\sigma \cdot 10^3$ |
|---|---|---|---|---|---|---|---|---|---|---|---|
| 349.15 | 3.387 | 927.9 | 173.63 | 374.9 | 485.5 | 2.590 | 2.906 | 1.907 | 1.794 | 110.6 | 1.66 |
| 351.15 | 3.524 | 912.4 | 184.51 | 378.4 | 484.6 | 2.599 | 2.901 | 2.000 | 1.932 | 106.2 | 1.45 |
| 353.15 | 3.666 | 895.8 | 196.46 | 382.0 | 483.4 | 2.607 | 2.896 | 2.117 | 2.106 | 101.4 | 1.26 |
| 355.15 | 3.813 | 878.0 | 209.72 | 385.7 | 482.0 | 2.619 | 2.890 | 2.268 | 2.330 | 96.3 | 1.07 |
| 357.15 | 3.964 | 858.6 | 224.59 | 389.7 | 480.4 | 2.629 | 2.883 | 2.472 | 2.627 | 90.7 | 0.89 |
| 358.15 | 4.041 | 848.2 | 232.75 | 391.7 | 479.5 | 2.634 | 2.879 | 2.602 | 2.815 | 87.7 | 0.80 |
| 359.15 | 4.119 | 837.1 | 241.50 | 393.8 | 476.4 | 2.640 | 2.875 | 2.790 | 3.041 | 82.6 | 0.71 |
| 361.15 | 4.280 | 812.7 | 261.14 | 397.3 | 476.0 | 2.652 | 2.867 | 3.202 | 3.656 | 77.7 | 0.54 |
| 363.15 | 4.445 | 783.9 | 284.65 | 403.3 | 472.9 | 2.665 | 2.856 | 3.961 | 4.662 | 69.6 | 0.38 |
| 365.15 | 4.615 | 747.3 | 314.13 | 409.2 | 469.0 | 2.680 | 2.844 | 5.565 | 6.575 | 59.8 | 0.24 |

**TABLE 38.** Thermodynamic Properties of Freon-22 in the Single-Phase Region

$T = 273{,}15$ K

| $p$ | $\rho$ | $z$ | $h$ | $s$ | $c_p$ | $w$ | $\mu$ | $\alpha \cdot 10^3$ |
|---|---|---|---|---|---|---|---|---|
| 0.01 | 0.38 | 0.9980 | 487.2 | 3.400 | 0.621 | 176.1 | 31.81 | 3.689 |
| 0.02 | 0.76 | 0.9959 | 487.0 | 3.333 | 0.623 | 175.9 | 31.80 | 3.716 |
| 0.03 | 1.15 | 0.9939 | 486.8 | 3.294 | 0.625 | 175.6 | 31.78 | 3.744 |
| 0.04 | 1.54 | 0.9919 | 486.6 | 3.265 | 0.627 | 175.4 | 31.76 | 3.773 |
| 0.05 | 1.92 | 0.9899 | 486.4 | 3.244 | 0.629 | 175.2 | 31.75 | 3.802 |
| 0.1 | 3.89 | 0.9796 | 485.4 | 3.174 | 0.638 | 174.0 | 31.68 | 3.949 |
| 0.2 | 7.94 | 0.9585 | 483.4 | 3.102 | 0.658 | 171.6 | 31.59 | 4.266 |
| 0.3 | 12.19 | 0.9367 | 481.3 | 3.057 | 0.679 | 169.1 | 31.54 | 4.617 |
| 0.4 | 16.66 | 0.9142 | 479.1 | 3.024 | 0.701 | 166.5 | 31.56 | 5.011 |
| 0.5 | 1283.25 | 0.0148 | 274.3 | 2.253 | 1.163 | 713.9 | —0.163 | 2.789 |
| 0.6 | 1283.62 | 0.0178 | 274.3 | 2.253 | 1.163 | 714.6 | —0.164 | 2.764 |
| 0.7 | 1284.00 | 0.0208 | 274.3 | 2.253 | 1.162 | 715.2 | —0.165 | 2.759 |
| 0.8 | 1284.37 | 0.0237 | 274.4 | 2.253 | 1.162 | 715.9 | —0.166 | 2.753 |
| 0.9 | 1285.75 | 0.0267 | 274.4 | 2.253 | 1.161 | 716.6 | —0.167 | 2.748 |
| 1.0 | 1285.12 | 0.0296 | 274.4 | 2.253 | 1.161 | 717.2 | —0.168 | 2.742 |
| 1.5 | 1286.96 | 0.0444 | 274.5 | 2.251 | 1.158 | 720.5 | —0.173 | 2.716 |
| 2.0 | 1289.00 | 0.0591 | 274.6 | 2.250 | 1.155 | 723.8 | —0.178 | 2.690 |
| 2.5 | 1290.58 | 0.0738 | 274.7 | 2.249 | 1.153 | 727.0 | —0.183 | 2.665 |
| 3.0 | 1292.36 | 0.0884 | 274.7 | 2.248 | 1.151 | 730.3 | —0.187 | 2.641 |
| 3.5 | 1294.11 | 0.1030 | 274.9 | 2.247 | 1.148 | 733.5 | —0.192 | 2.617 |
| 4.0 | 1296.85 | 0.1175 | 275.0 | 2.246 | 1.146 | 736.7 | —0.196 | 2.594 |
| 4.5 | 1298.55 | 0.1320 | 275.1 | 2.245 | 1.144 | 740.0 | —0.201 | 2.571 |
| 5.0 | 1299.25 | 0.1465 | 275.2 | 2.244 | 1.142 | 743.1 | —0.205 | 2.548 |
| 5.5 | 1300.91 | 0.1610 | 275.4 | 2.243 | 1.140 | 746.3 | —0.209 | 2.526 |
| 6.0 | 1302.58 | 0.1754 | 275.5 | 2.242 | 1.138 | 749.5 | —0.213 | 2.505 |
| 6.5 | 1304.20 | 0.1898 | 275.6 | 2.241 | 1.136 | 752.6 | —0.217 | 2.484 |
| 7.0 | 1305.83 | 0.2041 | 275.7 | 2.240 | 1.134 | 755.8 | —0.221 | 2.463 |
| 7.5 | 1307.41 | 0.2184 | 275.9 | 2.240 | 1.133 | 758.9 | —0.225 | 2.443 |
| 8.0 | 1309.00 | 0.2327 | 276.0 | 2.239 | 1.131 | 762.0 | —0.228 | 2.423 |
| 8.5 | 1310.55 | 0.2469 | 276.1 | 2.238 | 1.129 | 765.1 | —0.232 | 2.404 |
| 9.0 | 1312.11 | 0.2612 | 276.2 | 2.238 | 1.128 | 768.2 | —0.235 | 2.385 |
| 9.5 | 1313.63 | 0.2754 | 276.4 | 2.236 | 1.126 | 771.2 | —0.239 | 2.366 |
| 10.0 | 1315.15 | 0.2895 | 276.7 | 2.235 | 1.125 | 774.3 | —0.242 | 2.347 |
| 11.0 | 1318.12 | 0.3177 | 276.8 | 2.233 | 1.122 | 780.3 | —0.249 | 2.312 |
| 12.0 | 1321.03 | 0.3459 | 277.1 | 2.231 | 1.120 | 786.3 | —0.256 | 2.277 |
| 13.0 | 1323.89 | 0.3739 | 277.4 | 2.230 | 1.117 | 792.2 | —0.262 | 2.243 |
| 14.0 | 1326.68 | 0.4018 | 277.7 | 2.228 | 1.115 | 798.1 | —0.268 | 2.211 |
| 15.0 | 1329.42 | 0.4296 | 278.0 | 2.226 | 1.113 | 803.8 | —0.273 | 2.179 |
| 16.0 | 1332.11 | 0.4435 | 278.3 | 2.225 | 1.111 | 809.6 | —0.279 | 2.148 |
| 17.0 | 1334.74 | 0.4849 | 278.6 | 2.223 | 1.109 | 815.2 | —0.284 | 2.119 |
| 18.0 | 1337.33 | 0.5125 | 278.9 | 2.222 | 1.108 | 820.8 | —0.290 | 2.090 |
| 19.0 | 1339.87 | 0.5399 | 279.2 | 2.220 | 1.106 | 826.3 | —0.295 | 2.062 |
| 20.0 | 1342.36 | 0.5673 | 279.6 | 2.219 | 1.105 | 831.8 | —0.300 | 2.034 |

**TABLE 38.** Thermodynamic Properties of Freon-22 in the Single-Phase Region (*Continued*)

$$T = 283.15 \text{ K}$$

| $p$ | $\rho$ | $z$ | $h$ | $s$ | $c_p$ | $w$ | $\mu$ | $\alpha \cdot 10^3$ |
|------|--------|--------|--------|--------|--------|--------|--------|--------|
| 0.01 | 0.37 | 0.9982 | 493.0 | 3.431 | 0.633 | 179.0 | 28.381 | 3.555 |
| 0.02 | 0.74 | 0.9965 | 492.8 | 3.364 | 0.635 | 178.8 | 28.367 | 3.578 |
| 0.03 | 1.11 | 0.9947 | 492.6 | 3.325 | 0.637 | 178.6 | 28.353 | 3.602 |
| 0.04 | 1.48 | 0.9929 | 492.5 | 3.296 | 0.639 | 178.4 | 28.340 | 3.626 |
| 0.05 | 1.85 | 0.9911 | 492.3 | 3.274 | 0.640 | 178.2 | 28.327 | 3.650 |
| 0.1 | 3.74 | 0.9822 | 491.4 | 3.206 | 0.649 | 177.1 | 28.267 | 3.774 |
| 0.2 | 7.62 | 0.9639 | 489.5 | 3.134 | 0.668 | 175.0 | 28.175 | 4.038 |
| 0.3 | 11.66 | 0.9450 | 487.6 | 3.090 | 0.687 | 172.7 | 28.122 | 4.328 |
| 0.4 | 15.88 | 0.9255 | 485.6 | 3.057 | 0.708 | 170.4 | 28.112 | 4.648 |
| 0.5 | 20.29 | 0.9053 | 483.6 | 3.030 | 0.730 | 167.9 | 28.148 | 5.004 |
| 0.6 | 24.92 | 0.8823 | 481.5 | 3.007 | 0.754 | 165.3 | 28.237 | 5.406 |
| 0.7 | 1247.8 | 0.0206 | 286.5 | 2.303 | 1.182 | 659.0 | —0.106 | 2.976 |
| 0.8 | 1248.2 | 0.0235 | 285.5 | 2.303 | 1.181 | 659.7 | —0.108 | 2.969 |
| 0.9 | 1248.6 | 0.0265 | 285.5 | 2.303 | 1.181 | 660.3 | —0.110 | 2.962 |
| 1.0 | 1249.1 | 0.0294 | 285.6 | 2.303 | 1.180 | 660.9 | —0.111 | 2.954 |
| 1.5 | 1251.2 | 0.0440 | 285.6 | 2.302 | 1.176 | 664.2 | —0.118 | 2.919 |
| 2.0 | 1253.4 | 0.0586 | 285.7 | 2.300 | 1.173 | 667.3 | —0.125 | 2.885 |
| 2.5 | 1255.5 | 0.0731 | 285.8 | 2.299 | 1.169 | 670.5 | —0.131 | 2.851 |
| 3.0 | 1257.6 | 0.0876 | 285.9 | 2.298 | 1.166 | 673.6 | —0.138 | 2.819 |
| 3.5 | 1259.6 | 0.1021 | 285.9 | 2.297 | 1.163 | 676.7 | —0.144 | 2.789 |
| 4.0 | 1261.7 | 0.1165 | 286.0 | 2.297 | 1.159 | 679.7 | —0.150 | 2.757 |
| 4.5 | 1263.6 | 0.1308 | 286.1 | 2.295 | 1.156 | 682.8 | —0.156 | 2.728 |
| 5.0 | 1265.6 | 0.1451 | 286.2 | 2.294 | 1.154 | 685.7 | —0.162 | 2.699 |
| 5.5 | 1267.6 | 0.1594 | 286.3 | 2.293 | 1.151 | 688.7 | —0.167 | 2.671 |
| 6.0 | 1269.5 | 0.1736 | 286.4 | 2.292 | 1.148 | 691.7 | —0.173 | 2.643 |
| 6.5 | 1271.4 | 0.1878 | 286.5 | 2.291 | 1.146 | 694.6 | —0.178 | 2.616 |
| 7.0 | 1273.2 | 0.2019 | 286.6 | 2.290 | 1.143 | 697.5 | —0.183 | 2.590 |
| 7.5 | 1275.1 | 0.2160 | 286.7 | 2.289 | 1.141 | 700.4 | —0.188 | 2.565 |
| 8.0 | 1276.9 | 0.2301 | 286.8 | 2.288 | 1.139 | 703.2 | —0.193 | 2.540 |
| 8.5 | 1278.7 | 0.2442 | 286.9 | 2.287 | 1.136 | 708.0 | —0.198 | 2.515 |
| 9.0 | 1280.5 | 0.2582 | 287.0 | 2.286 | 1.134 | 708.8 | —0.203 | 2.492 |
| 9.5 | 1282.3 | 0.2721 | 287.2 | 2.285 | 1.132 | 711.6 | —0.207 | 2.468 |
| 10.0 | 1284.0 | 0.2861 | 287.3 | 2.284 | 1.130 | 714.4 | —0.212 | 2.445 |
| 11.0 | 1287.4 | 0.3138 | 287.5 | 2.282 | 1.127 | 719.8 | —0.221 | 2.401 |
| 12.0 | 1290.8 | 0.3415 | 287.8 | 2.280 | 1.123 | 725.1 | —0.229 | 2.358 |
| 13.0 | 1294.0 | 0.3640 | 288.0 | 2.278 | 1.120 | 730.4 | —0.237 | 2.317 |
| 14.0 | 1297.2 | 0.3964 | 288.3 | 2.276 | 1.117 | 735.5 | —0.245 | 2.278 |
| 15.0 | 1300.4 | 0.4237 | 288.6 | 2.275 | 1.114 | 740.6 | —0.253 | 2.240 |
| 16.0 | 1303.4 | 0.4609 | 288.7 | 2.273 | 1.111 | 745.6 | —0.260 | 2.203 |
| 17.0 | 1306.4 | 0.4779 | 289.2 | 2.271 | 1.109 | 750.4 | —0.267 | 2.167 |
| 18.0 | 1309.4 | 0.5049 | 289.5 | 2.270 | 1.107 | 755.3 | —0.273 | 2.132 |
| 19.0 | 1312.3 | 0.5318 | 289.8 | 2.268 | 1.105 | 760.0 | —0.280 | 2.099 |
| 20.0 | 1315.1 | 0.5586 | 290.1 | 2.266 | 1.103 | 764.6 | —0.286 | 2.066 |

**TABLE 38.** Thermodynamic Properties of Freon-22 in the Single-Phase Region (*Continued*)

$$T-293.15 \text{ K}$$

| $p$ | $\rho$ | $z$ | $k$ | $s$ | $c_p$ | $w$ | $\mu$ | $\varkappa \cdot 10^3$ |
|------|--------|--------|-------|-------|-------|-------|---------|-------------|
| 0.01 | 0.35 | 0.9985 | 499.0 | 3.459 | 0.645 | 181.8 | 25.260 | 3.431 |
| 0.02 | 0.71 | 0.9969 | 498.9 | 3.391 | 0.647 | 181.7 | 25.250 | 3.451 |
| 0.03 | 1.07 | 0.9954 | 498.7 | 3.352 | 0.649 | 181.5 | 25.240 | 3.471 |
| 0.04 | 1.43 | 0.9938 | 498.5 | 3.324 | 0.650 | 181.3 | 25.230 | 3.491 |
| 0.05 | 1.79 | 0.9922 | 498.4 | 3.302 | 0.652 | 181.1 | 25.220 | 3.511 |
| 0.1 | 3.60 | 0.9844 | 497.6 | 2.233 | 0.660 | 180.1 | 25.176 | 3.615 |
| 0.2 | 7.33 | 0.9683 | 495.9 | 3.163 | 0.677 | 178.2 | 25.107 | 3.836 |
| 0.3 | 11.18 | 0.9519 | 494.1 | 3.119 | 0.695 | 176.1 | 25.068 | 4.076 |
| 0.4 | 15.18 | 0.9349 | 492.4 | 3.087 | 0.713 | 174.0 | 25.058 | 4.337 |
| 0.5 | 19.33 | 0.9174 | 490.6 | 3.061 | 0.733 | 171.8 | 25.081 | 4.624 |
| 0.6 | 23.67 | 0.8994 | 488.7 | 3.039 | 0.754 | 169.5 | 25.140 | 4.942 |
| 0.7 | 28.20 | 0.8807 | 486.8 | 4.019 | 0.777 | 167.1 | 25.238 | 5.297 |
| 0.8 | 32.96 | 0.8612 | 484.8 | 3.001 | 0.801 | 164.6 | 25.379 | 5.698 |
| 0.9 | 37.97 | 0.8409 | 482.7 | 2.984 | 0.828 | 161.9 | 25.572 | 6.155 |
| 1.0 | 30.90 | 0.1148 | 409.8 | 2.696 | 1.272 | 540.8 | 12.974 | 1.398 |
| 1.5 | 1213.6 | 0.0438 | 297.2 | 2.348 | 1.208 | 609.5 | —0.041 | 3.206 |
| 2.0 | 1216.2 | 0.0583 | 297.2 | 2.347 | 1.203 | 612.9 | —0.051 | 3.158 |
| 2.5 | 1218.7 | 0.0728 | 297.2 | 2.345 | 1.198 | 616.2 | —0.060 | 3.113 |
| 3.0 | 1221.2 | 0.0872 | 297.3 | 2.344 | 1.193 | 619.4 | —0.069 | 3.069 |
| 3.5 | 1223.6 | 0.1015 | 297.3 | 2.343 | 1.189 | 622.6 | —0.077 | 3.026 |
| 4.0 | 1226.0 | 0.1157 | 297.4 | 2.342 | 1.184 | 625.8 | —0.086 | 2.986 |
| 4.5 | 1228.4 | 0.1300 | 297.3 | 2.341 | 1.180 | 628.9 | —0.094 | 2.946 |
| 5.0 | 1230.7 | 0.1441 | 297.5 | 2.339 | 1.176 | 631.9 | —0.102 | 2.908 |
| 5.5 | 1233.0 | 0.1589 | 297.5 | 2.338 | 1.172 | 634.9 | —0.109 | 2.871 |
| 6.0 | 1235.2 | 0.1723 | 297.6 | 2.337 | 1.169 | 637.9 | —0.117 | 2.835 |
| 6.5 | 1237.5 | 0.1863 | 297.7 | 2.336 | 1.165 | 640.8 | —0.124 | 2.801 |
| 7.0 | 1239.7 | 0.2003 | 297.8 | 2.335 | 1.162 | 643.7 | —0.131 | 2.767 |
| 7.5 | 1241.8 | 0.2143 | 297.8 | 2.334 | 1.158 | 646.6 | —0.138 | 2.734 |
| 8.0 | 1243.9 | 0.2282 | 297.9 | 2.333 | 1.155 | 649.4 | —0.145 | 2.703 |
| 8.5 | 1246.0 | 0.2420 | 298.0 | 2.331 | 1.152 | 652.1 | —0.151 | 2.672 |
| 9.0 | 1248.1 | 0.2558 | 298.1 | 2.330 | 1.149 | 654.8 | —0.157 | 2.642 |
| 9.5 | 1250.1 | 0.2696 | 298.2 | 2.329 | 1.146 | 657.5 | —0.163 | 2.612 |
| 10.0 | 1252.2 | 0.2833 | 298.3 | 2.328 | 1.144 | 660.2 | —0.169 | 2.584 |
| 11.0 | 1256.1 | 0.3107 | 298.5 | 2.326 | 1.139 | 665.3 | —0.181 | 2.528 |
| 12.0 | 1260.0 | 0.3379 | 298.7 | 2.324 | 1.134 | 670.4 | —0.192 | 2.476 |
| 13.0 | 1263.7 | 0.3650 | 298.9 | 2.322 | 1.129 | 675.9 | —0.202 | 2.426 |
| 14.0 | 1267.4 | 0.3919 | 299.1 | 2.320 | 1.125 | 680.1 | —0.213 | 2.377 |
| 15.0 | 1271.0 | 0.4187 | 299.4 | 2.319 | 1.121 | 684.7 | —0.222 | 2.331 |
| 16.0 | 1274.5 | 0.4454 | 299.6 | 2.317 | 1.118 | 689.2 | —0.231 | 2.287 |
| 17.0 | 1278.0 | 0.4719 | 299.9 | 2.315 | 1.114 | 693.6 | —0.240 | 2.244 |
| 18.0 | 1281.3 | 0.4984 | 300.2 | 2.313 | 1.111 | 697.9 | —0.249 | 2.202 |
| 19.0 | 1284.6 | 0.5247 | 300.5 | 2.312 | 1.108 | 702.1 | —0.257 | 2.162 |
| 20.0 | 1287.9 | 0.5509 | 300.8 | 2.309 | 1.105 | 706.2 | —0.265 | 2.123 |

**TABLE 38.** Thermodynamic Properties of Freon-22 in the Single-Phase Region (*Continued*)

$T=303.15$ K

| p | ρ | z | h | s | $c_p$ | w | μ | $\alpha \cdot 10^3$ |
|---|---|---|---|---|---|---|---|---|
| 0.01 | 0.34 | 0.9986 | 505.3 | 3,483 | 0,657 | 184,7 | 22,468 | 3.315 |
| 0.02 | 0.69 | 0.9973 | 505,1 | 3.416 | 0,658 | 184,5 | 22,462 | 3.332 |
| 0.03 | 1,03 | 0,9959 | 505,0 | 3,377 | 0,658 | 184,3 | 22,455 | 3.349 |
| 0.04 | 1.38 | 0.9945 | 504,9 | 3.349 | 0,661 | 181,1 | 22,449 | 3.366 |
| 0.05 | 1,73 | 0.9931 | 504.7 | 3,327 | 0,663 | 184.0 | 22,443 | 3.383 |
| 0,1 | 3,48 | 0.9862 | 504.0 | 3.259 | 0,670 | 183,1 | 22.416 | 3,471 |
| 0.2 | 7,06 | 0,9721 | 502,6 | 3.189 | 0,686 | 181,3 | 22.377 | 3,656 |
| 0.3 | 10,65 | 0.9577 | 500,9 | 3.146 | 0,702 | 179,5 | 22.358 | 3,855 |
| 0,4 | 14,55 | 0,9429 | 499,3 | 3.114 | 0,718 | 177,6 | 22,361 | 4,070 |
| 0.5 | 18,49 | 0,9277 | 497,7 | 3.089 | 0.736 | 175,6 | 22.386 | 4.304 |
| 0.6 | 22,57 | 0.9120 | 496.0 | 3,067 | 0,754 | 173,5 | 22.435 | 4,559 |
| 0.7 | 26.81 | 0.8958 | 494.3 | 3,048 | 0,774 | 171,4 | 22.511 | 4,840 |
| 0.8 | 31,22 | 0.8791 | 492.5 | 3,031 | 0,795 | 169,2 | 22.616 | 5.150 |
| 0.9 | 35,83 | 0.8619 | 490,7 | 3,015 | 0,818 | 166,9 | 22.752 | 5,497 |
| 1,0 | 40,65 | 0.8439 | 488,8 | 3,000 | 0,842 | 164,4 | 22.924 | 5,888 |
| 1,5 | 1173,1 | 0,0439 | 309,3 | 2.392 | 1,256 | 555,6 | 0.064 | 3.612 |
| 2,0 | 1176.2 | 0,0583 | 309,2 | 2.391 | 1,249 | 559,5 | 0.050 | 3.543 |
| 2,5 | 1179,3 | 0,0727 | 309,2 | 2.389 | 1,241 | 563,3 | 0.037 | 3.478 |
| 3,0 | 1182.3 | 0,0870 | 309,2 | 2.388 | 1,235 | 570,0 | 0.024 | 3.415 |
| 3,5 | 1185.3 | 0.1013 | 309,2 | 2.386 | 1,228 | 570,6 | 0.012 | 3.356 |
| 4,0 | 1188.2 | 0.1155 | 309,2 | 2.385 | 1,222 | 574,2 | 0.002 | 3.299 |
| 4.5 | 1191.0 | 0.1296 | 309,2 | 2.384 | 1,216 | 577,6 | —0,011 | 3.245 |
| 5,0 | 1193.8 | 0.1437 | 309,2 | 2.383 | 1,210 | 581,0 | —0,022 | 3.193 |
| 5,5 | 1196.5 | 0.1577 | 309,2 | 2.381 | 1,205 | 584,3 | —0,033 | 3,144 |
| 6,0 | 1199.2 | 0.1716 | 309,2 | 2.380 | 1,200 | 587,6 | —0,043 | 3.096 |
| 6,5 | 1201.9 | 0.1855 | 309,3 | 2.378 | 1,195 | 590,7 | —0,052 | 3.050 |
| 7,0 | 1204.4 | 0,1994 | 309,3 | 2.377 | 1,190 | 593,8 | —0,062 | 3.006 |
| 7,5 | 1207.0 | 0.2132 | 309,3 | 2.376 | 1,186 | 596,9 | —0,071 | 2.963 |
| 8,0 | 1209.5 | 0,2269 | 309,4 | 2.375 | 1,182 | 599,8 | —0,080 | 2.922 |
| 8,5 | 1212.0 | 0.2406 | 309,4 | 2.374 | 1,177 | 602,7 | —0,088 | 2.883 |
| 9,0 | 1214,4 | 0.2542 | 309,5 | 2.372 | 1,173 | 605,6 | —0,097 | 3.844 |
| 9,5 | 1216.8 | 0.2679 | 309,5 | 2.371 | 1,170 | 608,4 | —0,105 | 2.807 |
| 10,0 | 1219.1 | 0.2814 | 309,6 | 2.370 | 1,166 | 611,1 | —0,113 | 2.771 |
| 11,0 | 1223.7 | 0.3084 | 309,7 | 2.368 | 1,159 | 616,4 | —0,128 | 2.702 |
| 12,0 | 1228,2 | 0.3352 | 309,9 | 2.366 | 1,153 | 621,5 | —0,142 | 2,636 |
| 13,0 | 1232.6 | 0,3618 | 310,1 | 2.364 | 1,146 | 626,4 | —0,155 | 2.574 |
| 14,0 | 1236.8 | 0,3883 | 310,2 | 2.361 | 1,141 | 631,1 | —0,168 | 2.516 |
| 15,0 | 1241,0 | 0.4147 | 310,4 | 2.359 | 1,136 | 635,6 | —0,180 | 2,460 |
| 16,0 | 1245,0 | 0,4409 | 310,7 | 2.358 | 1,131 | 639,9 | —0,192 | 2.406 |
| 17,0 | 1249.0 | 0,4669 | 310,9 | 3.356 | 1,126 | 644,1 | —0,203 | 2.354 |
| 18,0 | 1252.8 | 0,4929 | 311,1 | 2.354 | 1,122 | 648,1 | —0,214 | 2.305 |
| 19,0 | 1256.6 | 0.5187 | 311,4 | 3.352 | 1,118 | 651,9 | —0,225 | 2.257 |
| 20,0 | 1260.3 | 0,5444 | 311,6 | 2,350 | 1,114 | 655,6 | —0,235 | 2,211 |

**TABLE 38.** Thermodynamic Properties of Freon-22 in the Single-Phase Region (*Continued*)

$T=313.15$ K

| $p$ | $\rho$ | $z$ | $h$ | $s$ | $c_p$ | $w$ | $\mu$ | $\alpha \cdot 10^3$ |
|---|---|---|---|---|---|---|---|---|
| 0.01 | 0.33 | 0.9988 | 511.8 | 3.506 | 0.668 | 187.4 | 19.999 | 3.207 |
| 0.02 | 0.67 | 0.9976 | 511.7 | 3.439 | 0.670 | 187.2 | 19.996 | 3.222 |
| 0.03 | 1.00 | 0.9964 | 511.5 | 3.400 | 0.671 | 187.1 | 19.993 | 3.236 |
| 0.04 | 1.34 | 0.9951 | 511.4 | 3.372 | 0.672 | 186.9 | 19.991 | 3.251 |
| 0.05 | 1.67 | 0.9939 | 511.3 | 3.351 | 0.674 | 186.8 | 19.988 | 3.265 |
| 0.1 | 3.36 | 0.9878 | 510.6 | 3.282 | 0.680 | 186.0 | 19.977 | 3.339 |
| 0.2 | 6.81 | 0.9753 | 509.2 | 4.212 | 0.694 | 184.3 | 19.965 | 3.495 |
| 0.3 | 10.35 | 0.9626 | 507.8 | 3.170 | 0.708 | 182.7 | 19.968 | 3.661 |
| 0.4 | 13.99 | 0.9496 | 506.4 | 3.139 | 0.723 | 180.9 | 19.985 | 3.839 |
| 0.5 | 17.74 | 0.9362 | 504.9 | 3.114 | 0.739 | 179.2 | 20.018 | 4.031 |
| 0.6 | 21.60 | 0.9225 | 503.4 | 3.093 | 0.755 | 177.3 | 20.068 | 4.238 |
| 0.7 | 25.59 | 0.9085 | 501.9 | 3.075 | 0.772 | 175.4 | 20.136 | 4.464 |
| 0.8 | 29.72 | 0.8940 | 500.3 | 3.058 | 0.790 | 173.5 | 20.223 | 4.709 |
| 0.9 | 34.00 | 0.8791 | 498.7 | 3.043 | 0.890 | 171.4 | 20.331 | 4.979 |
| 1.0 | 38.45 | 0.8637 | 497.0 | 3.029 | 0.829 | 169.3 | 20.460 | 5.277 |
| 1.5 | 64.06 | 0.7776 | 487.7 | 2.967 | 0.964 | 157.1 | 21.554 | 7.446 |
| 2.0 | 1132.3 | 0.0587 | 321.9 | 2.434 | 1.313 | 505.3 | 0.190 | 4.096 |
| 2.5 | 1136.2 | 0.0731 | 321.8 | 2.432 | 1.302 | 510.1 | 0.169 | 3.994 |
| 3.0 | 1140.0 | 0.0874 | 321.7 | 2.430 | 1.292 | 514.7 | 0.150 | 3.900 |
| 3.5 | 1143.7 | 0.1016 | 321.6 | 2.429 | 1.282 | 519.1 | 0.132 | 3.811 |
| 4.0 | 1147.3 | 0.1158 | 321.5 | 2.427 | 1.273 | 523.4 | 0.115 | 3.728 |
| 4.5 | 1150.8 | 0.1299 | 321.4 | 2.425 | 1.265 | 527.6 | 0.098 | 3.650 |
| 5.0 | 1154.2 | 0.1439 | 321.4 | 2.424 | 1.267 | 531.6 | 0.083 | 3.576 |
| 5.5 | 1157.5 | 0.1578 | 321.3 | 2.422 | 1.249 | 535.6 | 0.068 | 3.507 |
| 6.0 | 1160.7 | 0.1717 | 321.3 | 2.421 | 1.242 | 539.4 | 0.054 | 3.441 |
| 6.5 | 1163.9 | 0.1855 | 321.3 | 2.419 | 1.235 | 543.1 | 0.040 | 3.378 |
| 7.0 | 1167.0 | 0.1992 | 321.2 | 2.418 | 1.229 | 546.7 | 0.027 | 3.319 |
| 7.5 | 1170.1 | 0.2129 | 321.2 | 2.417 | 1.223 | 550.1 | 0.015 | 3.262 |
| 8.0 | 1173.1 | 0.2265 | 321.2 | 2.415 | 1.217 | 553.5 | 0.003 | 3.207 |
| 8.5 | 1176.0 | 0.2400 | 321.2 | 2.414 | 1.212 | 556.8 | —0.008 | 3.155 |
| 9.0 | 1178.9 | 0.2535 | 321.2 | 2.413 | 1.206 | 560.0 | —0.019 | 3.105 |
| 9.5 | 1181.7 | 0.2670 | 321.3 | 2.411 | 1.201 | 563.2 | —0.030 | 3.057 |
| 10.0 | 1184.4 | 0.2804 | 321.3 | 2.410 | 1.196 | 566.2 | —0.040 | 3.011 |
| 11.0 | 1189.8 | 0.3070 | 321.3 | 2.407 | 1.187 | 572.0 | —0.060 | 2.924 |
| 12.0 | 1195.1 | 0.3335 | 321.4 | 2.405 | 1.179 | 577.5 | —0.078 | 2.842 |
| 13.0 | 1200.1 | 0.3597 | 321.5 | 2.403 | 1.171 | 582.7 | —0.095 | 2.766 |
| 14.0 | 1205.1 | 0.3858 | 321.6 | 2.400 | 1.164 | 587.7 | —0.111 | 2.694 |
| 15.0 | 1209.8 | 0.4117 | 321.8 | 2.398 | 1.157 | 592.3 | —0.127 | 2.626 |
| 16.0 | 1214.5 | 0.4375 | 321.9 | 2.396 | 1.151 | 596.7 | —0.141 | 2.562 |
| 17.0 | 1219.0 | 0.4631 | 322.2 | 2.394 | 1.145 | 600.9 | —0.155 | 2.501 |
| 18.0 | 1223.5 | 0.4886 | 322.3 | 2.392 | 1.139 | 604.9 | —0.168 | 2.442 |
| 19.0 | 1227.8 | 0.5139 | 322.5 | 2.390 | 1.134 | 608.6 | —0.182 | 2.386 |
| 20.0 | 1232.0 | 0.5391 | 322.7 | 2.388 | 1.129 | 612.1 | —0.194 | 2.333 |

**TABLE 38.** Thermodynamic Properties of Freon-22 in the Single-Phase Region (*Continued*)

$$T=323.15 \text{ K}$$

| $p$ | $\rho$ | $z$ | $h$ | $s$ | $c_p$ | $w$ | $\mu$ | $\alpha \cdot 10^3$ |
|---|---|---|---|---|---|---|---|---|
| 0.01 | 0.32 | 0.9989 | 518.5 | 3.528 | 0.680 | 190.1 | 17.836 | 3.107 |
| 0.02 | 0.64 | 0.9978 | 518.4 | 3.461 | 0.681 | 189.9 | 17.835 | 3.119 |
| 0.03 | 0.97 | 0.9967 | 518.3 | 3.422 | 0.682 | 189.8 | 17.835 | 3.131 |
| 0.04 | 1.29 | 0.9957 | 518.1 | 3.394 | 0.683 | 189.7 | 17.835 | 3.143 |
| 0.05 | 1.62 | 0.9946 | 518.0 | 3.372 | 0.685 | 189.5 | 17.835 | 3.156 |
| 0.1 | 3.26 | 0.9891 | 517.4 | 3.304 | 0.691 | 188.8 | 17.837 | 3.219 |
| 0.2 | 6.58 | 0.9780 | 516.2 | 3.235 | 0.703 | 187.3 | 17.848 | 3.350 |
| 0.3 | 9.99 | 0.9697 | 514.9 | 3.193 | 0.715 | 185.8 | 17.869 | 3.490 |
| 0.4 | 13.47 | 0.9552 | 513.6 | 3.163 | 0.728 | 184.2 | 17.900 | 3.638 |
| 0.5 | 17.06 | 0.9434 | 512.3 | 3.138 | 0.742 | 182.6 | 17.942 | 3.797 |
| 0.6 | 20.73 | 0.9314 | 510.9 | 3.117 | 0.756 | 181.0 | 17.996 | 3.967 |
| 0.7 | 24.51 | 0.0191 | 509.6 | 3.099 | 0.771 | 179.3 | 18.061 | 4.150 |
| 0.8 | 28.41 | 0.9064 | 508.2 | 3.083 | 0.786 | 177.5 | 18.139 | 4.348 |
| 0.9 | 32.42 | 0.8935 | 506.7 | 3.069 | 0.802 | 175.7 | 18.231 | 4.562 |
| 1.0 | 36.57 | 0.9902 | 505.2 | 3.055 | 0.820 | 173.8 | 18.337 | 4.795 |
| 1.5 | 59.81 | 0.8072 | 497.1 | 2.997 | 0.927 | 163.4 | 19.135 | 6.377 |
| 2.0 | 1082.7 | 0.0595 | 335.4 | 2.477 | 1.406 | 447.7 | 0.392 | 4.942 |
| 2.5 | 1087.9 | 0.0740 | 335.1 | 2.475 | 1.387 | 454.1 | 0.358 | 4.765 |
| 3.0 | 1092.8 | 0.0883 | 334.9 | 2.473 | 1.370 | 460.1 | 0.326 | 4.606 |
| 3.5 | 1097.6 | 0.1026 | 334.7 | 2.471 | 1.355 | 466.0 | 0.297 | 4.461 |
| 4.0 | 1102.2 | 0.1168 | 334.4 | 2.469 | 1.341 | 471.5 | 0.270 | 4.329 |
| 4.5 | 1106.6 | 0.1309 | 334.3 | 2.467 | 1.328 | 476.9 | 0.245 | 4.208 |
| 5.0 | 1110.9 | 0.1449 | 334.2 | 2.465 | 1.317 | 482.0 | 0.221 | 4.096 |
| 5.5 | 1115.0 | 0.1587 | 334.0 | 2.463 | 1.306 | 486.9 | 0.199 | 3.992 |
| 6.0 | 1119.1 | 0.1726 | 333.9 | 2.461 | 1.296 | 491.7 | 0.179 | 3.896 |
| 6.5 | 1123.0 | 0.1863 | 333.8 | 2.4605 | 1.286 | 496.2 | 0.159 | 3.806 |
| 7.0 | 1126.8 | 0.2000 | 333.7 | 2.458 | 1.277 | 500.6 | 0.141 | 3.722 |
| 7.5 | 1130.5 | 0.2135 | 333.6 | 2.456 | 1.269 | 504.9 | 0.124 | 3.643 |
| 8.0 | 1134.1 | 0.2270 | 333.6 | 2.455 | 1.261 | 508.9 | 0.107 | 3.568 |
| 8.5 | 1137.6 | 0.2405 | 333.5 | 2.453 | 1.254 | 512.9 | 0.091 | 3.498 |
| 9.0 | 1141.0 | 0.2539 | 333.4 | 2.452 | 1.247 | 516.7 | 0.076 | 3.431 |
| 9.5 | 1144.4 | 0.2672 | 333.4 | 2.450 | 1.240 | 520.5 | 0.062 | 3.368 |
| 10.0 | 1147.7 | 0.2804 | 333.4 | 2.447 | 1.234 | 524.0 | 0.049 | 3.308 |
| 11.0 | 1154.0 | 0.3068 | 333.3 | 2.446 | 1.222 | 530.0 | 0.023 | 3.196 |
| 12.0 | 1160.2 | 0.3329 | 333.3 | 2.443 | 1.212 | 537.2 | 0.039 | 3.093 |
| 13.0 | 1166.1 | 0.3588 | 333.3 | 2.441 | 1.202 | 543.1 | 0.022 | 2.998 |
| 14.0 | 1171.8 | 0.3845 | 333.4 | 2.438 | 1.193 | 548.6 | 0.043 | 2.910 |
| 15.0 | 1177.3 | 0.4100 | 333.4 | 2.436 | 1.184 | 553.8 | 0.062 | 2.828 |
| 16.0 | 1182.7 | 0.4354 | 333.5 | 2.433 | 1.176 | 558.6 | 0.080 | 2.751 |
| 17.0 | 1187.8 | 0.4606 | 333.6 | 2.431 | 1.169 | 563.1 | 0.097 | 2.679 |
| 18.0 | 1193.0 | 0.4856 | 333.7 | 2.427 | 1.162 | 567.3 | 0.113 | 2.610 |
| 19.0 | 1197.9 | 0.5105 | 333.9 | 2.427 | 1.155 | 471.2 | 0.128 | 2.545 |
| 20.0 | 1202.7 | 0.5352 | 334.0 | 2.424 | 1.149 | 574.8 | 0.143 | 2.483 |

**TABLE 38.** Thermodynamic Properties of Freon-22 in the Single-Phase Region (*Continued*)

$T=333.15$ K

| $p$ | $\rho$ | $z$ | $h$ | $s$ | $c_p$ | $w$ | $\mu$ | $\alpha \cdot 10^8$ |
|------|--------|--------|-------|-------|-------|-------|--------|---------|
| 0.01 | 0.31 | 0.9990 | 525.4 | 3.549 | 0.691 | 192.8 | 15.951 | 3.012 |
| 0.02 | 0.63 | 0.9981 | 525.2 | 3.482 | 0.692 | 192.7 | 15.953 | 3.022 |
| 0.03 | 0.94 | 0.9971 | 525.1 | 3.442 | 0.693 | 192.5 | 15.955 | 3.033 |
| 0.04 | 1.25 | 0.9961 | 525.0 | 3.415 | 0.694 | 192.4 | 15.957 | 3.043 |
| 0.05 | 1.57 | 0.9952 | 524.9 | 3.393 | 0.695 | 192.3 | 15.960 | 3.054 |
| 0.1 | 3.15 | 0.9902 | 524.4 | 3.325 | 0.700 | 191.6 | 15.971 | 3.107 |
| 0.2 | 6.37 | 0.9803 | 523.2 | 3.256 | 0.711 | 190.2 | 15.999 | 3.219 |
| 0.3 | 9.65 | 0.9703 | 522.1 | 3.215 | 0.722 | 188.8 | 16.034 | 3.337 |
| 0.4 | 13.01 | 0.9600 | 520.9 | 3.185 | 0.734 | 187.4 | 16.077 | 3.462 |
| 0.5 | 16.44 | 0.9495 | 519.7 | 3.161 | 0.745 | 185.9 | 16.126 | 3.595 |
| 0.6 | 19.95 | 0.9389 | 518.5 | 3.140 | 0.758 | 184.4 | 16.183 | 3.736 |
| 0.7 | 23.55 | 0.9280 | 517.3 | 3.123 | 0.770 | 182.9 | 16.249 | 3.887 |
| 0.8 | 27.24 | 0.9169 | 516.0 | 3.107 | 0.784 | 181.3 | 16.322 | 4.048 |
| 0.9 | 31.03 | 0.9055 | 514.7 | 3.093 | 0.798 | 179.7 | 16.405 | 4.220 |
| 1.0 | 34.92 | 0.8939 | 513.4 | 3.080 | 0.812 | 178.0 | 16.497 | 4.406 |
| 1.5 | 56.34 | 0.8312 | 506.2 | 3.025 | 0.900 | 169.0 | 17.120 | 5.606 |
| 2.0 | 82.38 | 0.7579 | 497.8 | 2.977 | 1.031 | 158.2 | 18.120 | 7.621 |
| 2.5 | 1031.1 | 0.7569 | 349.6 | 2.519 | 1.519 | 391.6 | 0.656 | 6.085 |
| 3.0 | 1038.2 | 0.0902 | 349.1 | 2.516 | 1.486 | 400.2 | 0.595 | 5.758 |
| 3.5 | 1044.8 | 0.1046 | 348.7 | 2.513 | 1.459 | 408.3 | 0.542 | 5.479 |
| 4.0 | 1051.1 | 0.1188 | 348.4 | 2.511 | 1.434 | 415.9 | 0.494 | 5.237 |
| 4.5 | 1057.0 | 0.1329 | 348.0 | 2.508 | 1.413 | 423.1 | 0.451 | 5.024 |
| 5.0 | 1062.6 | 0.1469 | 347.7 | 2.506 | 1.394 | 429.9 | 0.412 | 4.836 |
| 5.5 | 1068.0 | 0.1608 | 347.4 | 2.504 | 1.377 | 436.4 | 0.377 | 4.667 |
| 6.0 | 1073.1 | 0.1745 | 347.2 | 2.502 | 1.362 | 442.5 | 0.345 | 4.514 |
| 6.5 | 1078.1 | 0.1882 | 347.0 | 2.500 | 1.348 | 448.4 | 0.315 | 4.376 |
| 7.0 | 1082.9 | 0.2018 | 346.8 | 2.498 | 1.335 | 454.0 | 0.287 | 4.250 |
| 7.5 | 1087.5 | 0.2153 | 346.6 | 2.496 | 1.324 | 459.4 | 0.262 | 4.133 |
| 8.0 | 1091.9 | 0.2267 | 346.4 | 2.494 | 1.313 | 464.6 | 0.238 | 4.026 |
| 8.5 | 1096.2 | 0.2420 | 346.3 | 2.492 | 1.303 | 469.5 | 0.216 | 3.927 |
| 9.0 | 1100.4 | 0.2553 | 346.1 | 2.490 | 1.294 | 474.2 | 0.195 | 3.834 |
| 9.5 | 1104.4 | 0.2685 | 346.0 | 2.489 | 1.285 | 478.8 | 0.175 | 3.748 |
| 10.0 | 1108.4 | 0.2816 | 345.9 | 2.487 | 1.277 | 483.2 | 0.156 | 3.667 |
| 11.0 | 1116.0 | 0.3077 | 345.7 | 2.484 | 1.262 | 491.4 | 0.122 | 3.519 |
| 12.0 | 1123.2 | 0.3335 | 345.6 | 2.481 | 1.248 | 499.1 | 0.091 | 3.386 |
| 13.0 | 1130.1 | 0.3591 | 345.5 | 2.478 | 1.236 | 536.1 | 0.063 | 3.267 |
| 14.0 | 1136.8 | 0.3844 | 345.4 | 2.475 | 1.225 | 512.7 | 0.038 | 3.158 |
| 15.0 | 1143.2 | 0.4096 | 345.4 | 2.472 | 1.215 | 518.7 | 0.014 | 3.059 |
| 16.0 | 1149.4 | 0.4346 | 345.4 | 2.469 | 1.205 | 524.3 | −0.008 | 2.967 |
| 17.0 | 1155.3 | 0.4593 | 345.4 | 2.467 | 1.197 | 529.5 | −0.029 | 2.881 |
| 18.0 | 1161.1 | 0.4839 | 345.5 | 2.464 | 1.188 | 534.2 | −0.048 | 2.801 |
| 19.0 | 1166.7 | 0.5083 | 345.5 | 2.462 | 1.181 | 538.5 | −0.067 | 2.726 |
| 20.0 | 1172.2 | 0.5326 | 345.6 | 2.460 | 1.173 | 542.5 | −0.084 | 2.655 |

**TABLE 38.** Thermodynamic Properties of Freon-22 in the Single-Phase Region (*Continued*)

$$T=343.15 \text{ K}$$

| $p$ | $\rho$ | $z$ | $h$ | $s$ | $c_p$ | $w$ | $\mu$ | $\alpha \cdot 10^{9}$ |
|------|------|------|------|------|------|------|------|------|
| 0.01 | 0.30 | 0.9991 | 532.3 | 3.569 | 0.702 | 195.4 | 14.320 | 2.923 |
| 0.02 | 0.61 | 0.9983 | 532.2 | 3.502 | 0.703 | 195.3 | 14.323 | 2.932 |
| 0.03 | 0.91 | 0.9974 | 532.1 | 3.463 | 0.704 | 195.2 | 14.327 | 2.941 |
| 0.04 | 1.22 | 0.9965 | 532.0 | 3.435 | 0.705 | 195.0 | 14.330 | 2.950 |
| 0.05 | 1.52 | 0.9956 | 531.9 | 3.414 | 0.706 | 194.9 | 14.334 | 2.959 |
| 0.1 | 3.06 | 0.9912 | 531.5 | 3.346 | 0.710 | 194.3 | 14.352 | 3.005 |
| 0.2 | 6.17 | 0.9823 | 530.4 | 3.277 | 0.720 | 193.0 | 14.392 | 3.100 |
| 0.3 | 9.34 | 0.9733 | 529.4 | 3.236 | 0.730 | 191.7 | 14.437 | 3.201 |
| 0.4 | 12.58 | 0.9641 | 528.3 | 3.206 | 0.739 | 190.4 | 14.487 | 3.307 |
| 0.5 | 15.87 | 0.9547 | 527.2 | 3.182 | 0.750 | 199.1 | 14.542 | 3.418 |
| 0.6 | 19.24 | 0.9452 | 526.1 | 3.162 | 0.760 | 187.7 | 14.602 | 3.537 |
| 0.7 | 22.68 | 0.9336 | 525.0 | 3.145 | 0.771 | 186.4 | 14.667 | 3.662 |
| 0.8 | 26.19 | 0.9257 | 523.9 | 3.130 | 0.783 | 184.9 | 14.737 | 3.795 |
| 0.9 | 29.80 | 0.9157 | 522.7 | 3.116 | 0.795 | 183.5 | 14.814 | 3.936 |
| 1.0 | 33.47 | 0.9054 | 521.5 | 3.104 | 0.807 | 182.0 | 14.896 | 4.087 |
| 1.5 | 53.43 | 0.8509 | 515.1 | 3.051 | 0.879 | 174.0 | 15.408 | 5.024 |
| 2.0 | 76.82 | 0.7891 | 507.8 | 3.007 | 0.979 | 164.9 | 16.134 | 6.451 |
| 2.5 | 105.79 | 0.7162 | 499.1 | 2.965 | 1.142 | 154.2 | 17.181 | 8.962 |
| 3.0 | 970.08 | 0.0937 | 365.0 | 2.563 | 1.711 | 329.7 | 10.924 | 8.200 |
| 3.5 | 980.76 | 0.1082 | 364.1 | 2.559 | 1.641 | 342.0 | 0.966 | 7.446 |
| 4.0 | 990.33 | 0.1224 | 363.4 | 2.555 | 1.586 | 353.2 | 0.863 | 6.866 |
| 4.5 | 999.04 | 0.1365 | 362.8 | 2.552 | 1.542 | 363.5 | 0.777 | 6.401 |
| 5.0 | 1007.0 | 0.1505 | 362.2 | 2.548 | 1.506 | 372.9 | 0.702 | 6.020 |
| 5.5 | 1014.5 | 0.1643 | 361.7 | 2.546 | 1.476 | 381.8 | 0.638 | 5.699 |
| 6.0 | 1021.5 | 0.1780 | 361.3 | 2.543 | 1.450 | 390.1 | 0.582 | 5.425 |
| 6.5 | 1028.0 | 0.1916 | 360.9 | 2.540 | 1.427 | 397.9 | 0.531 | 5.187 |
| 7.0 | 1034.3 | 0.2051 | 360.5 | 2.538 | 1.408 | 405.3 | 0.486 | 4.978 |
| 7.5 | 1040.2 | 0.2185 | 360.2 | 2.535 | 1.390 | 412.2 | 0.446 | 4.792 |
| 8.0 | 1045.8 | 0.2318 | 359.9 | 2.533 | 1.374 | 418.9 | 0.409 | 4.626 |
| 8.5 | 1051.2 | 0.2450 | 359.6 | 2.531 | 1.360 | 425.2 | 0.375 | 4.477 |
| 9.0 | 1056.4 | 0.2582 | 359.4 | 2.529 | 1.347 | 431.2 | 0.344 | 4.341 |
| 9.5 | 1061.4 | 0.2712 | 359.1 | 2.527 | 1.335 | 436.9 | 0.315 | 4.217 |
| 10.0 | 1066.2 | 0.2842 | 358.9 | 2.525 | 1.324 | 442.4 | 0.289 | 4.103 |
| 11.0 | 1075.4 | 0.3100 | 358.6 | 2.521 | 1.305 | 452.6 | 0.241 | 3.900 |
| 12.0 | 1084.0 | 0.3355 | 358.3 | 2.518 | 1.288 | 462.0 | 0.199 | 3.725 |
| 13.0 | 1092.2 | 0.3607 | 358.1 | 2.514 | 1.273 | 470.6 | 0.162 | 3.571 |
| 14.0 | 1100.0 | 0.3857 | 357.9 | 2.511 | 1.259 | 478.6 | 0.129 | 3.434 |
| 15.0 | 1107.4 | 0.4105 | 357.8 | 2.508 | 1.247 | 485.9 | 0.099 | 3.312 |
| 16.0 | 1114.5 | 0.4351 | 357.6 | 2.505 | 1.236 | 492.5 | 0.071 | 3.201 |
| 17.0 | 1121.3 | 0.4595 | 357.6 | 2.502 | 1.226 | 498.7 | 0.046 | 3.099 |
| 18.0 | 1127.9 | 0.4837 | 357.5 | 2.500 | 1.217 | 504.3 | 0.023 | 3.006 |
| 19.0 | 1134.3 | 0.5077 | 357.5 | 2.497 | 1.208 | 509.4 | 0.014 | 2.920 |
| 20.0 | 1140.5 | 0.5315 | 357.5 | 2.494 | 1.199 | 514.0 | 0.019 | 2.839 |

**TABLE 38.** Thermodynamic Properties of Freon-22 in the Single-Phase Region (*Continued*)

$T=353.15$ K

| $p$ | $\rho$ | $z$ | $h$ | $s$ | $c_p$ | $w$ | $\mu$ | $\alpha \cdot 10^3$ |
|---|---|---|---|---|---|---|---|---|
| 0.01 | 0.30 | 0.9992 | 539.5 | 3.589 | 0.713 | 198.0 | 12.914 | 2.839 |
| 0.02 | 0.59 | 0.9984 | 539.4 | 3.522 | 0.714 | 197.9 | 12.917 | 2.847 |
| 0.03 | 0.89 | 0.9976 | 539.3 | 3.483 | 0.715 | 197.8 | 12.921 | 2.855 |
| 0.04 | 1.18 | 0.9968 | 539.2 | 3.455 | 0.716 | 197.7 | 12.926 | 2.863 |
| 0.05 | 1.48 | 0.9960 | 539.1 | 3.434 | 0.716 | 197.5 | 12.930 | 2.870 |
| 0.1 | 2.97 | 0.9921 | 538.6 | 3.366 | 0.720 | 197.0 | 12.953 | 2.910 |
| 0.2 | 5.99 | 0.9840 | 537.7 | 3.297 | 0.728 | 195.8 | 13.001 | 2.992 |
| 0.3 | 9.05 | 0.9759 | 536.7 | 3.257 | 0.737 | 194.6 | 13.052 | 3.078 |
| 0.4 | 12.17 | 0.9676 | 535.8 | 3.227 | 0.745 | 193.4 | 13.106 | 3.168 |
| 0.5 | 15.35 | 0.9592 | 534.8 | 3.209 | 0.754 | 192.2 | 13.163 | 3.263 |
| 0.6 | 18.59 | 0.9507 | 533.8 | 3.184 | 0.764 | 190.9 | 13.224 | 3.363 |
| 0.7 | 21.88 | 0.9420 | 532.8 | 3.167 | 0.773 | 189.7 | 13.288 | 3.468 |
| 0.8 | 25.24 | 0.9333 | 531.7 | 3.152 | 0.783 | 188.4 | 13.355 | 3.579 |
| 0.9 | 28.67 | 0.9243 | 530.7 | 3.138 | 0.793 | 187.1 | 13.426 | 3.696 |
| 1.0 | 32.18 | 0.9152 | 529.6 | 3.126 | 0.804 | 185.7 | 13.501 | 3.820 |
| 1.5 | 50.93 | 0.8673 | 523.9 | 3.075 | 0.864 | 178.7 | 13.938 | 4.568 |
| 2.0 | 79.33 | 0.8143 | 517.5 | 3.034 | 0.943 | 170.8 | 14.496 | 5.632 |
| 2.5 | 106.73 | 0.7542 | 510.1 | 2.996 | 1.058 | 162.0 | 15.213 | 7.280 |
| 3.0 | 129.32 | 0.6832 | 501.1 | 2.958 | 1.252 | 151.7 | 16.156 | 10.238 |
| 3.5 | 174.49 | 0.5907 | 489.0 | 2.914 | 1.717 | 138.8 | 17.460 | 17.649 |
| 4.0 | 908.81 | 0.1296 | 380.8 | 2.604 | 1.963 | 276.6 | 1.664 | 1.124 |
| 4.5 | 924.97 | 0.1433 | 379.4 | 2.599 | 1.814 | 293.3 | 1.413 | 9.544 |
| 5.0 | 938.54 | 0.1569 | 378.2 | 2.594 | 1.714 | 307.8 | 1.226 | 8.419 |
| 5.5 | 950.33 | 0.1704 | 377.2 | 2.590 | 1.642 | 320.6 | 1.081 | 7.608 |
| 6.0 | 960.83 | 0.1839 | 376.4 | 2.586 | 1.587 | 332.3 | 0.962 | 6.989 |
| 6.5 | 970.33 | 0.1973 | 375.7 | 2.582 | 1.543 | 343.0 | 0.864 | 6.497 |
| 7.0 | 979.04 | 0.2106 | 375.1 | 2.579 | 1.508 | 352.9 | 0.781 | 6.096 |
| 7.5 | 987.11 | 0.2238 | 374.5 | 2.576 | 1.478 | 362.1 | 0.709 | 5.761 |
| 8.0 | 994.63 | 0.2369 | 374.0 | 2.573 | 1.452 | 370.8 | 0.646 | 5.476 |
| 8.5 | 1001.7 | 0.2499 | 373.6 | 2.570 | 1.430 | 378.9 | 0.591 | 5.229 |
| 9.0 | 1008.3 | 0.2628 | 373.2 | 2.568 | 1.410 | 393.8 | 0.542 | 5.013 |
| 9.5 | 1014.7 | 0.2757 | 372.8 | 2.566 | 1.393 | 400.7 | 0.497 | 4.823 |
| 10.0 | 1020.7 | 0.2885 | 372.5 | 2.563 | 1.378 | 413.4 | 0.457 | 4.465 |
| 11.0 | 1032.0 | 0.3139 | 371.9 | 2.559 | 1.351 | 425.0 | 0.387 | 4.362 |
| 12.0 | 1042.4 | 0.3390 | 371.4 | 2.555 | 1.329 | 435.6 | 0.328 | 4.120 |
| 13.0 | 1052.1 | 0.3639 | 371.0 | 2.551 | 1.310 | 445.2 | 0.278 | 3.915 |
| 14.0 | 1061.2 | 0.3885 | 370.7 | 2.547 | 1.294 | 454.1 | 0.233 | 3.739 |
| 15.0 | 1069.8 | 0.4129 | 370.4 | 2.544 | 1.279 | 462.2 | 0.194 | 3.585 |
| 16.0 | 1078.0 | 0.4371 | 370.2 | 2.541 | 1.266 | 469.6 | 0.159 | 3.448 |
| 17.0 | 1085.9 | 0.4610 | 370.0 | 2.537 | 1.255 | 476.4 | 0.128 | 3.327 |
| 18.0 | 1093.3 | 0.4848 | 369.9 | 2.534 | 1.244 | 482.6 | 0.100 | 3.216 |
| 19.0 | 1100.6 | 0.5084 | 369.8 | 2.532 | 1.234 | 488.2 | 0.074 | 3.116 |
| 20.0 | 1107.5 | 0.5318 | 369.7 | 2.529 | 1.225 | 493.5 | 0.050 | 3.024 |

**TABLE 38.** Thermodynamic Properties of Freon-22 in the Single-Phase Region (*Continued*)

$$T=363.15 \text{ K}$$

| $p$ | $\rho$ | $z$ | $h$ | $s$ | $c_p$ | $w$ | $\mu$ | $\alpha \cdot 10^3$ |
|---|---|---|---|---|---|---|---|---|
| 0.01 | 0.29 | 0.9993 | 546.7 | 3.608 | 0.724 | 200.6 | 11.704 | 2.760 |
| 0.02 | 0.57 | 0.9986 | 546.6 | 3.542 | 0.725 | 200.5 | 11.708 | 2.767 |
| 0.03 | 0.86 | 0.9978 | 546.5 | 3.503 | 0.725 | 200.4 | 11.713 | 2.774 |
| 0.04 | 1.15 | 0.9971 | 546.4 | 3.475 | 0.726 | 200.3 | 11.718 | 2.781 |
| 0.05 | 1.44 | 0.9964 | 546.4 | 3.453 | 0.727 | 200.1 | 11.723 | 2.787 |
| 0.1 | 2.89 | 0.9928 | 545.9 | 3.386 | 0.730 | 199.6 | 11.748 | 2.822 |
| 0.2 | 5.81 | 0.9855 | 545.1 | 3.317 | 0.737 | 198.5 | 11.800 | 2.893 |
| 0.3 | 8.78 | 0.9781 | 544.2 | 3.277 | 0.744 | 197.4 | 11.853 | 2.967 |
| 0.4 | 11.80 | 0.9706 | 543.3 | 3.247 | 0.752 | 196.3 | 11.909 | 3.045 |
| 0.5 | 14.87 | 0.9631 | 542.4 | 3.224 | 0.760 | 195.2 | 11.967 | 3.126 |
| 0.6 | 17.99 | 0.9554 | 541.5 | 3.205 | 0.768 | 194.1 | 12.026 | 3.211 |
| 0.7 | 21.16 | 0.9476 | 540.6 | 3.188 | 0.776 | 192.9 | 12.088 | 3.300 |
| 0.8 | 24.38 | 0.9397 | 539.6 | 3.173 | 0.784 | 191.7 | 12.152 | 3.393 |
| 0.9 | 27.66 | 0.9317 | 538.6 | 3.160 | 0.793 | 190.5 | 12.219 | 3.492 |
| 1.0 | 31.01 | 0.9236 | 537.7 | 3.148 | 0.802 | 189.3 | 12.287 | 3.595 |
| 1.5 | 48.75 | 0.8812 | 532.5 | 3.099 | 0.853 | 183.0 | 12.667 | 4.204 |
| 2.0 | 68.58 | 0.8351 | 526.8 | 3.059 | 0.917 | 176.2 | 13.113 | 5.025 |
| 2.5 | 91.28 | 0.7843 | 520.4 | 3.024 | 1.003 | 168.7 | 13.639 | 5.193 |
| 3.0 | 118.18 | 0.7270 | 513.0 | 2.990 | 1.130 | 160.4 | 14.258 | 7.998 |
| 3.5 | 151.94 | 0.6597 | 504.0 | 3.955 | 1.347 | 151.0 | 14.986 | 11.20 |
| 4.0 | 199.64 | 0.5738 | 492.1 | 2.914 | 1.849 | 139.8 | 15.817 | 18.83 |
| 4.5 | 791.42 | 0.1628 | 402.5 | 2.662 | 3.624 | 194.1 | 37.007 | 31.99 |
| 5.0 | 836.62 | 0.1714 | 398.0 | 2.648 | 2.449 | 225.5 | 26.050 | 17.43 |
| 5.5 | 862.53 | 0.1826 | 395.4 | 2.640 | 2.078 | 247.8 | 2.062 | 12.93 |
| 6.0 | 882.63 | 0.1947 | 393.6 | 2.633 | 1.885 | 266.0 | 1.715 | 10.61 |
| 6.5 | 898.95 | 0.2071 | 392.1 | 2.627 | 1.763 | 281.7 | 1.467 | 9.161 |
| 7.0 | 912.82 | 0.2196 | 391.0 | 2.623 | 1.679 | 295.6 | 1.281 | 8.157 |
| 7.5 | 924.97 | 0.2322 | 380.1 | 2.618 | 1.616 | 308.1 | 1.132 | 7.413 |
| 8.0 | 935.82 | 0.2448 | 389.1 | 2.615 | 1.566 | 319.5 | 1.011 | 6.836 |
| 8.5 | 945.67 | 0.2574 | 388.4 | 2.611 | 1.527 | 330.1 | 0.910 | 6.372 |
| 9.0 | 954.72 | 0.2700 | 387.7 | 2.608 | 1.494 | 339.9 | 0.824 | 5.991 |
| 9.5 | 963.09 | 0.2825 | 387.1 | 2.605 | 1.467 | 349.0 | 0.750 | 5.670 |
| 10.0 | 970.91 | 0.2950 | 385.6 | 2.602 | 1.443 | 357.7 | 0.685 | 5.395 |
| 11.0 | 985.26 | 0.3198 | 385.7 | 2.597 | 1.404 | 373.5 | 0.576 | 4.949 |
| 12.0 | 908.05 | 0.3443 | 385.0 | 2.592 | 1.374 | 387.6 | 0.489 | 4.600 |
| 13.0 | 1009.7 | 0.3687 | 384.4 | 2.587 | 1.349 | 400.5 | 0.417 | 4.317 |
| 14.0 | 1020.6 | 0.3929 | 383.9 | 2.583 | 1.329 | 412.2 | 0.356 | 4.082 |
| 15.0 | 1030.7 | 0.4168 | 383.4 | 2.579 | 1.311 | 422.8 | 0.304 | 3.883 |
| 16.0 | 1040.1 | 0.4405 | 383.1 | 2.576 | 1.296 | 432.6 | 0.258 | 3.712 |
| 17.0 | 1049.1 | 0.4641 | 382.8 | 2.572 | 1.282 | 441.5 | 0.218 | 3.562 |
| 18.0 | 1057.6 | 0.4874 | 382.5 | 2.569 | 1.270 | 449.6 | 0.183 | 3.430 |
| 19.0 | 1065.7 | 0.5106 | 382.3 | 2.556 | 1.259 | 457.1 | 0.151 | 3.312 |
| 20.0 | 1073.5 | 0.5335 | 382.1 | 2.563 | 1.249 | 463.8 | 0.123 | 3.206 |

**TABLE 38.** Thermodynamic Properties of Freon-22 in the Single-Phase Region (*Continued*)

$T = 373.15$ K

| $p$ | $\rho$ | $z$ | $h$ | $s$ | $c_p$ | $w$ | $\mu$ | $\alpha \cdot 10^4$ |
|------|--------|--------|-------|-------|--------|-------|--------|---------|
| 0.01 | 0.79 | 0.9993 | 554.0 | 3.628 | 0.734 | 203.1 | 10.667 | 2.686 |
| 0.02 | 0.58 | 0.9987 | 553.9 | 3.561 | 0.735 | 203.0 | 10.672 | 2.692 |
| 0.03 | 0.39 | 0.9980 | 553.9 | 3.522 | 0.736 | 202.9 | 10.677 | 2.698 |
| 0.04 | 1.12 | 0.9974 | 553.8 | 3.494 | 0.736 | 202.8 | 10.682 | 2.703 |
| 0.05 | 1.40 | 0.9967 | 553.7 | 3.472 | 0.737 | 202.7 | 10.688 | 2.709 |
| 0.1 | 2.81 | 0.9934 | 553.3 | 3.405 | 0.740 | 202.2 | 10.713 | 2.739 |
| 0.2 | 5.65 | 0.9868 | 552.5 | 3.337 | 0.746 | 201.2 | 10.766 | 2.802 |
| 0.3 | 8.53 | 0.9801 | 551.7 | 3.296 | 0.753 | 200.2 | 10.820 | 2.866 |
| 0.4 | 11.45 | 0.9733 | 550.9 | 3.267 | 0.755 | 199.2 | 10.875 | 2.933 |
| 0.5 | 14.42 | 0.9654 | 550.1 | 3.244 | 0.765 | 198.1 | 10.931 | 3.003 |
| 0.6 | 17.43 | 0.9595 | 549.2 | 3.225 | 0.772 | 197.1 | 10.989 | 3.076 |
| 0.7 | 20.48 | 0.9525 | 548.4 | 3.209 | 0.779 | 196.0 | 11.047 | 3.152 |
| 0.8 | 23.59 | 0.9453 | 547.5 | 3.194 | 0.786 | 195.0 | 11.107 | 3.232 |
| 0.9 | 26.74 | 0.9382 | 546.6 | 3.181 | 0.794 | 193.9 | 11.169 | 3.315 |
| 1.0 | 29.94 | 0.9307 | 545.7 | 3.169 | 0.802 | 192.8 | 11.231 | 3.402 |
| 1.5 | 46.81 | 0.8931 | 541.1 | 3.121 | 0.845 | 187.1 | 11.564 | 3.905 |
| 2.0 | 65.38 | 0.8526 | 535.9 | 3.083 | 0.897 | 181.7 | 11.931 | 4.556 |
| 2.5 | 86.15 | 0.8088 | 530.3 | 3.050 | 0.965 | 174.6 | 12.334 | 5.428 |
| 3.0 | 109.88 | 0.7609 | 624.0 | 3.019 | 1.057 | 167.7 | 12.772 | 6.654 |
| 3.5 | 137.89 | 0.7074 | 516.7 | 2.989 | 1.191 | 160.3 | 13.238 | 8.504 |
| 4.0 | 172.62 | 0.6458 | 508.0 | 2.957 | 1.411 | 152.1 | 13.710 | 11.63 |
| 4.5 | 219.84 | 0.5705 | 496.8 | 2.920 | 1.864 | 142.6 | 14.102 | 18.17 |
| 5.0 | 300.98 | 0.4630 | 479.5 | 2.868 | 3.496 | 132.0 | 13.941 | 42.00 |
| 5.5 | 625.52 | 0.2451 | 430.7 | 2.734 | 9.134 | 144.4 | 7.080 | 111.1 |
| 6.0 | 747.52 | 0.2237 | 417.2 | 2.696 | 3.285 | 184.1 | 3.993 | 28.96 |
| 6.5 | 793.82 | 0.2282 | 412.4 | 2.682 | 2.433 | 210.6 | 2.917 | 17.78 |
| 7.0 | 823.60 | 0.2369 | 409.5 | 2.672 | 3.088 | 231.6 | 2.320 | 13.37 |
| 7.5 | 845.96 | 0.2471 | 407.4 | 2.665 | 1.898 | 249.2 | 1.928 | 10.97 |
| 8.0 | 864.05 | 0.2580 | 405.8 | 2.659 | 1.775 | 264.7 | 1.647 | 9.449 |
| 8.5 | 879.35 | 0.2694 | 404.4 | 2.654 | 1.689 | 278.6 | 1.434 | 8.385 |
| 9.0 | 892.68 | 0.2810 | 403.3 | 2.650 | 1.624 | 291.1 | 1.265 | 7.596 |
| 9.5 | 904.53 | 0.2927 | 402.4 | 2.645 | 1.574 | 302.7 | 1.128 | 6.984 |
| 10.0 | 915.24 | 0.3045 | 401.5 | 2.642 | 1.534 | 313.4 | 1.014 | 6.494 |
| 11.0 | 934.08 | 0.3282 | 400.2 | 2.635 | 1.472 | 332.8 | 0.834 | 5.754 |
| 12.0 | 950.37 | 0.3519 | 399.0 | 2.629 | 1.428 | 349.9 | 0.698 | 5.218 |
| 13.0 | 964.83 | 0.3755 | 398.1 | 2.624 | 1.393 | 365.3 | 0.591 | 4.810 |
| 14.0 | 977.87 | 0.3990 | 397.4 | 2.619 | 1.366 | 379.1 | 0.504 | 4.486 |
| 15.0 | 989.81 | 0.4224 | 396.8 | 2.615 | 1.344 | 391.7 | 0.432 | 4.221 |
| 16.0 | 1000.8 | 0.4456 | 396.2 | 2.611 | 1.325 | 403.3 | 0.372 | 4.001 |
| 17.0 | 1011.1 | 0.4686 | 395.5 | 2.607 | 1.309 | 413.8 | 0.320 | 3.813 |
| 18.0 | 1020.8 | 0.4914 | 395.4 | 2.603 | 1.294 | 423.4 | 0.274 | 3.652 |
| 19.0 | 1030.0 | 0.5141 | 395.1 | 2.600 | 1.282 | 432.3 | 0.235 | 3.510 |
| 20.0 | 1038.7 | 0.5366 | 394.8 | 2.596 | 1.271 | 440.3 | 0.199 | 3.385 |

**TABLE 38.** Thermodynamic Properties of Freon-22 in the Single-Phase Region (*Continued*)

$T = 383.15$ K

| $p$ | $\rho$ | $z$ | $h$ | $s$ | $c_p$ | $w$ | $\mu$ | $a \cdot 10^3$ |
|------|--------|--------|-------|-------|-------|-------|--------|--------|
| 0.01 | 0.27 | 0.9994 | 561.5 | 3.647 | 0.745 | 205.6 | 9.782 | 2.615 |
| 0.02 | 0.54 | 0.9987 | 561.4 | 3.580 | 0.745 | 205.5 | 9.787 | 2.620 |
| 0.03 | 0.82 | 0.9982 | 561.3 | 3.541 | 0.746 | 205.4 | 9.792 | 2.626 |
| 0.04 | 1.09 | 0.9976 | 561.3 | 3.513 | 0.746 | 205.3 | 9.797 | 2.631 |
| 0.05 | 1.36 | 0.9970 | 561.2 | 3.492 | 0.747 | 205.2 | 9.802 | 2.636 |
| 0.1 | 2.73 | 0.9940 | 560.8 | 3.424 | 0.749 | 264.8 | 9.828 | 2.662 |
| 0.2 | 5.50 | 0.9879 | 560.1 | 3.356 | 0.755 | 203.8 | 9.879 | 2.717 |
| 0.3 | 8.30 | 0.9818 | 559.3 | 3.316 | 0.760 | 202.9 | 9.931 | 2.773 |
| 0.4 | 11.13 | 0.9756 | 558.6 | 3.287 | 0.766 | 201.9 | 9.984 | 2.832 |
| 0.5 | 14.00 | 0.9693 | 557.8 | 3.264 | 0.771 | 201.0 | 10.037 | 2.893 |
| 0.6 | 16.91 | 0.9630 | 557.0 | 3.245 | 0.777 | 200.0 | 10.091 | 2.956 |
| 0.7 | 19.86 | 0.8567 | 556.2 | 3.229 | 0.783 | 199.1 | 10.146 | 3.022 |
| 0.8 | 22.85 | 0.9502 | 555.4 | 3.215 | 0.789 | 198.1 | 10.201 | 3.090 |
| 0.9 | 25.89 | 0.9437 | 554.6 | 3.202 | 0.796 | 197.1 | 10.257 | 3.161 |
| 1.0 | 28.96 | 0.9371 | 553.8 | 3.190 | 0.802 | 196.1 | 10.314 | 3.236 |
| 1.5 | 45.08 | 0.9032 | 549.5 | 3.143 | 0.839 | 190.9 | 10.607 | 3.657 |
| 2.0 | 62.59 | 0.9673 | 544.9 | 3.107 | 0.883 | 185.6 | 10.913 | 4.184 |
| 2.5 | 81.84 | 0.8292 | 539.9 | 3.075 | 0.937 | 180.0 | 11.231 | 4.859 |
| 3.0 | 103.30 | 0.7882 | 534.3 | 3.046 | 1.007 | 174.1 | 11.556 | 5.748 |
| 3.5 | 127.70 | 0.7439 | 528.2 | 3.019 | 1.100 | 167.9 | 11.878 | 6.967 |
| 4.0 | 156.16 | 0.6953 | 521.2 | 2.991 | 1.233 | 161.3 | 12.178 | 8.731 |
| 4.5 | 190.67 | 0.6406 | 513.0 | 2.962 | 1.438 | 154.4 | 12.419 | 11.50 |
| 5.0 | 235.14 | 0.5772 | 503.0 | 2.930 | 1.798 | 147.2 | 12.507 | 16.41 |
| 5.5 | 299.01 | 0.4943 | 489.8 | 2.891 | 2.573 | 139.9 | 12.164 | 27.04 |
| 6.0 | 408.18 | 0.3990 | 470.4 | 2.836 | 4.454 | 135.7 | 10.457 | 52.23 |
| 6.5 | 561.85 | 0.3140 | 448.6 | 2.777 | 4.811 | 145.7 | 72.061 | 53.45 |
| 7.0 | 667.97 | 0.2849 | 436.1 | 2.742 | 3.463 | 166.5 | 49.876 | 32.68 |
| 7.5 | 726.83 | 0.2801 | 429.6 | 2.723 | 2.658 | 188.4 | 37.021 | 21.28 |
| 8.0 | 765.66 | 0.2836 | 425.6 | 2.711 | 2.247 | 208.2 | 29.154 | 15.70 |
| 8.5 | 794.04 | 0.2906 | 422.7 | 2.702 | 2.010 | 225.8 | 23.924 | 12.58 |
| 9.0 | 816.40 | 0.2992 | 420.7 | 2.695 | 1.859 | 241.5 | 20.201 | 10.61 |
| 9.5 | 834.90 | 0.3089 | 419.0 | 2.689 | 1.753 | 255.7 | 17.412 | 9.262 |
| 10.0 | 850.73 | 0.3191 | 417.6 | 2.684 | 1.676 | 268.8 | 15.240 | 8.281 |
| 11.0 | 876.99 | 0.3405 | 415.4 | 2.675 | 1.569 | 292.0 | 12.064 | 6.943 |
| 12.0 | 898.45 | 0.3625 | 413.7 | 2.668 | 1.498 | 312.3 | 0.984 | 6.069 |
| 13.0 | 916.72 | 0.3849 | 412.4 | 2.661 | 1.448 | 330.2 | 0.820 | 5.450 |
| 14.0 | 932.74 | 0.4074 | 411.3 | 2.656 | 1.410 | 346.4 | 0.692 | 4.986 |
| 15.0 | 947.05 | 0.4299 | 410.4 | 2.651 | 1.380 | 361.0 | 0.590 | 4.624 |
| 16.0 | 960.06 | 0.4524 | 409.7 | 2.646 | 1.356 | 374.3 | 0.507 | 4.333 |
| 17.0 | 972.01 | 0.4747 | 409.0 | 2.642 | 1.336 | 386.5 | 0.437 | 4.093 |
| 18.0 | 983.11 | 0.4970 | 408.5 | 2.637 | 1.319 | 397.9 | 0.378 | 3.891 |
| 19.0 | 993.50 | 0.5191 | 408.0 | 2.634 | 1.304 | 407.9 | 0.328 | 3.718 |
| 20.0 | 100.31 | 0.5411 | 407.6 | 2.630 | 1.291 | 417.3 | 0.283 | 3.568 |

**TABLE 38.** Thermodynamic Properties of Freon-22 in the Single-Phase Region (*Continued*)

$T = 393.15$ K

| $p$ | $\rho$ | $z$ | $h$ | $s$ | $c_p$ | $w$ | $\mu$ | $\alpha \cdot 10^2$ |
|---|---|---|---|---|---|---|---|---|
| 0.01 | 0.27 | 0.9994 | 569.0 | 3.666 | 0.755 | 208.1 | 9.027 | 2.548 |
| 0.02 | 0.53 | 0.9989 | 568.9 | 3.599 | 0.755 | 208.0 | 9.032 | 2.553 |
| 0.03 | 0.79 | 0.9983 | 568.9 | 3.560 | 0.756 | 207.9 | 9.037 | 2.557 |
| 0.04 | 1.06 | 0.9978 | 568.8 | 3.532 | 0.757 | 207.8 | 9.041 | 2.562 |
| 0.05 | 1.33 | 0.9972 | 568.8 | 3.510 | 0.757 | 207.7 | 9.046 | 2.567 |
| 0.1 | 2.66 | 0.9945 | 568.4 | 3.443 | 0.759 | 207.3 | 9.070 | 2.590 |
| 0.2 | 5.35 | 0.9889 | 567.7 | 3.376 | 0.763 | 206.4 | 9.119 | 2.638 |
| 0.3 | 8.07 | 0.9833 | 567.0 | 3.335 | 0.768 | 205.5 | 9.168 | 2.688 |
| 0.4 | 10.82 | 0.9776 | 566.3 | 3.306 | 0.773 | 204.7 | 9.217 | 2.740 |
| 0.5 | 13.61 | 0.9719 | 565.6 | 3.284 | 0.778 | 203.8 | 9.267 | 2.793 |
| 0.6 | 16.43 | 0.9662 | 564.9 | 3.265 | 0.783 | 202.9 | 9.317 | 2.848 |
| 0.7 | 19.28 | 0.9604 | 564.1 | 3.249 | 0.788 | 202.0 | 9.367 | 2.906 |
| 0.8 | 22.17 | 0.9545 | 563.4 | 3.235 | 0.793 | 201.1 | 9.417 | 2.965 |
| 0.9 | 25.10 | 0.9486 | 562.6 | 3.222 | 0.799 | 200.2 | 9.468 | 3.026 |
| 1.0 | 28.06 | 0.9426 | 561.9 | 3.210 | 0.804 | 100.3 | 9.519 | 2.090 |
| 1.5 | 43.51 | 0.9121 | 557.9 | 3.164 | 0.835 | 194.6 | 9.775 | 3.447 |
| 2.0 | 60.12 | 0.8800 | 553.7 | 3.129 | 0.872 | 189.8 | 10.032 | 3.882 |
| 2.5 | 78.14 | 0.8464 | 549.2 | 3.098 | 0.917 | 184.9 | 10.288 | 4.418 |
| 3.0 | 97.87 | 0.8108 | 544.2 | 3.071 | 0.972 | 179.7 | 10.536 | 5.092 |
| 3.5 | 119.74 | 0.7732 | 538.9 | 3.046 | 1.041 | 174.5 | 10.768 | 5.959 |
| 4.0 | 144.37 | 0.7329 | 533.0 | 3.021 | 1.132 | 169.0 | 10.971 | 7.106 |
| 4.5 | 172.63 | 0.6895 | 526.4 | 2.996 | 1.256 | 163.5 | 11.125 | 8.681 |
| 5.0 | 205.95 | 0.6422 | 518.9 | 2.971 | 1.433 | 157.8 | 11.194 | 10.95 |
| 5.5 | 246.65 | 0.5899 | 510.2 | 2.943 | 1.701 | 152.2 | 11.109 | 14.40 |
| 6.0 | 298.73 | 0.5313 | 499.8 | 2.912 | 2.131 | 147.1 | 10.729 | 19.92 |
| 6.5 | 368.28 | 0.4670 | 487.3 | 2.876 | 2.783 | 144.0 | 9.801 | 28.08 |
| 7.0 | 456.24 | 0.4058 | 473.3 | 2.837 | 3.346 | 145.8 | 8.196 | 34.37 |
| 7.5 | 545.81 | 0.3635 | 460.9 | 2.803 | 3.310 | 153.7 | 6.473 | 32.29 |
| 8.0 | 619.39 | 0.3417 | 451.9 | 2.778 | 2.945 | 166.2 | 6.096 | 26.19 |
| 8.5 | 674.52 | 0.3333 | 445.6 | 2.760 | 2.561 | 181.2 | 4.066 | 20.41 |
| 9.0 | 715.54 | 0.3327 | 441.2 | 2.747 | 2.264 | 196.9 | 3.309 | 16.18 |
| 9.5 | 747.10 | 0.3364 | 437.9 | 2.737 | 2.054 | 212.1 | 2.753 | 13.29 |
| 10.0 | 772.38 | 0.3425 | 433.4 | 2.729 | 1.905 | 226.55 | 2.335 | 11.28 |
| 11.0 | 811.26 | 0.3587 | 431.8 | 2.717 | 1.713 | 252.75 | 1.760 | 8.767 |
| 12.0 | 840.72 | 0.3776 | 429.2 | 2.707 | 1.597 | 275.78 | 1.389 | 7.286 |
| 13.0 | 864.54 | 0.3978 | 427.3 | 2.699 | 1.520 | 296.16 | 1.129 | 6.317 |
| 14.0 | 884.63 | 0.4186 | 425.7 | 2.692 | 1.465 | 314.45 | 0.937 | 5.634 |
| 15.0 | 802.08 | 0.4399 | 424.5 | 2.686 | 1.424 | 330.98 | 0.790 | 5.126 |
| 16.0 | 917.67 | 0.4613 | 423.4 | 2.681 | 1.392 | 346.03 | 0.674 | 4.733 |
| 17.0 | 931.56 | 0.4827 | 422.6 | 2.676 | 1.366 | 359.79 | 0.579 | 4.419 |
| 18.0 | 944.35 | 0.5042 | 421.8 | 2.671 | 1.345 | 372.43 | 0.501 | 4.161 |
| 19.0 | 956.17 | 0.5257 | 421.2 | 2.667 | 1.327 | 384.05 | 0.435 | 3.947 |
| 20.0 | 967.18 | 0.5470 | 420.7 | 2.663 | 1.312 | 394.75 | 0.378 | 3.764 |

**TABLE 38.** Thermodynamic Properties of Freon-22 in the Single-Phase Region (*Continued*)

$T = 403.15$ K

| $p$ | $\rho$ | $z$ | $h$ | $s$ | $c_p$ | $w$ | $\mu$ | $\alpha \cdot 10^3$ |
|---|---|---|---|---|---|---|---|---|
| 0.01 | 0.26 | 0.9995 | 576.7 | 3.685 | 0.765 | 210.5 | 8.384 | 2.484 |
| 0.02 | 0.52 | 0.9990 | 576.6 | 3.618 | 0.765 | 210.4 | 8.389 | 2.489 |
| 0.03 | 0.78 | 0.9985 | 576.5 | 3.579 | 0.766 | 210.3 | 8.393 | 2.493 |
| 0.04 | 1.03 | 0.9979 | 576.5 | 3.551 | 0.766 | 210.2 | 8.398 | 2.497 |
| 0.05 | 1.29 | 0.9975 | 576.4 | 3.530 | 0.766 | 210.2 | 8.402 | 2.501 |
| 0.1 | 2.59 | 0.9949 | 576.1 | 3.462 | 0.768 | 209.8 | 8.425 | 2.522 |
| 0.2 | 5.21 | 0.9898 | 575.4 | 3.395 | 0.772 | 204.0 | 8.470 | 2.565 |
| 0.3 | 7.86 | 0.9846 | 574.8 | 3.355 | 0.776 | 208.2 | 8.515 | 2.609 |
| 0.4 | 10.54 | 0.9794 | 574.1 | 3.326 | 0.780 | 207.3 | 8.560 | 2.655 |
| 0.5 | 13.24 | 0.9742 | 573.4 | 3.303 | 0.784 | 206.5 | 8.606 | 2.702 |
| 0.6 | 15.97 | 0.9689 | 572.8 | 3.284 | 0.789 | 205.7 | 8.651 | 2.751 |
| 0.7 | 18.74 | 0.9636 | 572.1 | 3.269 | 0.793 | 204.9 | 8.696 | 2.801 |
| 0.8 | 21.54 | 0.9583 | 571.4 | 3.254 | 0.797 | 204.1 | 8.741 | 2.853 |
| 0.9 | 24.36 | 0.9529 | 570.7 | 3.242 | 0.802 | 203.2 | 8.786 | 2.906 |
| 1.0 | 27.23 | 0.9475 | 570.0 | 3.230 | 0.807 | 202.4 | 8.831 | 2.961 |
| 1.5 | 42.07 | 0.9198 | 566.3 | 3.185 | 0.833 | 198.2 | 9.053 | 3.268 |
| 2.0 | 57.90 | 0.8910 | 562.4 | 3.150 | 0.864 | 193.8 | 9.269 | 3.631 |
| 2.5 | 74.90 | 0.8611 | 558.3 | 3.121 | 0.901 | 189.4 | 9.476 | 4.067 |
| 3.0 | 93.25 | 0.8299 | 553.9 | 3.095 | 0.946 | 184.9 | 9.668 | 4.595 |
| 3.5 | 113.24 | 0.7973 | 549.1 | 3.071 | 1.000 | 180.3 | 9.839 | 5.244 |
| 4.0 | 135.22 | 0.7631 | 544.0 | 3.049 | 1.067 | 175.7 | 9.981 | 6.053 |
| 4.5 | 159.65 | 0.7271 | 538.4 | 3.027 | 1.152 | 171.1 | 10.081 | 7.080 |
| 5.0 | 187.18 | 0.6891 | 532.4 | 3.004 | 1.262 | 166.4 | 10.123 | 8.413 |
| 5.5 | 218.72 | 0.6487 | 525.6 | 2.981 | 1.408 | 161.8 | 10.080 | 10.18 |
| 6.0 | 255.52 | 0.6058 | 518.1 | 2.957 | 1.604 | 157.6 | 9.910 | 12.55 |
| 6.5 | 299.22 | 0.5604 | 509.7 | 2.932 | 1.867 | 153.9 | 9.546 | 15.71 |
| 7.0 | 351.44 | 0.5137 | 500.3 | 2.905 | 2.193 | 151.7 | 8.906 | 19.51 |
| 7.5 | 411.95 | 0.4697 | 490.4 | 2.877 | 2.505 | 152.1 | 7.958 | 22.85 |
| 8.0 | 476.50 | 0.4331 | 480.8 | 2.850 | 2.666 | 155.7 | 6.837 | 24.02 |
| 8.5 | 538.35 | 0.4073 | 472.4 | 2.827 | 2.644 | 162.4 | 5.763 | 22.82 |
| 9.0 | 592.94 | 0.3916 | 465.5 | 2.808 | 2.511 | 171.5 | 4.842 | 20.36 |
| 9.5 | 638.89 | 0.3836 | 460.1 | 2.793 | 2.339 | 182.3 | 4.077 | 17.59 |
| 10.0 | 676.79 | 0.3812 | 455.9 | 2.780 | 2.169 | 194.2 | 3.453 | 15.05 |
| 11.0 | 734.13 | 0.3865 | 449.9 | 2.762 | 1.902 | 218.9 | 2.543 | 11.29 |
| 12.0 | 775.40 | 0.3992 | 445.8 | 2.748 | 1.729 | 242.6 | 1.949 | 8.961 |
| 13.0 | 807.12 | 0.4155 | 442.9 | 2.738 | 1.015 | 264.4 | 1.546 | 7.479 |
| 14.0 | 832.79 | 0.4357 | 440.7 | 2.730 | 1.536 | 284.2 | 1.260 | 6.477 |
| 15.0 | 854.38 | 0.4529 | 439.0 | 2.722 | 1.479 | 302.3 | 1.047 | 5.761 |
| 16.0 | 873.07 | 0.4728 | 437.6 | 2.716 | 1.436 | 318.8 | 0.883 | 5.226 |
| 17.0 | 889.60 | 0.4930 | 436.4 | 2.710 | 1.402 | 334.0 | 0.753 | 4.811 |
| 18.0 | 904.45 | 0.5134 | 435.5 | 2.705 | 1.375 | 347.9 | 0.648 | 4.480 |
| 19.0 | 918.98 | 0.5339 | 434.7 | 2.700 | 1.353 | 360.8 | 0.561 | 4.210 |
| 20.0 | 930.45 | 0.5545 | 433.9 | 2.696 | 1.334 | 372.7 | 0.488 | 3.985 |

**TABLE 38.** Thermodynamic Properties of Freon-22 in the Single-Phase Region (*Continued*)

$T = 413.15$ K

| $p$ | $\rho$ | $z$ | $h$ | $s$ | $c_p$ | $w$ | $\mu$ | $\alpha \cdot 10^8$ |
|---|---|---|---|---|---|---|---|---|
| 0.01 | 0.25 | 0.9995 | 584.4 | 3.703 | 0.775 | 212.9 | 7.839 | 2.424 |
| 0.02 | 0.50 | 0.9991 | 584.3 | 3.636 | 0.775 | 212.8 | 7.843 | 2.428 |
| 0.03 | 0.76 | 0.9986 | 584.3 | 3.598 | 0.775 | 212.7 | 7.847 | 2.431 |
| 0.04 | 1.01 | 0.9981 | 584.2 | 3.569 | 0.776 | 212.7 | 7.851 | 2.436 |
| 0.05 | 1.26 | 0.9976 | 584.1 | 3.548 | 0.776 | 212.6 | 7.855 | 2.439 |
| 0.1 | 2.53 | 0.9953 | 583.8 | 3.481 | 0.778 | 212.2 | 7.876 | 2.458 |
| 0.2 | 5.08 | 0.9906 | 583.2 | 3.414 | 0.781 | 211.5 | 7.917 | 2.496 |
| 0.3 | 7.66 | 0.9858 | 582.6 | 3.374 | 0.784 | 210.7 | 7.957 | 2.536 |
| 0.4 | 10.26 | 0.9811 | 582.0 | 3.345 | 0.788 | 210.0 | 7.998 | 2.577 |
| 0.5 | 12.89 | 0.9763 | 581.3 | 3.322 | 0.791 | 209.2 | 8.038 | 2.019 |
| 0.6 | 15.55 | 0.9714 | 580.7 | 3.304 | 0.795 | 208.5 | 8.078 | 2.662 |
| 0.7 | 18.23 | 0.9666 | 580.0 | 3.288 | 0.799 | 207.7 | 8.118 | 2.706 |
| 0.8 | 20.94 | 0.9617 | 579.4 | 3.274 | 0.802 | 207.0 | 8.158 | 2.752 |
| 0.9 | 23.68 | 0.9568 | 578.7 | 3.261 | 0.806 | 206.2 | 8.157 | 2.799 |
| 1.0 | 26.45 | 0.9516 | 578.0 | 3.250 | 0.810 | 205.4 | 8.236 | 2.848 |
| 1.5 | 40.75 | 0.9266 | 574.7 | 3.205 | 0.833 | 201.6 | 8.426 | 3.113 |
| 2.0 | 55.90 | 0.9006 | 571.0 | 3.171 | 0.859 | 197.6 | 8.606 | 3.421 |
| 2.5 | 72.02 | 0.8738 | 567.3 | 3.143 | 0.890 | 193.7 | 8.773 | 3.781 |
| 3.0 | 89.25 | 0.8462 | 563.2 | 3.118 | 0.926 | 189.7 | 8.922 | 4.205 |
| 3.5 | 107.76 | 0.8176 | 559.0 | 3.095 | 0.969 | 185.7 | 9.048 | 4.709 |
| 4.0 | 127.78 | 0.7880 | 554.5 | 3.074 | 1.021 | 181.7 | 9.148 | 5.310 |
| 4.5 | 149.58 | 0.7574 | 549.6 | 3.054 | 1.084 | 177.7 | 9.212 | 6.036 |
| 5.0 | 173.47 | 0.7256 | 544.5 | 3.034 | 1.161 | 173.7 | 9.239 | 6.920 |
| 5.5 | 199.91 | 0.6926 | 538.9 | 3.014 | 1.255 | 169.9 | 9.198 | 8.006 |
| 6.0 | 229.42 | 0.6584 | 532.9 | 2.994 | 1.372 | 166.2 | 9.089 | 9.345 |
| 6.5 | 262.62 | 0.6231 | 526.4 | 2.973 | 1.516 | 162.9 | 8.885 | 10.98 |
| 7.0 | 300.14 | 0.5871 | 519.4 | 2.952 | 1.689 | 160.3 | 8.553 | 12.92 |
| 7.5 | 342.35 | 0.5515 | 512.0 | 2.930 | 1.883 | 158.8 | 8.069 | 15.01 |
| 8.0 | 388.85 | 0.5179 | 504.3 | 2.908 | 2.070 | 159.0 | 7.419 | 16.88 |
| 8.5 | 438.00 | 0.4885 | 496.8 | 2.887 | 2.208 | 161.1 | 6.662 | 18.02 |
| 9.0 | 487.19 | 0.4650 | 489.7 | 2.867 | 2.272 | 165.2 | 5.879 | 18.17 |
| 9.5 | 533.98 | 0.4478 | 483.5 | 2.850 | 2.266 | 171.1 | 5.145 | 17.49 |
| 10.0 | 576.75 | 0.4365 | 478.1 | 2.834 | 2.211 | 178.4 | 4.490 | 16.28 |
| 11.0 | 648.18 | 0.4272 | 469.7 | 2.810 | 2.037 | 196.2 | 3.429 | 13.36 |
| 12.0 | 702.39 | 0.4301 | 463.8 | 2.792 | 1.861 | 216.6 | 2.638 | 10.77 |
| 13.0 | 743.86 | 0.4399 | 459.6 | 2.779 | 1.722 | 237.3 | 2.075 | 8.854 |
| 14.0 | 776.65 | 0.4538 | 459.5 | 2.768 | 1.620 | 257.3 | 1.669 | 7.503 |
| 15.0 | 803.51 | 0.4699 | 454.1 | 2.759 | 1.545 | 276.0 | 1.370 | 6.536 |
| 16.0 | 826.21 | 0.4875 | 452.2 | 2.752 | 1.488 | 293.4 | 1.144 | 5.824 |
| 17.0 | 845.87 | 0.5059 | 450.7 | 2.745 | 1.445 | 309.6 | 0.967 | 5.282 |
| 18.0 | 863.24 | 0.5249 | 449.4 | 2.739 | 1.410 | 324.6 | 0.827 | 4.856 |
| 19.0 | 878.82 | 0.5442 | 448.3 | 2.734 | 1.382 | 338.5 | 0.713 | 4.518 |
| 20.0 | 892.99 | 0.5638 | 447.4 | 2.727 | 1.360 | 351.4 | 0.619 | 4.239 |

**TABLE 38.** Thermodynamic Properties of Freon-22 in the Single-Phase Region (*Continued*)

$T = 423.15$ K

| $p$ | $\rho$ | $z$ | $h$ | $s$ | $c_p$ | $w$ | $\mu$ | $\alpha \cdot 10^3$ |
|---|---|---|---|---|---|---|---|---|
| 0.01 | 0.24 | 0.9996 | 592.2 | 3.722 | 0.784 | 215.3 | 7.377 | 2.367 |
| 0.02 | 0.49 | 0.9991 | 592.1 | 3.656 | 0.785 | 215.2 | 7.380 | 2.370 |
| 0.03 | 0.74 | 0.9987 | 592.1 | 3.616 | 0.785 | 215.1 | 7.384 | 2.373 |
| 0.04 | 0.99 | 0.9983 | 592.0 | 3.588 | 0.786 | 215.1 | 7.388 | 2.377 |
| 0.05 | 1.23 | 0.9978 | 592.0 | 3.567 | 0.786 | 215.0 | 7.391 | 2.380 |
| 0.1 | 2.47 | 0.9957 | 591.7 | 3.500 | 0.787 | 214.6 | 7.409 | 2.397 |
| 0.2 | 4.96 | 0.9913 | 591.1 | 3.432 | 0.789 | 214.0 | 7.445 | 2.432 |
| 0.3 | 7.47 | 0.9869 | 590.5 | 3.392 | 0.792 | 213.3 | 7.481 | 2.468 |
| 0.4 | 19.01 | 0.9825 | 589.9 | 3.364 | 0.795 | 212.6 | 7.517 | 2.504 |
| 0.5 | 12.56 | 0.9781 | 589.8 | 3.341 | 0.798 | 211.9 | 7.552 | 2.542 |
| 0.6 | 15.15 | 0.9736 | 588.7 | 3.323 | 0.801 | 211.2 | 7.587 | 2.581 |
| 0.7 | 17.75 | 0.9692 | 588.1 | 3.307 | 0.804 | 210.5 | 7.622 | 2.620 |
| 0.8 | 20.38 | 0.9647 | 587.5 | 3.293 | 0.808 | 209.8 | 7.656 | 2.661 |
| 0.9 | 23.04 | 0.9602 | 586.8 | 3.280 | 0.811 | 209.1 | 7.690 | 2.703 |
| 1.0 | 25.72 | 0.9557 | 586.2 | 3.269 | 0.815 | 208.4 | 7.723 | 2.746 |
| 1.5 | 39.53 | 0.9327 | 583.0 | 3.225 | 0.834 | 204.9 | 7.884 | 2.977 |
| 2.0 | 54.07 | 0.9091 | 579.6 | 3.192 | 0.856 | 201.3 | 8.031 | 3.242 |
| 2.5 | 69.43 | 0.8850 | 576.1 | 3.166 | 0.882 | 197.7 | 8.164 | 3.544 |
| 3.0 | 85.71 | 0.8603 | 572.4 | 3.140 | 0.912 | 194.2 | 8.278 | 3.892 |
| 3.5 | 103.02 | 0.8349 | 568.6 | 3.118 | 0.947 | 190.6 | 8.370 | 4.293 |
| 4.0 | 121.52 | 0.8090 | 564.5 | 3.098 | 0.988 | 187.1 | 8.438 | 4.757 |
| 4.5 | 141.45 | 0.7824 | 560.2 | 3.079 | 1.036 | 183.6 | 8.476 | 5.297 |
| 5.0 | 162.72 | 0.7552 | 555.7 | 3.061 | 1.093 | 180.1 | 8.480 | 5.928 |
| 5.5 | 185.85 | 0.7273 | 550.9 | 3.042 | 1.160 | 176.8 | 8.444 | 6.667 |
| 6.0 | 211.01 | 0.6989 | 545.9 | 3.025 | 1.240 | 173.7 | 8.360 | 7.530 |
| 6.5 | 238.47 | 0.6699 | 540.6 | 3.007 | 1.332 | 170.8 | 8.217 | 8.534 |
| 7.0 | 268.51 | 0.6407 | 535.0 | 2.989 | 1.440 | 168.3 | 8.004 | 9.678 |
| 7.5 | 301.34 | 0.6117 | 529.1 | 2.971 | 1.561 | 166.4 | 7.707 | 10.93 |
| 8.0 | 336.94 | 0.5835 | 523.0 | 2.953 | 1.688 | 165.5 | 7.317 | 12.20 |
| 8.5 | 374.96 | 0.5571 | 516.8 | 2.935 | 1.811 | 165.7 | 6.837 | 13.33 |
| 9.0 | 414.57 | 0.5336 | 510.6 | 2.917 | 1.914 | 167.2 | 6.289 | 14.15 |
| 9.5 | 454.60 | 0.5136 | 504.8 | 2.901 | 1.984 | 170.1 | 5.711 | 14.53 |
| 10.0 | 493.79 | 0.4977 | 499.4 | 2.885 | 2.018 | 174.3 | 5.144 | 14.47 |
| 11.0 | 566.01 | 0.4776 | 490.0 | 2.859 | 1.998 | 185.9 | 4.127 | 13.39 |
| 12.0 | 627.05 | 0.4703 | 482.7 | 2.838 | 1.909 | 200.7 | 3.299 | 11.69 |
| 13.0 | 676.69 | 0.4722 | 477.3 | 2.821 | 1.798 | 217.7 | 2.645 | 9.970 |
| 14.0 | 716.77 | 0.4801 | 473.1 | 2.808 | 1.696 | 235.7 | 2.138 | 8.509 |
| 15.0 | 749.53 | 0.4919 | 469.9 | 2.797 | 1.612 | 253.6 | 1.752 | 7.366 |
| 16.0 | 776.89 | 0.5062 | 467.4 | 2.788 | 1.545 | 271.0 | 1.455 | 6.490 |
| 17.0 | 800.24 | 0.5221 | 465.4 | 2.780 | 1.492 | 287.0 | 1.224 | 6.817 |
| 18.0 | 820.56 | 0.5391 | 463.7 | 2.773 | 1.450 | 302.9 | 1.041 | 5.290 |
| 19.0 | 838.55 | 0.5569 | 462.3 | 2.767 | 1.416 | 317.5 | 0.893 | 4.870 |
| 20.0 | 854.71 | 0.5751 | 461.2 | 2.762 | 1.388 | 331.2 | 0.773 | 4.530 |

**TABLE 38.** Thermodynamic Properties of Freon-22 in the Single-Phase Region (*Continued*)

$T = 433.15$ K

| $p$ | $\rho$ | $z$ | $h$ | $s$ | $c_p$ | $w$ | $\mu$ | $\alpha \cdot 10^3$ |
|------|--------|--------|-------|-------|-------|-------|-------|--------|
| 0.01 | 0.24 | 0.9996 | 600.1 | 3.740 | 0.794 | 217.6 | 6.986 | 2.312 |
| 0.02 | 0.48 | 0.9992 | 600.0 | 3.674 | 0.794 | 217.6 | 6.989 | 2.315 |
| 0.03 | 0.72 | 0.9988 | 599.9 | 3.635 | 0.794 | 217.5 | 6.992 | 2.318 |
| 0.04 | 0.96 | 0.9984 | 599.9 | 3.607 | 0.794 | 217.4 | 6.995 | 2.321 |
| 0.05 | 1.20 | 0.9980 | 599.9 | 3.585 | 0.795 | 217.4 | 6.998 | 2.324 |
| 0.1 | 2.41 | 0.9960 | 599.9 | 3.518 | 0.796 | 217.1 | 7.014 | 2.340 |
| 0.2 | 4.84 | 0.9920 | 599.0 | 3.451 | 0.798 | 216.4 | 7.045 | 2.372 |
| 0.3 | 7.29 | 0.9879 | 598.5 | 3.411 | 0.800 | 215.8 | 7.076 | 2.404 |
| 0.4 | 9.76 | 0.9839 | 597.9 | 3.382 | 0.803 | 215.1 | 7.107 | 2.437 |
| 0.5 | 12.25 | 0.9798 | 597.3 | 3.360 | 0.805 | 214.5 | 7.137 | 2.472 |
| 0.6 | 14.77 | 0.9767 | 596.7 | 3.341 | 0.808 | 213.9 | 7.167 | 2.506 |
| 0.7 | 17.30 | 0.9716 | 596.2 | 3.326 | 0.811 | 213.2 | 7.196 | 2.542 |
| 0.8 | 19.86 | 0.9674 | 595.6 | 3.312 | 0.813 | 212.6 | 7.225 | 2.578 |
| 0.9 | 22.43 | 0.9330 | 595.0 | 3.300 | 0.816 | 211.9 | 7.254 | 2.615 |
| 1.0 | 25.04 | 0.9591 | 594.4 | 3.288 | 0.819 | 211.3 | 7.282 | 2.653 |
| 1.5 | 38.39 | 0.9381 | 591.3 | 3.244 | 0.835 | 208.1 | 7.414 | 2.858 |
| 2.0 | 52.39 | 0.9167 | 588.2 | 3.212 | 0.854 | 204.8 | 7.533 | 3.087 |
| 2.5 | 67.08 | 0.8948 | 584.9 | 3.185 | 0.876 | 201.6 | 7.635 | 3.344 |
| 3.0 | 82.55 | 0.8726 | 581.5 | 3.161 | 0.901 | 198.4 | 7.720 | 3.634 |
| 3.5 | 98.87 | 0.8500 | 577.9 | 3.140 | 0.930 | 195.2 | 7.785 | 3.961 |
| 4.0 | 116.13 | 0.8270 | 574.3 | 3.121 | 0.963 | 192.0 | 7.827 | 4.330 |
| 4.5 | 134.44 | 0.9036 | 570.4 | 3.103 | 1.002 | 189.0 | 7.845 | 4.747 |
| 5.0 | 153.93 | 0.7799 | 566.4 | 3.085 | 1.046 | 185.9 | 7.835 | 5.220 |
| 5.5 | 174.71 | 0.7559 | 562.2 | 3.069 | 1.096 | 183.0 | 7.795 | 6.755 |
| 6.0 | 196.92 | 0.7316 | 557.8 | 3.052 | 1.154 | 180.2 | 7.721 | 6.359 |
| 6.5 | 220.72 | 0.7071 | 553.3 | 3.036 | 1.220 | 177.7 | 7.609 | 7.037 |
| 7.0 | 246.34 | 0.6826 | 548.6 | 3.021 | 1.293 | 175.4 | 7.451 | 7.788 |
| 7.5 | 273.57 | 0.6583 | 543.1 | 3.005 | 1.375 | 173.5 | 7.243 | 8.599 |
| 8.0 | 302.73 | 0.6345 | 538.6 | 2.990 | 1.463 | 172.2 | 6.978 | 9.443 |
| 8.5 | 333.61 | 0.6118 | 533.5 | 2.974 | 1.553 | 171.7 | 6.655 | 10.266 |
| 9.0 | 365.93 | 0.5905 | 528.3 | 2.908 | 1.639 | 172.0 | 6.277 | 10.999 |
| 9.5 | 399.19 | 0.5714 | 523.2 | 2.944 | 1.715 | 173.4 | 5.859 | 11.567 |
| 10.0 | 432.77 | 0.5548 | 518.3 | 2.929 | 1.775 | 175.7 | 5.418 | 11.918 |
| 11.0 | 498.33 | 0.5300 | 509.3 | 2.904 | 1.840 | 183.2 | 4.551 | 11.942 |
| 12.0 | 558.48 | 0.5159 | 501.6 | 2.882 | 1.839 | 193.6 | 3.781 | 11.273 |
| 13.0 | 611.18 | 0.5107 | 495.3 | 2.863 | 1.794 | 206.5 | 3.130 | 10.233 |
| 14.0 | 856.06 | 0.5124 | 490.3 | 2.848 | 1.728 | 221.0 | 2.590 | 9.087 |
| 15.0 | 693.66 | 0.5191 | 486.3 | 2.835 | 1.657 | 236.7 | 2.149 | 8.015 |
| 16.0 | 725.76 | 0.5293 | 483.1 | 2.825 | 1.592 | 252.7 | 1.795 | 7.099 |
| 17.0 | 752.88 | 0.5421 | 480.5 | 2.816 | 1.536 | 268.5 | 1.612 | 6.347 |
| 18.0 | 776.52 | 0.5566 | 478.4 | 2.808 | 1.490 | 283.9 | 1.285 | 5.740 |
| 19.0 | 797.19 | 0.5723 | 476.7 | 2.801 | 1.451 | 298.6 | 1.101 | 5.248 |
| 20.0 | 815.57 | 0.5888 | 475.2 | 2.794 | 1.418 | 312.6 | 0.950 | 4.846 |

**TABLE 38.** Thermodynamic Properties of Freon-22 in the Single-Phase Region (*Continued*)

$T = 443.15$ K

| $p$ | $\rho$ | $z$ | $h$ | $s$ | $c_p$ | $w$ | $\mu$ | $\alpha \cdot 10^3$ |
|------|--------|--------|-------|-------|-------|-------|-------|---------|
| 0.01 | 0.23 | 0.9996 | 608.1 | 3.759 | 0.807 | 219.9 | 6.656 | 2.259 |
| 0.02 | 0.47 | 0.9992 | 608.0 | 3.692 | 0.803 | 219.9 | 6.658 | 2.262 |
| 0.03 | 0.70 | 0.9989 | 607.9 | 3.653 | 0.803 | 219.8 | 6.661 | 2.265 |
| 0.04 | 0.94 | 0.9985 | 607.9 | 3.625 | 0.803 | 219.8 | 6.663 | 2.267 |
| 0.05 | 1.18 | 0.9981 | 607.8 | 3.604 | 0.804 | 219.7 | 6.666 | 2.271 |
| 0.1 | 2.36 | 0.9963 | 607.6 | 3.537 | 0.804 | 219.4 | 6.680 | 2.285 |
| 0.2 | 4.73 | 0.9925 | 607.0 | 3.469 | 0.806 | 218.8 | 6.706 | 2.314 |
| 0.3 | 7.12 | 0.9888 | 606.5 | 3.429 | 0.909 | 218.2 | 6.732 | 2.344 |
| 0.4 | 9.53 | 0.9851 | 605.9 | 3.401 | 0.810 | 217.7 | 6.758 | 2.374 |
| 0.5 | 11.96 | 0.9813 | 605.4 | 3.376 | 0.813 | 217.1 | 6.783 | 2.405 |
| 0.6 | 14.40 | 0.9775 | 604.8 | 3.360 | 0.815 | 216.5 | 6.808 | 2.437 |
| 0.7 | 16.87 | 0.9737 | 604.3 | 3.344 | 0.817 | 215.9 | 6.832 | 2.469 |
| 0.8 | 19.36 | 0.9699 | 603.7 | 3.331 | 0.819 | 215.3 | 6.856 | 2.501 |
| 0.9 | 21.86 | 0.9661 | 603.2 | 3.318 | 0.822 | 214.7 | 6.880 | 2.535 |
| 1.0 | 24.40 | 0.9623 | 602.6 | 3.307 | 0.824 | 214.1 | 6.902 | 2.570 |
| 1.5 | 37.33 | 0.9430 | 599.7 | 3.264 | 0.838 | 211.2 | 7.009 | 2.751 |
| 2.0 | 50.83 | 0.9234 | 596.7 | 3.231 | 0.854 | 208.2 | 7.101 | 2.952 |
| 2.5 | 64.93 | 0.9036 | 593.6 | 3.205 | 0.872 | 205.3 | 7.177 | 3.174 |
| 3.0 | 79.69 | 0.8835 | 590.5 | 3.182 | 0.893 | 202.4 | 7.237 | 3.419 |
| 3.5 | 95.16 | 0.8632 | 587.2 | 3.161 | 0.917 | 199.5 | 7.278 | 3.690 |
| 4.0 | 111.40 | 0.8426 | 583.8 | 3.143 | 0.945 | 196.7 | 7.299 | 3.990 |
| 4.5 | 128.50 | 0.8219 | 580.3 | 3.125 | 0.976 | 193.9 | 7.301 | 4.322 |
| 5.0 | 146.51 | 0.8009 | 576.7 | 3.109 | 1.010 | 191.2 | 7.279 | 4.688 |
| 5.5 | 165.51 | 0.7799 | 572.9 | 3.093 | 1.050 | 188.6 | 7.234 | 5.093 |
| 6.0 | 185.59 | 0.7587 | 569.1 | 3.078 | 1.094 | 186.1 | 7.164 | 5.539 |
| 6.5 | 206.83 | 0.7375 | 565.1 | 3.063 | 1.143 | 183.8 | 7.087 | 6.027 |
| 7.0 | 229.29 | 0.7165 | 561.0 | 3.049 | 1.197 | 181.7 | 6.939 | 6.555 |
| 7.5 | 255.02 | 0.6957 | 556.8 | 3.035 | 1.256 | 180.0 | 6.777 | 7.118 |
| 8.0 | 278.03 | 0.6753 | 552.5 | 3.021 | 1.320 | 178.6 | 6.579 | 7.704 |
| 8.5 | 304.26 | 0.6556 | 548.1 | 3.007 | 1.386 | 177.8 | 6.343 | 8.290 |
| 9.0 | 331.69 | 0.6370 | 543.7 | 2.994 | 1.452 | 177.7 | 6.068 | 8.848 |
| 9.5 | 359.78 | 0.6197 | 539.3 | 2.980 | 1.515 | 178.1 | 5.760 | 9.343 |
| 10.0 | 388.53 | 0.6040 | 535.0 | 2.968 | 1.573 | 179.4 | 5.426 | 9.741 |
| 11.0 | 446.20 | 0.5786 | 526.8 | 2.944 | 1.662 | 184.2 | 4.728 | 10.17 |
| 12.0 | 501.72 | 0.5613 | 519.4 | 2.922 | 1.708 | 191.8 | 4.057 | 10.10 |
| 13.0 | 553.01 | 0.5517 | 512.9 | 2.903 | 1.716 | 201.4 | 3.459 | 9.664 |
| 14.0 | 599.02 | 0.5485 | 507.5 | 2.887 | 1.696 | 212.9 | 2.940 | 8.997 |
| 15.0 | 639.47 | 0.5505 | 502.9 | 2.873 | 1.658 | 225.7 | 2.495 | 8.226 |
| 16.0 | 674.64 | 0.5566 | 499.2 | 2.861 | 1.611 | 239.4 | 2.119 | 7.453 |
| 17.0 | 705.15 | 0.5658 | 496.1 | 2.851 | 1.564 | 253.7 | 1.803 | 6.744 |
| 18.0 | 731.72 | 0.5773 | 493.5 | 2.842 | 1.520 | 268.1 | 1.542 | 6.125 |
| 19.0 | 755.06 | 0.5905 | 491.3 | 2.834 | 1.480 | 282.3 | 1.325 | 5.599 |
| 20.0 | 775.76 | 0.6205 | 489.5 | 2.827 | 1.446 | 296.1 | 1.146 | 5.157 |

**TABLE 38.** Thermodynamic Properties of Freon-22 in the Single-Phase Region (*Continued*)

$T=453.15$ K

| $p$ | $\rho$ | $z$ | $h$ | $s$ | $c_p$ | $w$ | $\mu$ | $\alpha \cdot 10^3$ |
|------|--------|--------|-------|-------|-------|-------|-------|---------|
| 0.01 | 0.23 | 0.9997 | 616.1 | 3.777 | 0.812 | 222.3 | 6.378 | 2.209 |
| 0.02 | 0.46 | 0.9993 | 616.1 | 3.710 | 0.812 | 222.2 | 6.380 | 2.212 |
| 0.03 | 0.69 | 0.9990 | 616.0 | 3.671 | 0.812 | 222.2 | 6.382 | 2.214 |
| 0.04 | 0.92 | 0.9986 | 616.0 | 3.643 | 0.812 | 222.1 | 6.384 | 2.217 |
| 0.05 | 1.15 | 0.9983 | 615.9 | 3.622 | 0.812 | 222.0 | 6.386 | 2.220 |
| 0.1 | 2.30 | 0.9966 | 615.7 | 3.555 | 0.813 | 221.8 | 6.397 | 2.233 |
| 0.2 | 4.62 | 0.9931 | 615.2 | 3.487 | 0.815 | 221.2 | 6.419 | 2.260 |
| 0.3 | 6.96 | 0.9896 | 614.6 | 3.448 | 0.816 | 220.7 | 6.441 | 2.288 |
| 0.4 | 9.31 | 0.9862 | 614.7 | 3.419 | 0.818 | 229.2 | 6.462 | 2.315 |
| 0.5 | 11.68 | 0.9827 | 613.5 | 3.397 | 0.820 | 219.6 | 6.482 | 2.234 |
| 0.6 | 14.06 | 0.9792 | 613.0 | 3.378 | 0.822 | 219.1 | 6.502 | 2.373 |
| 0.7 | 16.47 | 0.9757 | 612.5 | 3.363 | 0.824 | 218.5 | 6.522 | 2.402 |
| 0.8 | 18.89 | 0.9722 | 611.9 | 3.349 | 0.826 | 218.0 | 6.541 | 2.432 |
| 0.9 | 21.32 | 0.9687 | 611.4 | 3.337 | 0.828 | 217.5 | 6.559 | 2.462 |
| 1.0 | 23.78 | 0.9652 | 610.9 | 3.326 | 0.830 | 216.9 | 6.577 | 2.493 |
| 1.5 | 36.34 | 0.9475 | 608.1 | 3.282 | 0.841 | 214.2 | 6.659 | 2.656 |
| 2.0 | 49.38 | 0.9295 | 605.3 | 3.250 | 0.855 | 211.5 | 6.727 | 2.833 |
| 2.5 | 62.95 | 0.9116 | 602.3 | 3.224 | 0.870 | 208.9 | 6.780 | 3.026 |
| 3.0 | 77.08 | 0.8932 | 599.4 | 3.202 | 0.888 | 206.2 | 6.817 | 3.236 |
| 3.5 | 91.82 | 0.9748 | 596.3 | 3.182 | 0.908 | 203.8 | 6.838 | 3.465 |
| 4.0 | 107.20 | 0.8564 | 593.1 | 3.164 | 0.930 | 201.1 | 6.842 | 3.713 |
| 4.5 | 123.28 | 0.8378 | 589.9 | 3.147 | 0.955 | 198.6 | 6.828 | 3.982 |
| 5.0 | 140.10 | 0.8191 | 586.6 | 3.131 | 0.984 | 196.1 | 6.797 | 4.275 |
| 5.5 | 157.70 | 0.8004 | 583.2 | 3.116 | 1.016 | 193.8 | 6.746 | 4.591 |
| 6.0 | 176.15 | 0.7817 | 579.8 | 3.102 | 1.050 | 191.5 | 6.677 | 4.933 |
| 6.5 | 195.48 | 0.7632 | 576.2 | 3.088 | 1.088 | 189.4 | 6.587 | 5.300 |
| 7.0 | 215.72 | 0.7447 | 572.6 | 3.075 | 1.130 | 187.5 | 6.476 | 5.690 |
| 7.5 | 236.90 | 0.7266 | 568.9 | 3.062 | 1.175 | 185.9 | 6.341 | 6.101 |
| 8.0 | 269.02 | 0.7088 | 565.2 | 3.049 | 1.222 | 184.5 | 6.181 | 6.526 |
| 8.5 | 282.04 | 0.6917 | 561.4 | 3.036 | 1.272 | 183.6 | 5.996 | 6.954 |
| 9.0 | 305.89 | 0.6753 | 557.5 | 3.025 | 1.323 | 183.1 | 5.783 | 7.372 |
| 9.5 | 330.44 | 0.6598 | 553.7 | 3.013 | 1.374 | 183.2 | 5.546 | 7.762 |
| 10.0 | 355.51 | 0.6456 | 549.9 | 3.001 | 1.422 | 183.8 | 5.288 | 8.107 |
| 11.0 | 406.34 | 0.6213 | 542.6 | 2.979 | 1.507 | 187.1 | 4.730 | 8.600 |
| 12.0 | 456.51 | 0.6033 | 535.7 | 2.959 | 1.569 | 192.6 | 4.165 | 8.791 |
| 13.0 | 504.41 | 0.5915 | 529.5 | 2.941 | 1.605 | 200.2 | 3.637 | 8.705 |
| 14.0 | 549.95 | 0.5853 | 524.1 | 2.924 | 1.618 | 209.3 | 3.163 | 8.406 |
| 15.0 | 589.55 | 0.5839 | 519.3 | 2.910 | 1.612 | 219.6 | 2.745 | 7.962 |
| 16.0 | 626.04 | 0.5866 | 515.2 | 2.897 | 1.591 | 231.1 | 2.376 | 7.434 |
| 17.0 | 658.55 | 0.5926 | 511.7 | 2.886 | 1.562 | 243.4 | 2.058 | 6.880 |
| 18.0 | 687.38 | 0.6010 | 508.7 | 2.876 | 1.530 | 256.3 | 1.782 | 6.341 |
| 19.0 | 712.98 | 0.6116 | 506.2 | 2.868 | 1.497 | 269.3 | 1.546 | 6.848 |
| 20.0 | 735.80 | 0.6238 | 504.1 | 2.860 | 1.465 | 282.4 | 1.345 | 5.406 |

**TABLE 38.** Thermodynamic Properties of Freon-22 in the Single-Phase Region (*Continued*)

$$T = 463.15 \text{ K}$$

| $p$ | $\rho$ | $z$ | $h$ | $s$ | $c_p$ | $w$ | $\mu$ | $a \cdot 10^3$ |
|------|--------|--------|-------|-------|-------|-------|-------|-------|
| 0.01 | 0.23 | 0.9997 | 624.3 | 3.794 | 0.820 | 224.5 | 6.144 | 2.162 |
| 0.02 | 0.45 | 0.9994 | 624.2 | 3.728 | 0.821 | 224.5 | 6.146 | 2.164 |
| 0.03 | 0.67 | 0.9990 | 624.2 | 3.689 | 0.821 | 224.5 | 6.147 | 2.167 |
| 0.04 | 0.90 | 0.9987 | 624.1 | 3.661 | 0.821 | 224.4 | 6.149 | 2.169 |
| 0.05 | 1.13 | 0.9984 | 624.0 | 3.640 | 0.821 | 224.4 | 6.151 | 2.171 |
| 0.1 | 2.25 | 0.9968 | 623.8 | 3.573 | 0.822 | 224.1 | 6.159 | 2.184 |
| 0.2 | 4.52 | 0.9936 | 623.3 | 3.505 | 0.823 | 223.6 | 6.177 | 2.209 |
| 0.3 | 6.80 | 0.9904 | 622.8 | 3.466 | 0.824 | 223.1 | 6.194 | 2.234 |
| 0.4 | 9.10 | 0.9872 | 622.3 | 3.437 | 0.826 | 222.6 | 6.210 | 2.259 |
| 0.5 | 11.41 | 0.9840 | 621.8 | 3.415 | 0.827 | 222.1 | 6.226 | 2.286 |
| 0.6 | 13.74 | 0.9808 | 621.2 | 3.396 | 0.829 | 221.6 | 6.242 | 2.313 |
| 0.7 | 16.08 | 0.0776 | 620.7 | 3.381 | 0.830 | 221.1 | 6.264 | 2.339 |
| 0.8 | 18.44 | 0.9743 | 620.2 | 3.367 | 0.832 | 220.8 | 6.271 | 2.380 |
| 0.9 | 20.81 | 0.9711 | 619.7 | 3.350 | 0.834 | 220.1 | 6.285 | 2.395 |
| 1.0 | 23.20 | 0.9678 | 619.2 | 3.344 | 0.835 | 219.7 | 6.299 | 2.423 |
| 1.5 | 35.40 | 0.9515 | 616.5 | 3.301 | 0.845 | 217.2 | 6.358 | 2.570 |
| 2.0 | 48.03 | 0.9351 | 613.8 | 3.289 | 0.856 | 214.8 | 6.403 | 2.728 |
| 2.5 | 61.12 | 0.9186 | 611.0 | 3.243 | 0.869 | 212.3 | 6.435 | 2.897 |
| 3.0 | 74.69 | 0.9020 | 608.2 | 3.221 | 0.884 | 209.9 | 6.453 | 3.079 |
| 3.5 | 88.78 | 0.8853 | 605.3 | 3.202 | 0.901 | 207.6 | 6.456 | 3.274 |
| 4.0 | 103.41 | 0.8685 | 602.4 | 3.184 | 0.920 | 205.2 | 6.445 | 3.483 |
| 4.5 | 118.63 | 0.8518 | 599.4 | 3.168 | 0.941 | 203.0 | 6.419 | 3.706 |
| 5.0 | 134.46 | 0.8350 | 596.3 | 3.153 | 0.964 | 200.7 | 6.378 | 3.945 |
| 5.5 | 150.93 | 0.8183 | 593.2 | 3.138 | 0.990 | 198.6 | 6.322 | 4.198 |
| 6.0 | 168.08 | 0.8016 | 590.1 | 3.125 | 1.018 | 196.5 | 6.252 | 4.468 |
| 6.5 | 185.91 | 0.7851 | 586.9 | 3.112 | 1.048 | 194.6 | 6.186 | 4.753 |
| 7.0 | 204.47 | 0.7688 | 583.6 | 3.099 | 1.081 | 192.6 | 6.063 | 5.052 |
| 7.5 | 223.75 | 0.7527 | 580.3 | 3.087 | 1.116 | 191.3 | 5.944 | 5.363 |
| 8.0 | 243.74 | 0.7370 | 577.0 | 3.075 | 1.159 | 190.0 | 5.808 | 5.682 |
| 8.5 | 264.42 | 0.7219 | 573.6 | 3.064 | 1.191 | 189.0 | 5.653 | 6.004 |
| 9.0 | 285.73 | 0.7073 | 570.3 | 3.053 | 1.231 | 188.4 | 5.479 | 6.321 |
| 9.5 | 307.60 | 0.6935 | 566.9 | 3.042 | 1.271 | 188.2 | 5.287 | 6.624 |
| 10.0 | 329.90 | 0.6807 | 563.5 | 3.031 | 1.311 | 188.5 | 5.079 | 6.903 |
| 11.0 | 375.28 | 0.6582 | 557.0 | 3.011 | 1.386 | 190.7 | 4.627 | 7.352 |
| 12.0 | 420.62 | 0.0406 | 550.8 | 2.992 | 1.447 | 194.9 | 4.154 | 7.618 |
| 13.0 | 464.72 | 0.6282 | 545.0 | 2.975 | 1.493 | 200.9 | 3.696 | 7.696 |
| 14.0 | 506.66 | 0.6205 | 539.8 | 2.959 | 1.523 | 208.4 | 3.273 | 7.612 |
| 15.0 | 545.88 | 0.6170 | 535.0 | 2.945 | 1.538 | 217.0 | 2.892 | 7.400 |
| 16.0 | 582.06 | 0.6173 | 530.7 | 2.932 | 1.539 | 226.6 | 2.551 | 7.094 |
| 17.0 | 615.10 | 0.6206 | 527.2 | 2.920 | 1.530 | 237.1 | 2.247 | 6.726 |
| 18.0 | 645.07 | 0.6266 | 524.0 | 2.910 | 1.514 | 248.2 | 1.977 | 6.326 |
| 19.0 | 672.13 | 0.6348 | 521.2 | 2.900 | 1.492 | 259.9 | 1.737 | 5.922 |
| 20.0 | 696.55 | 0.6562 | 518.8 | 2.892 | 1.469 | 271.8 | 1.528 | 5.534 |

**TABLE 38.** Thermodynamic Properties of Freon-22 in the Single-Phase Region (*Continued*)

$T = 473.15$ K

| $p$ | $\rho$ | $z$ | $h$ | $s$ | $c_p$ | $w$ | $\mu$ | $\alpha \cdot 10^{J}$ |
|------|--------|--------|-------|-------|-------|-------|-------|-------|
| 0.01 | 0.22 | 0.9997 | 632.5 | 3.812 | 0.829 | 226.8 | 5.948 | 2.116 |
| 0.02 | 0.44 | 0.9994 | 632.4 | 3.746 | 0.829 | 226.8 | 5.949 | 2.118 |
| 0.03 | 0.66 | 0.9991 | 632.4 | 3.707 | 0.829 | 226.7 | 5.950 | 2.120 |
| 0.04 | 0.88 | 0.9988 | 632.3 | 3.679 | 0.829 | 226.7 | 5.952 | 2.123 |
| 0.05 | 1.10 | 0.9985 | 632.3 | 3.658 | 0.829 | 226.6 | 5.953 | 2.125 |
| 0.1 | 2.21 | 0.9970 | 632.0 | 3.591 | 0.830 | 226.4 | 5.959 | 2.137 |
| 0.2 | 4.42 | 0.9941 | 631.5 | 3.523 | 0.831 | 226.0 | 5.972 | 2.160 |
| 0.3 | 6.65 | 0.9911 | 631.0 | 3.483 | 0.832 | 225.5 | 5.985 | 2.184 |
| 0.4 | 8.90 | 0.9882 | 630.6 | 3.455 | 0.833 | 225.1 | 5.997 | 2.207 |
| 0.5 | 11.16 | 0.9852 | 630.1 | 3.433 | 0.834 | 224.6 | 6.009 | 2.223 |
| 0.6 | 13.43 | 0.9822 | 629.5 | 3.415 | 0.836 | 224.2 | 6.020 | 2.256 |
| 0.7 | 15.71 | 0.9793 | 629.0 | 3.399 | 0.837 | 223.7 | 6.031 | 2.281 |
| 0.8 | 18.01 | 0.9763 | 628.5 | 3.385 | 0.838 | 223.3 | 6.041 | 2.306 |
| 0.9 | 20.33 | 0.9733 | 628.0 | 3.373 | 0.840 | 222.8 | 6.051 | 2.332 |
| 1.0 | 22.65 | 0.9703 | 627.5 | 3.362 | 0.841 | 222.4 | 6.060 | 2.358 |
| 1.5 | 34.52 | 0.9559 | 625.0 | 3.319 | 0.849 | 220.1 | 6.098 | 2.491 |
| 2.0 | 46.76 | 0.9402 | 622.3 | 3.288 | 0.859 | 217.9 | 6.124 | 2.633 |
| 2.5 | 59.40 | 0.9251 | 619.7 | 3.262 | 0.870 | 215.7 | 6.137 | 2.784 |
| 3.0 | 72.48 | 0.9099 | 617.0 | 3.240 | 0.882 | 213.5 | 6.138 | 2.943 |
| 3.5 | 85.99 | 0.8947 | 614.3 | 3.221 | 0.897 | 211.4 | 6.125 | 3.112 |
| 4.0 | 99.98 | 0.8797 | 611.5 | 3.204 | 0.912 | 209.2 | 6.100 | 3.289 |
| 4.5 | 114.45 | 0.8642 | 608.7 | 3.188 | 0.930 | 207.1 | 6.063 | 3.477 |
| 5.0 | 129.44 | 0.8490 | 605.9 | 3.173 | 0.949 | 205.1 | 6.014 | 3.675 |
| 5.5 | 144.96 | 0.8339 | 603.0 | 3.160 | 0.970 | 203.1 | 5.952 | 3.882 |
| 6.0 | 161.04 | 0.8190 | 600.1 | 3.146 | 0.993 | 201.2 | 5.879 | 4.100 |
| 6.5 | 177.68 | 0.8041 | 697.2 | 3.134 | 1.017 | 199.4 | 5.794 | 4.327 |
| 7.0 | 194.89 | 0.7895 | 594.2 | 3.122 | 1.044 | 197.8 | 5.697 | 4.563 |
| 7.5 | 212.68 | 0.7751 | 591.2 | 3.111 | 1.072 | 196.3 | 5.588 | 4.805 |
| 8.0 | 231.03 | 0.7611 | 588.2 | 3.099 | 1.101 | 195.0 | 5.465 | 5.052 |
| 8.5 | 249.93 | 0.7476 | 585.2 | 3.089 | 1.132 | 194.0 | 5.330 | 5.300 |
| 9.0 | 269.32 | 0.7345 | 582.2 | 3.078 | 1.164 | 193.4 | 5.181 | 5.545 |
| 9.5 | 289.15 | 0.7222 | 579.2 | 3.068 | 1.196 | 193.0 | 5.020 | 5.781 |
| 10.0 | 309.35 | 0.7105 | 576.2 | 3.058 | 1.228 | 193.1 | 4.845 | 6.003 |
| 11.0 | 350.45 | 0.6899 | 570.3 | 3.039 | 1.290 | 194.6 | 4.467 | 6.383 |
| 12.0 | 391.73 | 0.6733 | 564.7 | 3.022 | 1.347 | 197.9 | 4.067 | 6.648 |
| 13.0 | 432.30 | 0.6610 | 559.4 | 3.006 | 1.393 | 202.8 | 3.671 | 6.787 |
| 14.0 | 471.43 | 0.6528 | 554.4 | 2.990 | 1.430 | 209.1 | 3.297 | 6.810 |
| 15.0 | 508.52 | 0.6483 | 550.0 | 2.976 | 1.455 | 216.4 | 2.954 | 6.734 |
| 16.0 | 543.55 | 0.6470 | 545.9 | 2.964 | 1.470 | 224.7 | 2.644 | 6.580 |
| 17.0 | 676.06 | 0.6487 | 542.2 | 2.952 | 1.477 | 233.7 | 2.364 | 6.369 |
| 18.0 | 606.12 | 0.6528 | 538.9 | 2.942 | 1.475 | 253.6 | 2.111 | 6.101 |
| 19.0 | 633.74 | 0.6590 | 536.0 | 2.932 | 1.487 | 264.2 | 1.882 | 5.811 |
| 20.0 | 659.04 | 0.6671 | 533.4 | 2.923 | 1.454 | 275.2 | 1.676 | 5.508 |

**TABLE 39.** Transport Properties of Freon-22 on the Saturation Line

| $T$ | $p$ | $\eta' \cdot 10^6$ | $\eta'' \cdot 10^5$ | $\nu' \cdot 10^6$ | $\nu'' \cdot 10^6$ | $\lambda' \cdot 10$ | $\lambda'' \cdot 10^6$ | $a' \cdot 10^6$ | $a'' \cdot 10^6$ | $Pr'$ | $Pr''$ |
|---|---|---|---|---|---|---|---|---|---|---|---|
| 233.15 | 0.1054 | 338.7 | 10.19 | 0.2392 | 2.0839 | 115.0 | 6.66 | 0.0722 | 2.3181 | 3.311 | 0.887 |
| 235.15 | 0.1155 | 328.1 | 10.28 | 0.2327 | 1.9287 | 114.1 | 6.80 | 0.0715 | 2.1770 | 3.253 | 0.887 |
| 237.15 | 0.1264 | 318.6 | 10.37 | 0.2270 | 1.7910 | 113.1 | 6.94 | 0.0709 | 2.0179 | 3.200 | 0.887 |
| 238.15 | 0.1321 | 314.2 | 10.42 | 0.2244 | 1.7251 | 112.6 | 7.01 | 0.0706 | 1.9538 | 3.178 | 0.884 |
| 239.15 | 0.1380 | 309.8 | 10.47 | 0.2217 | 1.6645 | 112.2 | 7.08 | 0.0703 | 1.8887 | 3.151 | 0.881 |
| 241.15 | 0.1505 | 301.9 | 10.56 | 0.2171 | 1.5461 | 111.2 | 7.23 | 0.0698 | 1.7584 | 3.106 | 0.879 |
| 243.15 | 0.1639 | 294.5 | 10.66 | 0.2127 | 1.4406 | 110.2 | 7.37 | 0.0694 | 1.6380 | 3.061 | 0.879 |
| 245.15 | 0.1782 | 287.1 | 10.75 | 0.2084 | 1.3438 | 109.3 | 7.52 | 0.0690 | 1.5285 | 3.018 | 0.879 |
| 247.15 | 0.1934 | 280.2 | 10.85 | 0.2043 | 1.2558 | 108.3 | 7.67 | 0.0686 | 1.4295 | 2.976 | 0.879 |
| 248.15 | 0.2014 | 276.8 | 10.90 | 0.2023 | 1.2138 | 107.9 | 7.74 | 0.0684 | 1.3791 | 2.955 | 0.879 |
| 249.15 | 0.2097 | 273.4 | 10.95 | 0.2003 | 1.1736 | 107.4 | 7.82 | 0.0682 | 1.3347 | 2.933 | 0.879 |
| 251.15 | 0.2269 | 266.9 | 11.04 | 0.1965 | 1.0985 | 106.4 | 7.97 | 0.0679 | 1.2489 | 2.892 | 0.879 |
| 253.15 | 0.2453 | 260.4 | 11.15 | 0.1927 | 1.0305 | 105.5 | 8.12 | 0.0676 | 1.1689 | 2.849 | 0.881 |
| 255.15 | 0.2647 | 254.3 | 11.24 | 0.1891 | 0.9665 | 104.5 | 8.28 | 0.0573 | 1.0970 | 2.808 | 0.882 |
| 257.15 | 0.2854 | 248.3 | 11.34 | 0.1856 | 0.9079 | 103.6 | 8.43 | 0.0671 | 1.0273 | 2.766 | 0.883 |
| 258.15 | 0.2961 | 245.3 | 11.39 | 0.1838 | 0.8809 | 103.1 | 8.51 | 0.0669 | 0.9958 | 2.746 | 0.884 |
| 259.15 | 0.3072 | 242.5 | 11.45 | 0.1822 | 0.8551 | 102.7 | 8.59 | 0.0667 | 0.9648 | 2.725 | 0.886 |
| 261.15 | 0.3303 | 236.8 | 11.55 | 0.1788 | 0.8049 | 101.7 | 8.75 | 0.0665 | 0.9061 | 2.686 | 0.888 |
| 263.15 | 0.3548 | 231.4 | 11.66 | 0.1755 | 0.7591 | 100.8 | 8.91 | 0.0662 | 0.8518 | 2.649 | 0.891 |
| 265.15 | 0.3805 | 226.2 | 11.76 | 0.1726 | 0.7158 | 99.9 | 9.07 | 0.0659 | 0.8012 | 2.618 | 0.893 |
| 267.15 | 0.4077 | 221.3 | 11.87 | 0.1697 | 0.6764 | 98.9 | 9.24 | 0.0655 | 0.7543 | 2.589 | 0.896 |
| 268.15 | 0.4218 | 218.9 | 11.93 | 0.1683 | 0.6583 | 98.4 | 9.32 | 0.0653 | 0.7322 | 2.577 | 0.898 |
| 269.15 | 0.4363 | 216.5 | 11.98 | 0.1669 | 0.6396 | 97.9 | 9.41 | 0.0651 | 0.7106 | 2.564 | 0.900 |
| 271.15 | 0.4664 | 212.0 | 12.08 | 0.1643 | 0.6049 | 96.8 | 9.58 | 0.0646 | 0.6700 | 2.543 | 0.903 |
| 273.15 | 0.4981 | 208.0 | 12.21 | 0.1621 | 0.5738 | 95.7 | 9.75 | 0.0641 | 0.6320 | 2.528 | 0.907 |
| 275.15 | 0.5313 | 204.0 | 12.32 | 0.1599 | 0.5439 | 94.7 | 9.93 | 0.0636 | 0.5969 | 2.511 | 0.911 |

**TABLE 39.** Transport Properties of Freon-22 on the Saturation Line (Continued)

| T | p | $\eta' \cdot 10^6$ | $\eta'' \cdot 10^6$ | $v' \cdot 10^6$ | $v'' \cdot 10^6$ | $\lambda' \cdot 10^6$ | $\lambda'' \cdot 10^6$ | $a' \cdot 10^6$ | $a'' \cdot 10^6$ | Pr' | Pr'' |
|---|---|---|---|---|---|---|---|---|---|---|---|
| 277.15 | 0.5662 | 200.0 | 12.43 | 0.1576 | 0.5158 | 93.8 | 10.10 | 0.0632 | 0.5631 | 2.493 | 0.915 |
| 278.15 | 0.5842 | 198.2 | 12.48 | 0.1566 | 0.5024 | 93.4 | 10.19 | 0.0630 | 0.5476 | 2.485 | 0.917 |
| 279.15 | 0.6027 | 196.0 | 12.54 | 0.1553 | 0.4897 | 93.0 | 10.28 | 0.0628 | 0.5323 | 2.475 | 0.919 |
| 281.15 | 0.6410 | 192.1 | 12.65 | 0.1534 | 0.4651 | 92.1 | 10.46 | 0.0623 | 0.5032 | 2.460 | 0.924 |
| 283.15 | 0.6811 | 188.4 | 12.74 | 0.1510 | 0.4420 | 91.2 | 10.65 | 0.0618 | 0.4762 | 2.442 | 0.928 |
| 285.15 | 0.7231 | 184.6 | 12.88 | 0.1488 | 0.4206 | 90.4 | 10.83 | 0.0613 | 0.4503 | 2.430 | 0.934 |
| 287.15 | 0.7669 | 180.9 | 13.00 | 0.1467 | 0.4006 | 89.5 | 11.02 | 0.0608 | 0.4264 | 2.416 | 0.939 |
| 288.15 | 0.7895 | 179.0 | 13.06 | 0.1456 | 0.3910 | 89.1 | 11.12 | 0.0605 | 0.4151 | 2.405 | 0.942 |
| 289.15 | 0.8127 | 177.2 | 13.12 | 0.1446 | 0.3816 | 88.7 | 11.21 | 0.0602 | 0.4037 | 2.397 | 0.946 |
| 291.15 | 0.8604 | 173.6 | 13.24 | 0.1425 | 0.3638 | 87.8 | 11.41 | 0.0596 | 0.3827 | 2.389 | 0.951 |
| 293.15 | 0.9103 | 170.0 | 13.37 | 0.1404 | 0.3473 | 86.6 | 11.61 | 0.0589 | 0.3628 | 2.380 | 0.957 |
| 295.15 | 0.9622 | 166.4 | 13.49 | 0.1383 | 0.3315 | 85.7 | 11.81 | 0.0582 | 0.3440 | 2.370 | 0.963 |
| 297.15 | 1.0163 | 162.8 | 13.63 | 0.1362 | 0.3168 | 84.9 | 12.02 | 0.0577 | 0.3263 | 2.361 | 0.970 |
| 298.15 | 1.0441 | 161.0 | 13.69 | 0.1352 | 0.3095 | 84.5 | 12.12 | 0.0574 | 0.3177 | 2.355 | 0.974 |
| 299.15 | 1.0726 | 159.2 | 13.76 | 0.1341 | 0.3028 | 84.1 | 12.23 | 0.0570 | 0.3096 | 2.349 | 0.977 |
| 301.15 | 1.1311 | 155.6 | 13.90 | 0.1320 | 0.2896 | 83.3 | 12.44 | 0.0564 | 0.2936 | 2.337 | 0.986 |
| 303.15 | 1.1920 | 152.0 | 14.04 | 0.1298 | 0.2771 | 82.4 | 12.66 | 0.0558 | 0.2787 | 2.326 | 0.994 |
| 305.15 | 1.2553 | 148.4 | 14.19 | 0.1276 | 0.2655 | 81.7 | 12.88 | 0.0552 | 0.2644 | 2.312 | 1.004 |
| 307.15 | 1.3210 | 144.9 | 14.34 | 0.1255 | 0.2543 | 80.9 | 13.11 | 0.0545 | 0.2509 | 2.302 | 1.014 |
| 308.15 | 1.3548 | 143.1 | 14.42 | 0.1244 | 0.2491 | 80.4 | 13.22 | 0.0541 | 0.2444 | 2.294 | 1.019 |
| 309.15 | 1.3892 | 141.3 | 14.50 | 0.1233 | 0.2439 | 80.0 | 13.34 | 0.0538 | 0.2381 | 2.286 | 1.024 |
| 311.15 | 1.4600 | 137.7 | 14.66 | 0.1211 | 0.2340 | 79.2 | 13.58 | 0.0531 | 0.2260 | 2.279 | 1.035 |
| 313.15 | 1.5335 | 134.2 | 14.83 | 0.1189 | 0.2280 | 78.4 | 13.82 | 0.0524 | 0.2176 | 2.271 | 1.048 |
| 315.15 | 1.6096 | 130.7 | 15.00 | 0.1167 | 0.2155 | 77.3 | 14.07 | 0.0515 | 0.2032 | 2.264 | 1.061 |
| 317.15 | 1.6885 | 127.4 | 15.18 | 0.1147 | 0.2070 | 76.5 | 14.32 | 0.0508 | 0.1924 | 2.257 | 1.075 |
| 318.15 | 1.7290 | 125.8 | 15.28 | 0.1137 | 0.2030 | 76.1 | 14.45 | 0.0504 | 0.1867 | 2.253 | 1.084 |
| 319.15 | 1.7702 | 124.2 | 15.37 | 0.1128 | 0.1990 | 75.7 | 14.58 | 0.0501 | 0.1822 | 2.249 | 1.092 |
| 321.15 | 1.8548 | 121.2 | 15.57 | 0.1111 | 0.1914 | 75.0 | 14.85 | 0.0494 | 0.1725 | 2.245 | 1.109 |

**TABLE 39.** Transport Properties of Freon-22 on the Saturation Line *(Continued)*

| $T$ | $p$ | $\eta'\cdot10^6$ | $\eta''\cdot10^6$ | $\nu'\cdot10^6$ | $\nu''\cdot10^2$ | $\lambda'\cdot10^5$ | $\lambda''\cdot10^5$ | $a'\cdot10^5$ | $a''\cdot10^6$ | $Pr'$ | $Pr''$ |
|---|---|---|---|---|---|---|---|---|---|---|---|
| 323.15 | 1.9424 | 118.4 | 15.78 | 0.1094 | 0.1841 | 74.2 | 15.13 | 0.0487 | 0.1632 | 2.247 | 1.128 |
| 325.15 | 2.033 | 115.8 | 16.00 | 0.1080 | 0.1772 | 73.5 | 15.42 | 0.0480 | 0.1542 | 2.250 | 1.150 |
| 327.15 | 2.127 | 113.5 | 16.22 | 0.1069 | 0.1706 | 72.7 | 15.71 | 0.0472 | 0.1454 | 2.262 | 1.173 |
| 328.15 | 2.175 | 112.4 | 16.34 | 0.1063 | 0.1671 | 72.3 | 15.86 | 0.0468 | 0.1412 | 2.271 | 1.186 |
| 329.15 | 2.224 | 111.3 | 16.46 | 0.1058 | 0.1643 | 71.9 | 16.01 | 0.0464 | 0.1370 | 2.279 | 1.199 |
| 331.15 | 2.324 | 109.2 | 16.72 | 0.1049 | 0.1583 | 71.1 | 16.33 | 0.0456 | 0.1290 | 2.299 | 1.228 |
| 333.15 | 2.427 | 107.2 | 16.99 | 0.1041 | 0.1526 | 70.3 | 16.66 | 0.0447 | 0.1210 | 2.324 | 1.262 |
| 335.15 | 2.534 | 105.0 | 17.28 | 0.1031 | 0.1472 | 69.4 | 17.00 | 0.0438 | 0.1133 | 2.351 | 1.299 |
| 337.15 | 2.645 | 102.7 | 17.58 | 0.1020 | 0.1421 | 68.5 | 17.35 | 0.0428 | 0.1059 | 2.377 | 1.342 |
| 338.15 | 2.701 | 101.4 | 17.74 | 0.1013 | 0.1394 | 68.0 | 17.53 | 0.0423 | 0.1021 | 2.393 | 1.366 |
| 339.15 | 2.759 | 100.1 | 17.90 | 0.1006 | 0.1369 | 67.6 | 17.72 | 0.0418 | 0.0985 | 2.403 | 1.390 |
| 341.15 | 2.877 | 97.4 | 18.25 | 0.0991 | 0.1322 | 65.7 | 18.11 | 0.0407 | 0.0913 | 2.431 | 1.447 |
| 343.15 | 2.998 | 94.7 | 18.63 | 0.0976 | 0.1276 | 65.8 | 18.52 | 0.0396 | 0.0843 | 2.461 | 1.514 |
| 345.15 | 3.124 | 92.1 | 19.04 | 0.0963 | 0.1233 | 61.9 | 18.95 | 0.0384 | 0.0774 | 2.506 | 1.592 |
| 347.15 | 3.253 | 89.4 | 19.48 | 0.0948 | 0.1190 | 63.9 | 19.41 | 0.0370 | 0.0706 | 2.563 | 1.686 |
| 348.15 | 3.319 | 88.0 | 19.71 | 0.0941 | 0.1169 | 63.4 | 19.65 | 0.0363 | 0.0673 | 2.590 | 1.739 |
| 349.15 | 3.387 | 86.6 | 19.96 | 0.0933 | 0.1150 | 62.9 | 19.89 | 0.0355 | 0.0639 | 2.626 | 1.830 |
| 351.15 | 3.524 | 83.8 | 20.50 | 0.0918 | 0.1111 | 62.0 | 20.4 | 0.0339 | 0.0572 | 2.703 | 1.941 |
| 353.15 | 3.666 | 80.9 | 21.10 | 0.0903 | 0.1074 | 61.9 | 21.0 | 0.0321 | 0.0508 | 2.812 | 2.116 |
| 355.15 | 3.813 | 77.9 | 21.70 | 0.0887 | 0.1035 | 59.6 | 21.6 | 0.0299 | 0.0442 | 2.964 | 2.341 |
| 357.15 | 3.964 | 74.7 | 22.40 | 0.0870 | 0.0997 | 58.3 | 22.3 | 0.0274 | 0.0378 | 3.167 | 2.639 |
| 358.15 | 4.041 | 73.1 | 22.80 | 0.0862 | 0.0981 | 57.6 | 22.7 | 0.0261 | 0.0347 | 3.302 | 2.827 |
| 359.15 | 4.119 | 71.4 | 23.20 | 0.0853 | 0.0961 | 57.0 | 23.1 | 0.0244 | 0.0315 | 3.495 | 3.054 |
| 361.15 | 4.280 | 67.7 | 24.20 | 0.0833 | 0.0927 | 55.3 | 24.0 | 0.0212 | 0.0251 | 3.920 | 3.686 |
| 363.15 | 4.445 | 63.6 | 25.40 | 0.0811 | 0.0892 | 53.3 | 25.0 | 0.0171 | 0.0188 | 4.726 | 4.737 |
| 365.15 | 4.615 | 58.8 | 26.80 | 0.0787 | 0.0853 | 51.0 | 26.3 | 0.0122 | 0.0127 | 6.416 | 6.700 |

**TABLE 40.** Transport Properties of Freon-22 in the Single-Phase Region

| | T=273.15 K | | | | T=283.15 K | | | |
|---|---|---|---|---|---|---|---|---|
| $p$ | $\eta \cdot 10^6$ | $\lambda \cdot 10^6$ | $a \cdot 10^6$ | Pr | $\eta \cdot 10^6$ | $\lambda \cdot 10^6$ | $a \cdot 10^6$ | Pr |
| 0.1 | 11.9 | 9.1 | 3.6890 | 0.830 | 12.3 | 9.8 | 4.0290 | 0.818 |
| 0.5 | 208.0 | 95.8 | 0.0642 | 2.525 | 12.6 | 10.3 | 0.6954 | 0.893 |
| 1.0 | 210.0 | 96.2 | 0.0645 | 2.534 | 189.2 | 91.5 | 0.0621 | 2.440 |
| 2.0 | 212.0 | 97.0 | 0.0651 | 2.524 | 191.6 | 92.3 | 0.0628 | 2.435 |
| 3.0 | 214.0 | 97.8 | 0.0658 | 2.518 | 194.0 | 93.2 | 0.0635 | 2.427 |
| 4.0 | 217.0 | 98.6 | 0.0664 | 2.522 | 196.4 | 94.1 | 0.0643 | 2.419 |
| 5.0 | 219.0 | 99.4 | 0.0670 | 2.524 | 198.7 | 95.0 | 0.0650 | 2.414 |
| 6.0 | 222.0 | 100.2 | 0.0676 | 2.526 | 201.0 | 95.8 | 0.0657 | 2.409 |
| 8.0 | 226.0 | 101.7 | 0.0687 | 2.526 | 206.0 | 97.4 | 0.0670 | 2.409 |
| 10.0 | 231.0 | 102.9 | 0.0696 | 2.526 | 210.0 | 98.7 | 0.0680 | 2.404 |
| 12.0 | 235.0 | 104.0 | 0.0703 | 2.531 | 214.0 | 99.9 | 0.0689 | 2.406 |
| 14.0 | 240.0 | 104.8 | 0.0708 | 2.553 | 218.0 | 100.7 | 0.0695 | 2.418 |
| 16.0 | 244.0 | 105.4 | 0.0712 | 2.572 | 223.0 | 101.4 | 0.0700 | 2.443 |
| 18.0 | 249.0 | 105.8 | 0.0714 | 2.608 | 227.0 | 101.9 | 0.0703 | 2.466 |
| 20.0 | 253.0 | 106.3 | 0.0717 | 2.630 | 231.0 | 102.3 | 0.0705 | 2.491 |

**TABLE 40.** Transport Properties of Freon-22 in the Single-Phase Region (*Continued*)

| | T=293.15 K | | | | T=303.15 K | | | |
|---|---|---|---|---|---|---|---|---|
| $p$ | $\eta \cdot 10^6$ | $\lambda \cdot 10^6$ | $a \cdot 10^6$ | Pr | $\eta \cdot 10^6$ | $\lambda \cdot 10^6$ | $a \cdot 10^6$ | Pr |
| 0.1 | 12.7 | 10.4 | 4.3760 | 0.807 | 13.1 | 11.0 | 4.736 | 0.797 |
| 0.5 | 12.0 | 10.9 | 0.773 | 0.804 | 13.4 | 11.5 | 0.847 | 0.854 |
| 1.0 | 170.2 | 86.8 | 2.208 | 2.494 | 13.8 | 12.3 | 0.369 | 0.945 |
| 2.0 | 172.7 | 87.7 | 0.0599 | 2.369 | 154.4 | 83.2 | 0.0566 | 2.318 |
| 3.0 | 175.2 | 88.6 | 0.0608 | 2.359 | 157.2 | 84.3 | 0.0627 | 2.303 |
| 4.0 | 180.0 | 90.6 | 0.0624 | 2.352 | 159.8 | 85.3 | 0.0588 | 2.289 |
| 5.0 | 182.4 | 91.5 | 0.0635 | 2.349 | 162.4 | 86.3 | 0.0597 | 2.278 |
| 6.0 | 185.0 | 93.2 | 0.0646 | 2.346 | 164.9 | 87.3 | 0.0607 | 2.267 |
| 8.0 | 187.0 | 93.2 | 0.0656 | 2.337 | 169.7 | 89.2 | 0.0624 | 2.249 |
| 10.0 | 191.4 | 93.7 | 0.0664 | 2.327 | 174.3 | 90.8 | 0.0639 | 2.238 |
| 12.0 | 195.8 | 95.9 | 0.0671 | 2.315 | 178.7 | 92.1 | 0.0650 | 2.230 |
| 14.0 | 200.0 | 96.9 | 0.0680 | 2.322 | 183.0 | 93.2 | 0.0660 | 2.240 |
| 16.0 | 204.0 | 97.7 | 0.0685 | 2.334 | 187.1 | 94.0 | 0.0668 | 2.251 |
| 18.0 | 208.0 | 98.2 | 0.0690 | 2.353 | 191.2 | 94.7 | 0.0674 | 2.265 |
| 20.0 | 212.0 | 98.7 | 0.0693 | 2.373 | 195.2 | 95.3 | 0.0679 | 2.282 |

**TABLE 40.** Transport Properties of Freon-22 in the Single-Phase Region (*Continued*)

| | | $T=313,15$ K | | | | $T=323,15$ K | | |
|---|---|---|---|---|---|---|---|---|
| $p$ | $\eta \cdot 10^6$ | $\lambda \cdot 10^6$ | $a \cdot 10^6$ | Pr | $\eta \cdot 10^6$ | $\lambda \cdot 10^6$ | $a \cdot 10^6$ | Pr |
| 0.1 | 13.5 | 11.67 | 5.101 | 0.789 | 13.9 | 12.3 | 5.4686 | 0.784 |
| 0.5 | 13.8 | 12.16 | 0.927 | 0.839 | 14.2 | 12.8 | 1.009 | 0.824 |
| 1.0 | 14.2 | 12.86 | 0.404 | 0.914 | 14.5 | 13.4 | 0.448 | 0.888 |
| 2.0 | 135.8 | 78.8 | 0.0530 | 2.263 | 118.6 | 74.4 | 0.0488 | 2.241 |
| 3.0 | 139.1 | 79.9 | 0.0542 | 2.249 | 122.2 | 75.6 | 0.0505 | 2.214 |
| 4.0 | 142.2 | 81.0 | 0.0555 | 2.235 | 125.6 | 76.8 | 0.0520 | 2.193 |
| 5.0 | 145.1 | 82.2 | 0.0566 | 2.221 | 128.8 | 78.0 | 0.0533 | 2.175 |
| 6.0 | 147.9 | 83.2 | 0.0577 | 2.208 | 131.8 | 79.2 | 0.0546 | 2.158 |
| 8.0 | 153.1 | 85.2 | 0.0597 | 2.187 | 137.4 | 81.4 | 0.0569 | 2.129 |
| 10.0 | 158.0 | 87.0 | 0.0614 | 2.172 | 142.6 | 83.2 | 0.0587 | 2.115 |
| 12.0 | 162.7 | 88.4 | 0.0627 | 2.170 | 147.5 | 84.9 | 0.0604 | 2.106 |
| 14.0 | 167.1 | 89.6 | 0.0639 | 2.171 | 152.1 | 86.2 | 0.0616 | 2.105 |
| 16.0 | 171.4 | 90.6 | 0.0648 | 2.177 | 156.6 | 87.3 | 0.0627 | 2.109 |
| 18.0 | 175.5 | 91.4 | 0.0656 | 2.187 | 160.8 | 88.2 | 0.0636 | 2.118 |
| 20.0 | 179.5 | 92.0 | 0.0661 | 2.203 | 164.9 | 88.9 | 0.0643 | 2.131 |

**TABLE 40.** Transport Properties of Freon-22 in the Single-Phase Region (*Continued*)

| | | $T=333,15$ K | | | | $T=343,15$ K | | |
|---|---|---|---|---|---|---|---|---|
| $p$ | $\eta \cdot 10^6$ | $\lambda \cdot 10^6$ | $a \cdot 10^6$ | Pr | $\eta \cdot 10^6$ | $\lambda \cdot 10^6$ | $a \cdot 10^6$ | Pr |
| 0.1 | 14.3 | 12.93 | 5.8570 | 0.776 | 14.7 | 13.56 | 6.2450 | 0.772 |
| 0.5 | 14.6 | 13.38 | 1.0920 | 0.811 | 15.0 | 13.99 | 1.1750 | 0.802 |
| 1.0 | 14.9 | 14.01 | 0.4941 | 0.864 | 15.3 | 14.59 | 0.5402 | 0.845 |
| 2.0 | 16.1 | 15.64 | 0.1841 | 1.060 | 16.3 | 16.08 | 0.2138 | 0.993 |
| 3.0 | 109.3 | 71.2 | 0.0462 | 2.281 | 94.7 | 66.6 | 0.0401 | 2.433 |
| 4.0 | 112.5 | 72.6 | 0.0482 | 2.222 | 99.1 | 68.0 | 0.0433 | 2.311 |
| 5.0 | 115.5 | 73.8 | 0.0498 | 2.182 | 102.1 | 69.4 | 0.0458 | 2.216 |
| 6.0 | 118.4 | 75.1 | 0.0514 | 2.147 | 106.4 | 70.7 | 0.0477 | 2.182 |
| 8.0 | 123.9 | 77.4 | 0.0540 | 2.102 | 112.8 | 73.2 | 0.0509 | 2.117 |
| 10.0 | 129.0 | 79.5 | 0.0562 | 2.072 | 118.3 | 75.3 | 0.0533 | 2.080 |
| 12.0 | 133.9 | 81.2 | 0.0579 | 2.057 | 122.8 | 77.3 | 0.0554 | 2.046 |
| 14.0 | 138.5 | 82.7 | 0.0594 | 2.052 | 127.1 | 78.9 | 0.0570 | 2.028 |
| 16.0 | 142.9 | 83.9 | 0.0606 | 2.052 | 131.3 | 80.3 | 0.0583 | 2.021 |
| 18.0 | 147.2 | 84.9 | 0.0615 | 2.060 | 135.4 | 81.4 | 0.0593 | 2.024 |
| 20.0 | 151.4 | 85.8 | 0.0624 | 2.084 | 139.3 | 82.3 | 0.0602 | 2.029 |

**TABLE 40.** Transport Properties of Freon-22 in the Single-Phase Region (*Continued*)

| | T=353.15 K | | | | T=363.15 K | | | |
|---|---|---|---|---|---|---|---|---|
| p | $\eta \cdot 10^6$ | $\lambda \cdot 10^6$ | $a \cdot 10^6$ | Pr | $\eta \cdot 10^6$ | $\lambda \cdot 10^6$ | $a \cdot 10^6$ | Pr |
| 0.1 | 15.1 | 14.19 | 6.6380 | 0.767 | 15.5 | 14.82 | 7.037 | 0.764 |
| 0.5 | 14.3 | 14.61 | 1.2620 | 0.740 | 15.7 | 15.23 | 1.348 | 0.784 |
| 1.0 | 15.6 | 15.18 | 0.5867 | 0.828 | 16.0 | 15.78 | 0.634 | 0.813 |
| 2.0 | 16.6 | 16.56 | 0.2214 | 0.944 | 16.9 | 17.06 | 0.271 | 0.906 |
| 3.0 | 18.4 | 18.54 | 0.1145 | 1.242 | 18.4 | 18.78 | 0.141 | 1.107 |
| 4.0 | 83.2 | 61.0 | 0.0342 | 2.677 | 21.6 | 21.7 | 0.0588 | 1.838 |
| 5.0 | 88.8 | 63.5 | 0.0395 | 2.397 | 71.3 | 55.9 | 0.0473 | 2.168 |
| 6.0 | 93.2 | 65.4 | 0.0439 | 2.262 | 79.0 | 59.6 | 0.0358 | 2.499 |
| 8.0 | 100.4 | 68.3 | 0.0473 | 2.134 | 88.6 | 63.9 | 0.0436 | 2.171 |
| 10.0 | 106.6 | 70.6 | 0.0502 | 2.081 | 95.7 | 66.9 | 0.0477 | 2.064 |
| 12.0 | 112.3 | 72.6 | 0.0524 | 2.056 | 101.6 | 69.4 | 0.0506 | 2.012 |
| 14.0 | 117.2 | 74.3 | 0.0541 | 2.041 | 107.0 | 71.2 | 0.0525 | 1.997 |
| 16.0 | 122.0 | 75.9 | 0.0556 | 2.035 | 112.1 | 73.0 | 0.0542 | 1.990 |
| 18.0 | 126.2 | 77.4 | 0.0569 | 2.028 | 117.0 | 74.6 | 0.0555 | 1.992 |
| 20.0 | 130.4 | 78.7 | 0.0580 | 2.030 | 121.5 | 76.1 | 0.0568 | 1.994 |

**TABLE 40.** Transport Properties of Freon-22 in the Single-Phase Region (*Continued*)

| | T=373.15 K | | | | T=383.15 K | | | |
|---|---|---|---|---|---|---|---|---|
| p | $\eta \cdot 10^6$ | $\lambda \cdot 10^6$ | $a \cdot 10^6$ | Pr | $\eta \cdot 10^6$ | $\lambda \cdot 10^6$ | $a \cdot 10^6$ | Pr |
| 0.1 | 15.9 | 15.4 | 7.4410 | 0.761 | 16.3 | 16.08 | 7.8610 | 0.758 |
| 0.5 | 16.1 | 15.8 | 1.4360 | 0.777 | 16.5 | 16.46 | 1.5250 | 0.771 |
| 1.0 | 16.4 | 16.4 | 0.6817 | 0.801 | 16.7 | 16.97 | 0.7306 | 0.790 |
| 2.0 | 17.2 | 17.6 | 0.2998 | 0.876 | 17.5 | 18.12 | 0.3279 | 0.851 |
| 3.0 | 18.5 | 19.1 | 0.1647 | 1.022 | 18.6 | 19.53 | 0.1877 | 0.962 |
| 4.0 | 20.8 | 21.4 | 0.0879 | 1.371 | 20.6 | 21.4 | 0.1111 | 1.187 |
| 5.0 | 26.5 | 26.3 | 0.0541 | 2.619 | 23.9 | 24.3 | 0.0575 | 1.768 |
| 6.0 | 59.1 | 50.2 | 0.0204 | 3.867 | 32.1 | 31.6 | 0.0174 | 4.524 |
| 8.0 | 76.2 | 58.7 | 0.0383 | 2.304 | 61.9 | 52.1 | 0.0303 | 2.670 |
| 10.0 | 85.2 | 62.8 | 0.0447 | 2.081 | 74.4 | 58.4 | 0.0410 | 2.135 |
| 12.0 | 91.9 | 65.8 | 0.0485 | 1.994 | 82.5 | 62.1 | 0.0461 | 1.990 |
| 14.0 | 97.5 | 68.1 | 0.0510 | 1.956 | 88.8 | 64.9 | 0.0493 | 1.929 |
| 16.0 | 102.6 | 70.1 | 0.0528 | 1.939 | 94.2 | 67.2 | 0.0516 | 1.901 |
| 18.0 | 107.4 | 71.9 | 0.0544 | 1.933 | 99.0 | 69.2 | 0.0534 | 1.887 |
| 20.0 | 112.1 | 73.5 | 0.0558 | 1.938 | 103.6 | 71.0 | 0.0548 | 1.884 |

**TABLE 40.** Transport Properties of Freon-22 in the Single-Phase Region (*Continued*)

| | $T=393,15$ K | | | | $T=403,15$ K | | | |
|---|---|---|---|---|---|---|---|---|
| $p$ | $\eta \cdot 10^6$ | $\lambda \cdot 10^6$ | $a \cdot 10^6$ | Pr | $\eta \cdot 10^6$ | $\lambda \cdot 10^6$ | $a \cdot 10^6$ | Pr |
| 0.1 | 16.6 | 16.71 | 8.2730 | 0.756 | 17.0 | 17.34 | 8.7040 | 0.753 |
| 0.5 | 16.8 | 17.08 | 1.6130 | 0.767 | 17.2 | 17.70 | 1.7050 | 0.761 |
| 1.0 | 17.1 | 17.57 | 0.7788 | 0.781 | 17.4 | 18.18 | 0.8273 | 0.774 |
| 2.0 | 17.8 | 18.67 | 0.3561 | 0.823 | 18.1 | 19.23 | 0.3844 | 0.813 |
| 3.0 | 18.8 | 19.97 | 0.2099 | 0.917 | 19.1 | 20.4 | 0.2313 | 0.884 |
| 4.0 | 20.5 | 21.6 | 0.1321 | 1.074 | 20.5 | 21.9 | 0.1518 | 0.999 |
| 5.0 | 23.0 | 23.8 | 0.0807 | 1.385 | 22.6 | 23.8 | 0.1007 | 1.198 |
| 6.0 | 27.2 | 27.5 | 0.0432 | 2.108 | 25.6 | 26.4 | 0.0644 | 1.555 |
| 8.0 | 46.2 | 43.3 | 0.0237 | 3.142 | 36.5 | 36.1 | 0.0284 | 2.696 |
| 10.0 | 63.1 | 53.2 | 0.0361 | 2.260 | 52.0 | 47.4 | 0.0352 | 2.360 |
| 12.0 | 73.2 | 58.2 | 0.0433 | 2.009 | 63.9 | 54.0 | 0.0403 | 2.046 |
| 14.0 | 80.5 | 61.6 | 0.0475 | 1.914 | 72.3 | 58.3 | 0.0456 | 1.905 |
| 16.0 | 86.4 | 64.3 | 0.0503 | 1.870 | 78.9 | 61.4 | 0.0490 | 1.845 |
| 18.0 | 91.4 | 66.5 | 0.0524 | 1.849 | 84.3 | 63.9 | 0.0514 | 1.814 |
| 20.0 | 96.0 | 68.5 | 0.0540 | 1.839 | 89.1 | 66.0 | 0.0532 | 1.801 |

**TABLE 40.** Transport Properties of Freon-22 in the Single-Phase Region (*Continued*)

| | $T=413,15$ K | | | | $T=423,15$ K | | | |
|---|---|---|---|---|---|---|---|---|
| $p$ | $\eta \cdot 10^6$ | $\lambda \cdot 10^6$ | $a \cdot 10^6$ | Pr | $\eta \cdot 10^6$ | $\lambda \cdot 10^6$ | $a \cdot 10^6$ | Pr |
| 0.1 | 17.4 | 17.97 | 9.1290 | 0.752 | 17.7 | 18.6 | 9.5720 | 0.751 |
| 0.5 | 17.5 | 18.32 | 1.7970 | 0.758 | 17.9 | 18.9 | 1.8860 | 0.756 |
| 1.0 | 17.8 | 18.78 | 0.8766 | 0.767 | 18.1 | 19.4 | 0.9250 | 0.762 |
| 2.0 | 18.4 | 19.79 | 0.4121 | 0.799 | 18.7 | 20.4 | 0.4408 | 0.785 |
| 3.0 | 19.3 | 20.9 | 0.2529 | 0.856 | 19.6 | 21.4 | 0.2738 | 0.834 |
| 4.0 | 20.6 | 22.3 | 0.1709 | 0.943 | 20.7 | 22.7 | 0.1891 | 0.901 |
| 5.0 | 22.4 | 23.9 | 0.1186 | 1.088 | 22.3 | 24.2 | 0.1361 | 1.007 |
| 6.0 | 24.7 | 26.0 | 0.0766 | 1.303 | 24.3 | 26.0 | 0.0994 | 1.159 |
| 8.0 | 32.2 | 32.6 | 0.0405 | 2.045 | 30.1 | 30.9 | 0.0543 | 1.644 |
| 10.0 | 43.4 | 42.0 | 0.0329 | 2.285 | 38.2 | 38.2 | 0.0383 | 2.018 |
| 12.0 | 55.2 | 49.7 | 0.0380 | 2.067 | 47.9 | 45.6 | 0.0381 | 2.005 |
| 14.0 | 64.5 | 54.8 | 0.0436 | 1.907 | 57.2 | 51.3 | 0.0422 | 1.891 |
| 16.0 | 71.7 | 58.4 | 0.0475 | 1.827 | 64.9 | 55.4 | 0.0461 | 1.810 |
| 18.0 | 77.6 | 61.2 | 0.0503 | 1.788 | 71.2 | 58.6 | 0.0492 | 1.762 |
| 20.0 | 82.7 | 63.6 | 0.0524 | 1.768 | 76.6 | 61.2 | 0.0516 | 1.737 |

**TABLE 40.** Transport Properties of Freon-22 in the Single-Phase Region (*Continued*)

| | T=433.15 K | | | | T=443.15 K | | | |
|---|---|---|---|---|---|---|---|---|
| p | $\eta \cdot 10^6$ | $\lambda \cdot 10^6$ | $a \cdot 10^6$ | Pr | $\eta \cdot 10^6$ | $\lambda \cdot 10^6$ | $a \cdot 10^6$ | Pr |
| 0.1 | 18.1 | 19.2 | 10.0200 | 0.749 | 18.4 | 19.9 | 10.4800 | 0.747 |
| 0.5 | 18.3 | 19.6 | 1.9850 | 0.751 | 18.6 | 20.2 | 2.0770 | 0.749 |
| 1.0 | 18.5 | 20.0 | 0.9752 | 0.756 | 18.8 | 20.6 | 1.0250 | 0.753 |
| 2.0 | 19.0 | 20.9 | 0.4671 | 0.778 | 19.4 | 21.5 | 0.4953 | 0.769 |
| 3.0 | 19.8 | 22.0 | 0.2958 | 0.812 | 20.1 | 22.5 | 0.3162 | 0.798 |
| 4.0 | 20.9 | 23.1 | 0.2066 | 0.871 | 21.1 | 23.6 | 0.2242 | 0.845 |
| 5.0 | 22.3 | 24.5 | 0.1522 | 0.952 | 22.4 | 24.8 | 0.1676 | 0.912 |
| 6.0 | 24.0 | 26.0 | 0.1144 | 1.065 | 23.9 | 26.3 | 0.1293 | 0.994 |
| 8.0 | 28.8 | 30.2 | 0.0682 | 1.395 | 28.0 | 29.8 | 0.0812 | 1.240 |
| 10.0 | 35.2 | 35.9 | 0.0467 | 1.740 | 33.2 | 34.4 | 0.0563 | 1.518 |
| 12.0 | 42.8 | 42.3 | 0.0412 | 1.861 | 39.4 | 39.9 | 0.0466 | 1.687 |
| 14.0 | 51.0 | 48.0 | 0.0423 | 1.836 | 46.3 | 45.2 | 0.0445 | 1.737 |
| 16.0 | 58.6 | 52.6 | 0.0455 | 1.774 | 53.2 | 49.8 | 0.0458 | 1.721 |
| 18.0 | 65.2 | 56.0 | 0.0484 | 1.735 | 59.7 | 53.6 | 0.0482 | 1.693 |
| 20.0 | 70.8 | 58.9 | 0.0509 | 1.704 | 65.4 | 56.6 | 0.0504 | 1.671 |

**TABLE 40.** Transport Properties of Freon-22 in the Single-Phase Region (*Continued*)

| | T=453.15 K | | | | T=463.15 K | | | |
|---|---|---|---|---|---|---|---|---|
| p | $\eta \cdot 10^6$ | $\lambda \cdot 10^6$ | $a \cdot 10^6$ | Pr | $\eta \cdot 10^6$ | $\lambda \cdot 10^6$ | $a \cdot 10^6$ | Pr |
| 0.1 | 18.8 | 20.5 | 10.9400 | 0.746 | 19.1 | 21.1 | 11.5500 | 0.746 |
| 0.5 | 19.0 | 20.8 | 2.1720 | 0.747 | 19.3 | 21.4 | 2.2680 | 0.746 |
| 1.0 | 19.2 | 21.2 | 1.0710 | 0.750 | 19.5 | 21.8 | 1.1250 | 0.747 |
| 2.0 | 19.7 | 22.1 | 0.5234 | 0.761 | 20.0 | 22.7 | 0.5521 | 0.754 |
| 3.0 | 20.4 | 23.0 | 0.3360 | 0.788 | 20.7 | 23.6 | 0.3574 | 0.775 |
| 4.0 | 21.3 | 24.1 | 0.2417 | 0.822 | 21.5 | 24.6 | 0.2586 | 0.804 |
| 5.0 | 22.5 | 25.3 | 0.1835 | 0.875 | 22.6 | 25.7 | 0.1982 | 0.848 |
| 6.0 | 23.9 | 26.6 | 0.1438 | 0.943 | 23.9 | 26.9 | 0.1572 | 0.904 |
| 8.0 | 27.5 | 29.7 | 0.0938 | 1.131 | 27.2 | 29.7 | 0.1057 | 1.056 |
| 10.0 | 32.0 | 33.6 | 0.0665 | 1.354 | 31.2 | 33.2 | 0.0768 | 1.232 |
| 12.0 | 37.2 | 38.3 | 0.0535 | 1.524 | 35.6 | 37.2 | 0.0611 | 1.385 |
| 14.0 | 42.8 | 43.0 | 0.0484 | 1.610 | 40.4 | 41.4 | 0.0536 | 1.486 |
| 16.0 | 48.9 | 47.4 | 0.0476 | 1.641 | 45.6 | 45.5 | 0.0508 | 1.542 |
| 18.0 | 54.9 | 51.3 | 0.0488 | 1.637 | 51.0 | 49.2 | 0.0504 | 1.569 |
| 20.0 | 60.6 | 54.5 | 0.0506 | 1.629 | 56.3 | 52.5 | 0.0513 | 1.575 |

**TABLE 40.** Transport Properties of Freon-22 in the Single-Phase Region (*Continued*)

| | | | | T=473.15 K | | | | | |
|---|---|---|---|---|---|---|---|---|---|
| p | $\eta \cdot 10^6$ | $\lambda \cdot 10^6$ | $a \cdot 10^6$ | Pr | p | $\eta \cdot 10^5$ | $\lambda \cdot 10^6$ | $a \cdot 10^6$ | Pr |
| 0.1 | 19.5 | 21.8 | 11.9100 | 0.742 | 8.0 | 26.9 | 29.9 | 0.1176 | 0.990 |
| 0.5 | 19.6 | 22.1 | 2.3740 | 0.741 | 10.0 | 30.5 | 33.0 | 0.0869 | 1.135 |
| 1.0 | 19.8 | 22.4 | 1.1760 | 0.744 | 12.0 | 34.5 | 36.5 | 0.0692 | 1.273 |
| 2.0 | 20.3 | 23.3 | 0.5801 | 0.748 | 14.0 | 38.7 | 40.3 | 0.0598 | 1.373 |
| 3.0 | 21.0 | 24.2 | 0.3786 | 0.765 | 16.0 | 43.2 | 44.0 | 0.0551 | 1.443 |
| 4.0 | 21.8 | 25.1 | 0.2753 | 0.792 | 18.0 | 47.9 | 47.5 | 0.0531 | 1.487 |
| 5.0 | 22.8 | 26.1 | 0.2125 | 0.829 | 20.0 | 52.7 | 50.7 | 0.0529 | 1.511 |
| 6.0 | 24.0 | 27.3 | 0.1600 | 0.873 | | | | | |

# FOUR

## THERMOPHYSICAL PROPERTIES OF FREON-23

### 4.1. A REVIEW OF PUBLISHED DATA ABOUT THERMODYNAMIC PROPERTIES

At the present time, Freon-23 (trifluoromethane or fluoroform) is considered to be a thoroughly investigated substance (Tables 41 and 42). Meanwhile, information about its thermodynamic properties is absent in the reference books [0.8–0.10]. Review reference [0.34] gives only summaries of articles from journals published up to 1967.

### Experimental Data in the Single-Phase Region

Table 41 and Fig. 32 indicate that the density (compressibility) of the gas and the liquid at elevated pressures was measured in the region $T = 203–473$ K, $p = 0.1–30$ MPa. However, part of the experimental data is represented in graphical or analytical forms [4.18, 4.32]. Therefore, the upper limit of the thermodynamic tables of the present work is 20 MPa.

Hou and Martin [4.25] were the first to investigate in detail the $pvT$ variables of Freon-23. Their experimental data cover the saturation curve and the region of liquid and superheated vapor. The saturation pressure was investigated statistically in a condensing thermometer and the single-phase region was investigated using the nonballast constant-volume-piezometer method. The temperature in the experiments was measured by a mercury thermometer and the pressure by a standard spring manometer. The working volumes in the apparatus were determined during calibration tests with water. The analysis of this work

**TABLE 41.** Experimental Studies of Thermodynamic Properties of Freon-23 in the Single-Phase Region

| Year | Authors | Measured property | Temperature, K | Pressure, MPa | Phase[a] | Number of experimental points | Reference |
|---|---|---|---|---|---|---|---|
| 1956 | Vandercoy and Devris | $c_v$ | 300 | 0.0 | G | — | [1.7] |
| 1959 | Hou and Martin | $\varrho$ | 222—392 | 0.4—13.8 | G, L | 79 | [4.25] |
| 1964 | Dymond and Smith | $B_1$ | 273—333 | 0.1 | G | 6 | [4.21] |
| 1966 | Thomas and Zander | $\varrho$ | 240—300 | 0.1—30 | L | Graphic | [4.32] |
| 1968 | Wagner | $\varrho$ | 203—353 | 0.2—3.3 | G | 84 | [4.34] |
| 1970 | Hajjar and MacWood | $B_1$ | 313—403 | 0.1 | G | 4 | [3.45] |
| 1970 | Suther and Cole | $B_1$ | 323—416 | 0.1 | G | 3 | [3.61] |
| 1970 | Lange and Stein | $\varrho$ | 243—368 | 0.7—4.5 | G | 25 | [4.26] |
| 1970 | Wenzel and Balaban | $c_p$ | 295—330 | 0.1—9 | G | Graphic | [4.35] |
| 1971 | Haworth and Sutton | $B_1$ | 298—328 | 0.1 | G | 3 | [3.46] |
| 1975, 1977 | Raskazov et al. | $\varrho$ | 258—393 | 0.7—19.2 | G | 226 | [4.3, 4.14] |
| 1975 | Raskazov et al. | $\delta_T$ | 243—373 | 0.5—5.0 | G | 119 | [4.4] |
| 1975, 1976 | Raskazov et al. | $\varrho$ | 238—343 | 0.9—20.0 | L | 98 | [4.6, 4.14] |
| 1975 | Gruzdev and Shumskaya | $c_p$ | 299—453 | 0.1—2.0 | G | 48 | [2.9] |
| 1975 | Timoshenko et al. | $B_1, B_2$ | 243—363 | — | G | 28 | [4.13] |
| 1976 | Belzile et al. | $\varrho$ | 298 | 0.4—16.5 | G | 32 | [4.18] |
|  |  | $B_1, B_2$ | 273—473 | — | G | 15 |  |

[a] L, Liquid; G, gas.

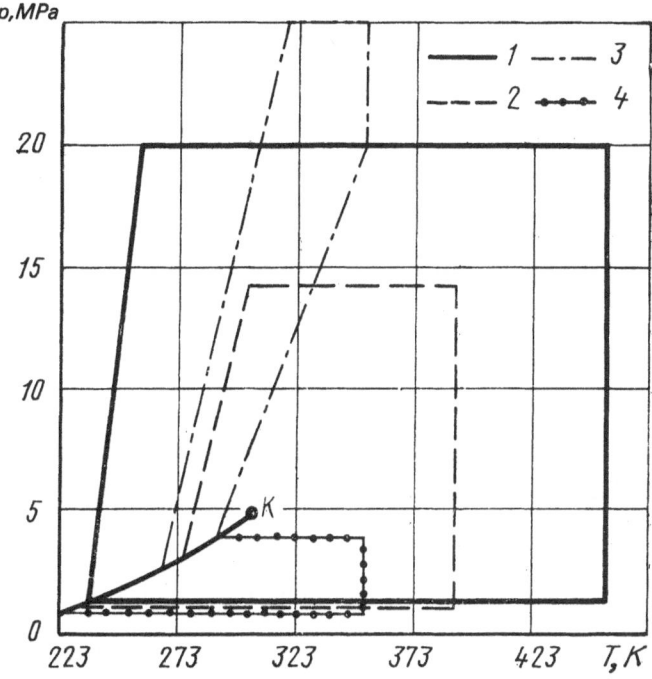

**FIG. 32.** Regions of experimental investigation of thermal properties of Freon-23 in references: 1, Raskazov et al. [4.3, 4.6]; 2, Hou and Martin [4.25]; 3, Thomas and Zander [4.32]; 4, Wagner [4.34].

resulted in an estimate of the error in measuring the saturated pressure of 0.5% and measuring the specific volume at 1%.

Wagner [4.34] experimentally investigated the $pvT$ variables of superheated vapor of Freon-23; the tests were carried out at constant volume. The researcher estimated the error in pressure measurement at 0.03 atm and the temperature (measured by a chromel alumel thermocouple) at 0.1 K. In the overlapping region of state, the discrepancy of experimental data in [4.25] and [4.34] amounts to 0.5–1.2% [4.14].

A more thorough and detailed investigation of compressibility of Freon-23 in the gaseous and liquid phase was carried out at MEI under the supervision of D. C. Raskazov and E. K. Petrov [4.3, 4.6, 4.14]. For the determination of compressibility of the gas, the researchers employed the method of successive expansion on the isotherm. Unlike the classical Burnett method, the expansion of the substance was carried out in a previously graduated piezometer (volume $V^{II}$). The quantity of the substance in the piezometer was determined by weighing after freezing out the freon into a separating container. The formula for the calculation of the specific volume is

$$v_{j-1} = \frac{V^{II}_{p,\,t}}{m^{II}_{j-1}} \tag{4.1}$$

where $V_{p,t}^{II}$, is the volume of the graduated piezometer, taking into account the deformation due to temperature and pressure; $m_{j-1}^{II}$ is the mass of the investigated substance in the graduated piezometer after the passages (j = 1, 2, 3, . . . , j is the number of passages).

Given the mass $m$ during the successive expansion of the substance from the volume of the main piezometer $V^I$ to the evacuated volume $V^{II}$, the geometrical constant of the instrument can be evaluated from the relation

$$N = \frac{m_{j-1}^{II}}{m_j^{II}} = \frac{V^I + V^{II}}{V^I} \tag{4.2}$$

Then the mass $m$ in the volume $V^{II}$ is determined from the relation (4.2) without the need for weighing after each passage.

The temperature during the tests was measured by a 10 ohm resistance thermometer coupled with a potentiometer (accuracy: class 0.002). The temperature regime was held stable so that the deviation would not exceed ±0.005 K. The error in measuring the temperature was not more than 0.02 K. The pressure was measured by a dead-weight manometer (accuracy: class 0.05). To separate the investigated substance from the pressure-measuring oil system a differential mercury manometer, having accuracy of 2 mm water column, was employed. Weighing was done on an analytical balance having sensitivity of 0.1 mg.

To determine the density of liquid freon, the method of a nonballast constant-volume piezometer with membrane-mercury separator was used. The mass of the substance in the piezometer was determined by weighing after freezing out Freon-23 in the separating container. The temperature and pressure were measured using the same instruments as in Rcf. [4.3]. Experiments were conducted along the isochores. The maximum error of measuring specific volume fell within the range 0.1–0.2%.

Purity of the investigated Freon-23, according to chromatography, amounted to 99.98% (moisture, 0.0042%, low-boiling substances, 0.01). Prior to commencement of experiments, the freon was additionally purified to remove low-boiling impurities.

To reveal the probable systematic error, the regions of study in Refs. [4.3, 4.6] overlapped. The results of specific volume measurements by two different pieces of apparatus correspond, with a deviation in the region of experimental errors.

The main purpose of the work by Lange and Stein [4.26] was the investigation of thermal properties of gaseous mixtures of Freon-23. For pure Freon-23 a small number of points were obtained (Table 41). The measurements were carried out with the Burnett method (the so-called classical version) using a membrane zero indicator of pressure having a sensitivity of 0.7 mbar. The investigated Freon-23 was 99.98% pure.

Belzille et al. [4.18] used a modified version of the Burnett method with three piezometers; two of them had a volume of 550 cm³, and the third had a volume of 183 cm³. This allowed them to carry out the experiments with considerably differing values of the geometric constants of the measuring cell ($N_I \approx$ 1.18, $N_{II} \approx 1.36$). In paper [4.18] the experimental values of $\rho$ are given on the 298 K isotherm only. For the other isotherms, the coefficients obtained for the following expansion from $pvT$ data are given

$$z = 1 + B^{(p)} p + C^{(p)} p^2 + \ldots \qquad (4.3)$$

The recurrent formula, linking the coefficients of the power series (4.3) with the virial coefficients of Eq. (0.7), is presented in Ref. [0.5]. In this case

$$B^{(p)} = B_1/RT; \; C^{(p)} = (B_2 - B_1^2) \, (RT)^2. \qquad (4.4)$$

The calculated values of $B_1$ from the data in [4.18] using the relationships (4.4) agree fairly well with the values obtained in the works of MEI [4.3, 4.5, 4.13]. The temperature variation, however, is clearly distorted (Fig. 33). It is essential to note that the values of $B_1$ and $B_2$ correspond in a wide range of temperature dependence. These values are obtained from the measurements of $pvT$ relations

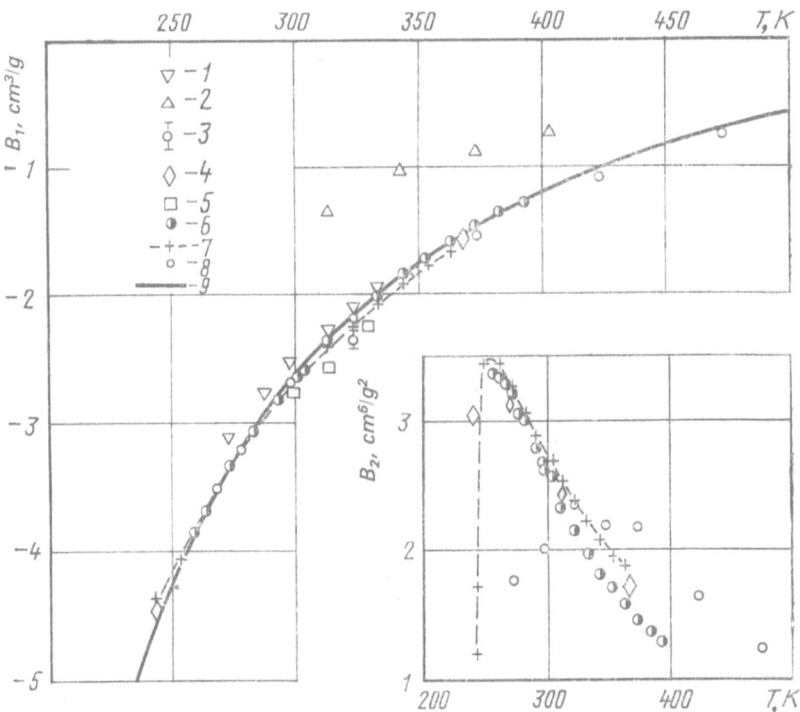

**FIG. 33.** Second and third virial coefficients of Freon-23. Experimental data: 1, Dymond and Smith [4.21]; 2, Hajjar and MacWood [3.45]; 3, Suther and Cole [3.61]; 4, Lange and Stein [4.26]; 5, Haworth and Sutton [3.46]; 6, Raskazov et al. [4.3]; 7, Timoshenko et al. [4.13]; 8, Belzile et al. [4.18]. Calculated results: 9, present work.

[4.3, 4.26], isothermal Joule–Thomson effect [4.5], and refractive index [4.13]. The values of $B_1$, determined from the experimental $pvT$ data of Wagner [4.34] were found to be low by 5–11% [4.5, 4.13]. It is also noted in Ref. [4.5] that the thermal equation of state, obtained from the data in [4.34], gives values of $\delta_T = -(\partial h/\partial p)_T$ practically independent of pressure. This does not correspond with reality. Therefore, the $pvT$ data in Ref. [4.34] should be considered unreliable.

D. C. Raskazov and L. A. Krukov [4.4] carried out detailed measurements of $\delta_T$ in the gaseous phase using the method of a flowing compensating calorimeter with a throttle-valve device. The Freon-23 used in the tests was 99.94–99.98% pure (moisture, 0.0015%, low-boiling substances, 0.02–0.06%). As estimated by the researchers, the error in the experimental data is 0.7–1.2%. The results of the measurements are approximated by the equations

$$\delta_T/\delta_T^0 = 1 + \sum_{i=1}^{4} \sum_{j=0}^{7} a_{ij}\, p^i/T^j \tag{4.5}$$

$$\delta_T^0 = \sum_{k=0}^{5} c_k/T^k \tag{4.6}$$

the coefficients of which are presented in Ref. [4.5]. Presented there also are the tables of $\delta_T$, $h$, $c_p$ calculated from these equations in the interval $T = $ 253–373 K and $p = $ 0.1–5 MPa.

Constant-pressure heat capacity of gaseous Freon-23 was measured in two studies (see Table 41). But numerical results are given only in a paper by V. A. Gruzdev and A. I. Shumskaya [2.9]. These researchers point out the good agreement of experimental values of $c_p$ in the interval $T = $ 300–360 K with the calculated values in Ref. [4.5]. The data by Barho [0.37] covering the heat capacity in the ideal gas state turned out to be high. The discrepancy exceeds the sum of experimental and calculation errors. In the present work, the initial values of thermodynamic functions in the ideal gas state are taken from the data in Ref. [1.39].

## Experimental Data on the Saturation Line

Table 42 shows that the saturation vapor pressure and orthobaric density of vapor and liquid have been measured repeatedly. We adopted experimental data of MEI [4.3] as reference values of $p_s$. These data were obtained statistically for a high-purity substance and with an error estimated by the authors at 0.1–0.2%. The results of comparison with the data of other researchers are shown in Fig. 34.

Discrepancies between different groups of experimental data covering the density of Freon-23 on the saturation line are relatively high (Fig. 35). These discrepancies can be explained in some cases. The measured values of $\rho'$ by the

TABLE 42. Experimental Studies of Thermodynamic Properties of Freon-23 on the Saturation Line

| Year | Authors | Measured property | Temperature, K | Phase[a] | Number of experimental points | Reference |
|------|---------|-------------------|----------------|----------|-------------------------------|-----------|
| 1959 | Hou and Martin | $p_s$, $\varrho$ | 141—298 | L | 47 | [4.25] |
| 1962 | Valentine et al. | $p_s$ | 145—191 | V | 13 | [4.33] |
| | | $c_p$ | 123—189 | L | 9 | |
| | | $r$ | 191 | L–V | 3 | |
| 1964 | Elchardus and Maestre | $p_s$ | 173—273 | V | 6 | [4.23] |
| 1966 | Thomas and Zander | $p_s$, $\varrho$ | 253—299 | L | 11 | [4.32] |
| 1968 | Wagner | $p_s$, $\varrho$ | 195—283 | V | 10 | [4.34] |
| 1975 | Raskazov et al. | $p_s$ | 198—298 | V | 15 | [4.3] |
| 1975 | Timoshenko et al. | $p_s$ | 243—283 | V | 5 | [4.12] |
| 1975 | Shavandrin et al. | $\varrho$ | 158—299 | L | 29 | [4.17] |
| | | $\varrho$ | 208—299 | V | 18 | |
| 1977 | Khodeieva and Gubochkina | $\varrho$ | 298,5—299 | L–V | Graphic | [4.15] |

[a] L, Liquid; V, vapor.

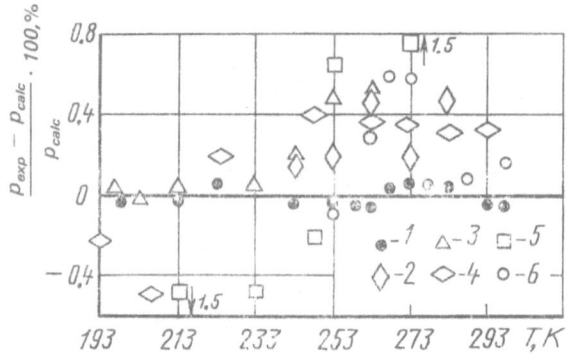

**FIG. 34.** Deviation of experimental values of saturation vapor pressure for Freon-23 from those adopted in the present work: 1, Raskazov et al. [4.3, 4.14]; 2, Timoshenko et al. [4.12]; 3, Wagner [4.34]; 4, Hou and Martin [4.25]; 5, Elchardus [4.23]; 6, Thomas and Zander [4.32].

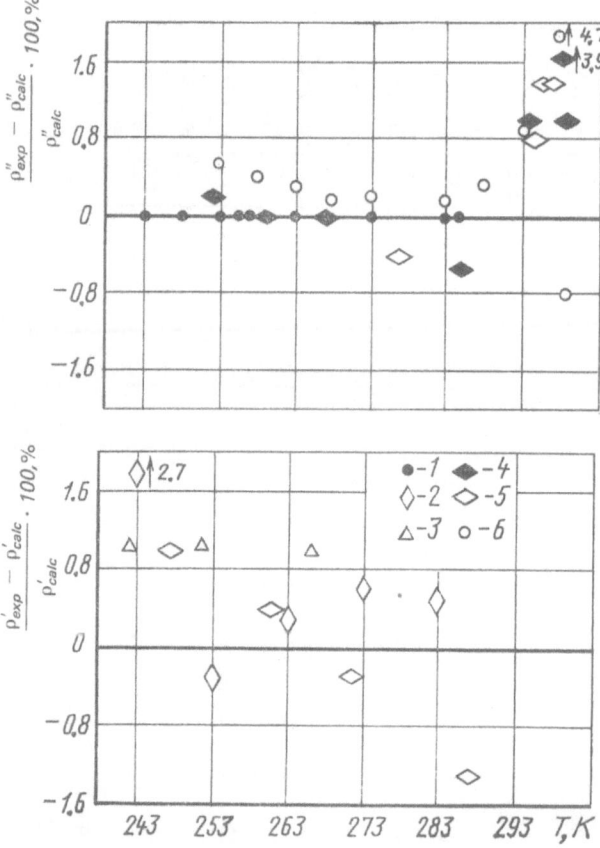

**FIG. 35.** Deviation of experimental values of density on the saturation line for Freon-23 from the adopted values in the present work: 1, Raskazov et al. [4.3, 4.6]; 2, Timoshenko et al. [4.12]; 3, Wagner [4.34]; 4 and 5, Hou and Martin [4.25]; 6, Thomas and Zander [4.32].

pycnometer method in Ref. [4.25] agree well with the experimental data of MEI, which are recommended in the present work. The data obtained by distant extrapolation of experimental isochoric lines on the vapor-pressure curve deviate from the calculated results by 4%. Considerable discrepancies are noticed in a wide zone around the critical point. Due to this fact, the experimental data of L. M. Shavandrin et al. [4.17] are of special interest. These data were obtained by the quasistatistical thermograph method for a high-purity substance (99.98%) with an estimated error of 0.03–0.15%. The experimental data obtained are approximated by equations of type (0.26) and, for computational procedure, the following equations were obtained

$$\tilde{\rho}' = 1.14125 \, (1 - \tau)^{0.355} - 0.12084 \, (1 - \tau)^{0.670} \qquad (4.7)$$

$$\tilde{\rho}'' = 1.13007 \, (1 - \tau)^{0.355} - 0.10334 \, (1 - \tau)^{0.670}, \qquad (4.8)$$

where $\tilde{\rho} = \rho - \rho_{cr} - b(1 - \tau)$; $\rho_{cr} = 0.5255$ g/cm³, $b = 0.54958$ g/cm, $T_{cr} = 298.98$ K.

The values of the critical parameters for Freon-23 recommended in experimental and analytical references differ considerably (Table 43). From our point of view, the more reliable results are obtained in the experimental investigations in [4.15, 4.17, 4.25]. The values of temperature at the triple and normal boiling points are adopted from the work of Valentine et al. [4.33]. Incidentally, this work is the only investigation in which experimental data about heat capacity and heat of vaporization for liquid Freon-23 were obtained (Table 42).

To determine the surface tension, the generalized equation (1.9) with the following coefficients is recommended:

$$n = 1.249, B = 4.6109 \cdot 10^{-3} \, N/m$$

## Equations of State and Tables

Several versions of equations of state of type (0.2)–(0.5) are known. The coefficients of these equations were determined from $pvT$ data of Hou and Martin [4.25] covering the region $\tau = 0.74$–1.3, $\omega = 0.03$–1.9.

It is supposed that Eq. (0.2) is most suitable for the calculations at moderate density [0.42]. Thus, at $\omega \leq 1.5$ Eq. (0.2) having 11 coefficients approximates the experimental data in [4.25] with a standard deviation of 0.45%, and at $\omega > 1.5$ the deviation rises to 4.14%. Adding the term $(A_6 + B_6 T) \exp(- av)$ [3.50] to the righthand side of Eq. (0.2), as noted in Ref. [4.26], does not practically change the standard deviation, but the maximum deviation decreases to 1%.

Comparative calculations from the generalized equations of state (0.3) and (0.5), which were carried out in Refs. [0.22, 0.23, 1.12], indicated that the standard deviation from the experimental data in [4.25] stays on the level 1.2–1.4%.

Morsy [0.42, 0.43, 4.28], using Eq. (0.4) and a computer method for de-

**TABLE 43.** Parameters of Fundamental Points on the Saturation Line for Freon-23[a]

| Year | Authors | $T_{cr}$, K | $p_{cr}$, MPa | $\rho_{cr}$, kg/m³ | $T_{NBP}$, K | $T_0$, K | Reference |
|------|---------|-------------|---------------|---------------------|---------------|----------|-----------|
| 1936 | Ruff et al. | 305.9 | — | 516 | 188.7 | 113 | [4.30] |
| 1937 | Henne | — | — | — | 190.9 | 110 | [4.24] |
| 1942 | Seger | 288.15 | 4.75 | 516 | 183.1 | — | [4.31] |
| 1948 | Whitney | 300.15 | — | 550 | — | — | [0.34] |
| 1956 | Plank | 302.15 | 4.54 | 496 | — | — | [0.45] |
| 1957 | Riedel | 302.15 | 5.06 | 530 | — | — | [4.29] |
| 1959 | Hou and Martin | 299.06 | 4.836 | 525 | — | — | [4.25] |
| 1962 | Valentine et al. | — | — | — | 191.3 | 117.97 | [4.33] |
| 1964 | Elchardus and Maestre | 288.15 | 4.75 | — | 190.97 | 118.15 | [4.23] |
| 1965 | Eiseman | — | — | — | 189.97 | — | [4.22] |
| 1965 | Michel | — | — | — | 191.15 | 117.97 | [4.27] |
| 1966 | Morsy | 299.06 | 4.836 | 525 | 191.05 | — | [4.28] |
| 1967 | Tsiklis and Prokhodov | 299.05 | — | 516 | 191.02 | — | [4.16] |
| 1968 | Wagner | 299.45 | 4.87 | 527 | — | — | [4.34] |
| 1970 | Phillips and Murphy | — | — | — | 191.05 | — | [1.50] |
| 1971 | Mermin | 300.15 | — | 550 | — | — | [4.17] |
| 1974 | Fomichev and Khokhlachev | 299.06 | — | 525 | — | — | [4.17] |
| 1975 | Shavandrin et al. | 298.98 | — | 525.5 | — | — | [4.17] |
| 1976 | Perelshtein and Parushin | 299.45 | 4.818 | 528 | — | — | [1.15] |
| 1977 | Khodeieva and Gubochkina | 298.955 | — | 526 | — | — | [4.15] |

[a] The adopted values of the parameters in this book are presented in Table 3.

**TABLE 44.** Coefficients of Equation of State (0.15) for Freon-23 ($\rho$ in g/cm³; $\tau = T/100$)

| $i$ | Values $b_{ij}$ at $j$ | | | |
|---|---|---|---|---|
| | 0 | 1 | 2 | 3 |
| 1 | $0.118261891 \cdot 10^1$ | $-0.905353810 \cdot 10^1$ | $0.699906826 \cdot 10^1$ | $-0.474556397 \cdot 10^0$ |
| 2 | $0.292086834 \cdot 10^1$ | $-0.171728480 \cdot 10^2$ | $0.147680242 \cdot 10^2$ | $0.983951291 \cdot 10^2$ |
| 3 | $0.286742894 \cdot 10^0$ | $-0.478164350 \cdot 10^1$ | $0.271065494 \cdot 10^2$ | $0.616892171 \cdot 10^1$ |
| 4 | $0.116222185 \cdot 10^1$ | $-0.260959815 \cdot 10^1$ | $-0.266431581 \cdot 10^2$ | $-0.822805725 \cdot 10^2$ |
| 5 | $0.633125551 \cdot 10^0$ | $0.289659151 \cdot 10^1$ | $0.153725797 \cdot 10^2$ | $0.323704664 \cdot 10^2$ |
| 6 | $-0.134528266 \cdot 10^1$ | $0.285487378 \cdot 10^1$ | $0.531991934 \cdot 10^1$ | $-0.292706129 \cdot 10^2$ |

**TABLE 44.** Coefficients of Equation of State (0.15) for Freon-23 ($\rho$ in g/cm³; $\tau = T/100$) (*Continued*)

| $i$ | Values $b_{ij}$ at $j$ | | | |
|---|---|---|---|---|
| | 4 | 5 | 6 | 7 |
| 1 | $-0.163433562 \cdot 10^3$ | $-0.443533446 \cdot 10^2$ | $0.725293982 \cdot 10^3$ | $-0.728147824 \cdot 10^3$ |
| 2 | $0.100907421 \cdot 10^2$ | $-0.117687673 \cdot 10^3$ | $-0.187312399 \cdot 10^3$ | $0.565148313 \cdot 10^3$ |
| 3 | $0.580641377 \cdot 10^2$ | $0.102534046 \cdot 10^3$ | $0.172624399 \cdot 10^3$ | $-0.292705520 \cdot 10^3$ |
| 4 | $-0.123143917 \cdot 10^3$ | $-0.219968266 \cdot 10^3$ | $-0.408125954 \cdot 10^2$ | $-0.802026686 \cdot 10^2$ |
| 5 | $0.341771583 \cdot 10^1$ | $0.358257395 \cdot 10^3$ | $-0.724264269 \cdot 10^2$ | $0.135670487 \cdot 10^3$ |
| 6 | $-0.104350620 \cdot 10^2$ | $0.230834361 \cdot 10^2$ | $-0.117554935 \cdot 10^3$ | |

**TABLE 45.** Coefficients of Auxiliary Interpolation Equations (0.18)–(0.21) for Freon-23 (at $\bar\tau = T/100$)

| $j$ | $A_{sj}$ | $\alpha_j$ | $\beta_j$ | $\tau_j$ |
|---|---|---|---|---|
| 0 | $-0.13014035\cdot10^3$ | $0.12304738\cdot10^2$ | $0.14214576\cdot10^2$ | $0.55790852\cdot10^2$ |
| 1 | $0.16664575\cdot10^3$ | $-0.34681429\cdot10^2$ | $-0.33622323\cdot10^2$ | $-0.20198257\cdot10^3$ |
| 2 | $-0.46097185\cdot10^2$ | $0.20967974\cdot10^3$ | $0.75219442\cdot10^0$ | $0.69988531\cdot10^3$ |
| 3 | $-0.74769300\cdot10^1$ | $-0.34240717\cdot10^3$ | $0.13114625\cdot10^3$ | $-0.12195008\cdot10^4$ |
| 4 | $-0.72273341\cdot10^1$ | $0.22003892\cdot10^3$ | $-0.14437032\cdot10^3$ | $0.82729148\cdot10^3$ |
| 5 | $0.79581501\cdot10^1$ | — | — | — |
| 6 | $-0.12492154\cdot10^1$ | — | — | — |
| 7 | $-0.25364268\cdot10^0$ | — | — | — |
| 8 | $0.62947726\cdot10^{-1}$ | — | — | — |

termination of the coefficients, lowered the standard deviation to 0.28%. From this equation of state and with the interpolation formulas (0.11) and (0.14), tables of $v$, $h$, $s$ on the saturation line and $h$, lg $p$ diagram in the range $T = 133-453$ K and $p \leqslant 20$ MPa were computed.

Döring and Löffler [0.39] also used Eq. (0.4), but the coeffficients were determined from the experimental data of Wagner [4.34] for superheated vapor. Since, as discussed previously, the results in [4.34] differ considerably from other known $pvT$ measurements, the reference data in [0.39] have lost their practical significance.

Experimental data obtained at MEI about thermodynamic properties of Freon-23 were used to obtain several local equations of state of type (0.7) and to compute thermodynamic tables. Raskazov et al. [4.3] compiled tables of $\rho$, $h$, $s$, $c_p$, and $w$ in the range $T = 258-393$ K, $p = 0.1-15$ MPa. Another version of the tables is compiled in Ref. [4.13]. In the present work, a single thermal equation of state is used for liquid and gas.

## 4.2. SYSTEM OF EQUATIONS FOR THE CALCULATION OF THERMODYNAMIC PROPERTIES

When deriving the equation of state of Freon-23, a computer method for the joint analysis of experimental data covering thermal properties of liquid and gas is employed [4.3]. For the initial data, the experimental $pvT$ variations of MEI [4.3, 4.6, 4.14] are adopted. These data cover the region $T = 233-393$ K, $p = 0.7-20$ MPa in the gaseous and liquid phases with about 330 measured values of $z(\rho, T)$ and $p_s(T)$. The derived equation of state has the form (0.15) with coefficients shown in Table 44.

Table 45 shows the coefficients of interpolation formulas (0.18)–(0.21) for temperature dependence of saturated vapor pressure and for three functions in the ideal gas state (enthalpy, heat capacity, and entropy).

The standard deviation in the approximation of initial experimental compressibility data [4.3, 4.6] by the equation of state amounts to 0.18%. Figure 36 shows the histogram of deviation.

The calculation of thermodynamic properties on the saturation line (the boiling and condensation line) was carried out using an independent equation for the vapor pressure curve. The standard deviation in the approximation of initial experimental data about $p_s$, $T$ variables by the polynomial (0.21) is equal to 0.04%.

The values of $\rho'$, $\rho''$, and $B_1$ calculated by using the equation of state agree well with the most reliable experimental data (see Figs. 33 and 35). The deviation of calculated values of $\delta_T$ from the experimental values reaches 5% in the region of low pressures. When the pressure increases, the discrepancy drops (Table 46).

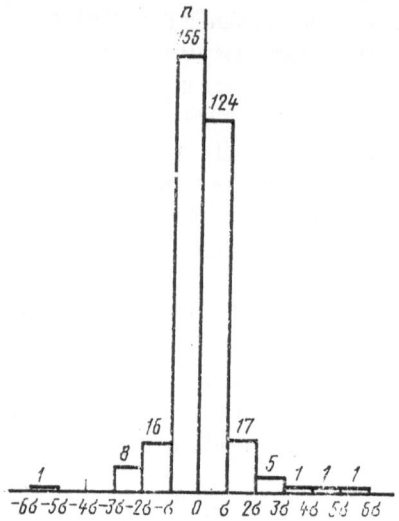

**FIG. 36.** Histogram of deviation of calculated values of compressibility for Freon-23, by Eq. (0.15), from the experimental data in [4.3, 4.6].

The agreement between the values of $c_p$, calculated from the equation, and the experimental data in [2.9] is good. The larger part of the data has a deviation of 1.5–2%, increasing up to 0.3–0.8% with increasing temperature (Table 47).

## 4.3. A REVIEW OF PUBLISHED DATA ABOUT VISCOSITY AND THERMAL CONDUCTIVITY

Numerical data about the viscosity of Freon-23 is lacking in the existing handbooks, although experimental information about $\eta_G$ and $\eta_L$ appeared at the beginning of the 1960s (Table 48).

Detailed investigation of viscosity of liquid and gaseous Freon-23 over a wide temperature and pressure region was conducted recently by C. I. Ivanchenko [0.15] and by D. C. Raskazov et al. [4.7–4.10]. In these works capillary viscometers are used: in the work of OTIFI a viscometer designed by I. F. Golubev (variant V) was used, and in the works of MEI a modified Rankine viscometer was used. The results of the indicated investigations correspond well with each other. The discrepancies do not exceed the sum of their errors, with the exception of some experimental data on the isotherm 273 K, where the deviation is 6%. It is essential to note that experimental data about viscosity of gaseous Freon-23 at low pressures, obtained in the works [2.30, 2.38, 4.11] and [4.7–4.9], are extremely close. For the determination of viscosity of gaseous Freon-23 at atmospheric pressure it is recommended that Eq. (1.18) be used, with the following values for the constants:

**TABLE 46.** Comparison of Calculation Results with Experimental Data [4.4] for the Isothermal Joule–Thomson Effect for Freon-23

| $T$, K | $p$, MPa | $\delta_T^{exp}$, cm³/g | $\delta_T^{calc}$, cm³/g | $\delta(\delta_T)$, % |
|---|---|---|---|---|
| 252.98 | 1.32 | 23.02 | 23.24 | −0.97 |
| 253.26 | 1.15 | 20.82 | 21.18 | −1.7 |
| 253.09 | 1.01 | 19.24 | 19.94 | −3.5 |
| 253.18 | 0.85 | 17.99 | 18.73 | −4.0 |
| 253.14 | 0.58 | 16.26 | 17.13 | −5.0 |
| 253.13 | 0.51 | 15.92 | 16.79 | −5.2 |
| 273.15 | 2.34 | 22.70 | 25.11 | −9.6 |
| 273.15 | 2.21 | 23.80 | 23.25 | 2.3 |
| 273.16 | 1.68 | 18.02 | 18.05 | −0.17 |
| 273.15 | 1.50 | 16.76 | 17.00 | −1.4 |
| 273.16 | 1.25 | 15.35 | 15.76 | −2.6 |
| 273.16 | 1.07 | 14.55 | 15.04 | −3.3 |
| 293.17 | 2.87 | 17.93 | 18.21 | −1.6 |
| 293.16 | 2.85 | 17.83 | 18.10 | −1.5 |
| 293.12 | 2.60 | 16.49 | 16.59 | −0.6 |
| 293.15 | 2.19 | 14.54 | 14.77 | −1.5 |
| 293.14 | 1.21 | 11.67 | 12.01 | −2.9 |

$$\eta_{T_{cr}} = 14.9 \ \mu Pa \cdot s;$$
$$a_0 = 0.78981; \qquad a_2 = -0.15669;$$
$$a_1 = 0.31640; \qquad a_3 = 0.02565.$$

When deriving this equation, the initial $\eta_T$, $T$ data sets included 30 experimental values for $\eta_T$, measured in works [0.15, 2.30, 2.38, 4.7–4.9] in the temperature range 203–473 K.

To calculate the variation of viscosity of gaseous and liquid Freon-23 with

**TABLE 47.** Comparison of Calculation Results with Experimental Data [2.9] on Constant-Pressure Heat Capacity $c_p$ for Freon-23

| $T$, K | $p$, MPa | $c_p^{exp}$, kJ/(kJ · K) | $c_p^{calc}$, kJ/(kJ · K) | $\delta c_p$, % |
|---|---|---|---|---|
| 300.05 | 0.251 | 0.751 | 0.753 | −0.32 |
| 300.35 | 0.932 | 0.810 | 0.820 | −1.2 |
| 300.35 | 1.300 | 0.850 | 0.864 | −1.6 |
| 300.35 | 1.671 | 0.894 | 0.915 | −2.3 |
| 300.45 | 1.950 | 0.933 | 0.960 | −2.9 |
| 323.15 | 1.675 | 0.875 | 0.893 | −2.1 |
| 323.25 | 1.325 | 0.847 | 0.861 | −1.6 |
| 323.25 | 1.036 | 0.826 | 0.837 | −1.3 |
| 323.35 | 0.637 | 0.800 | 0.808 | −0.9 |
| 323.35 | 0.259 | 0.778 | 0.787 | −1.16 |
| 345.15 | 0.288 | 0.807 | 0.811 | −0.5 |

**TABLE 47.** Comparison of Calculation Results with Experimental Data [2.9] on Constant-Pressure Heat Capacity $c_p$ for Freon-23 (*Continued*)

| $T$, K | $p$, MPa | $c_p^{exp}$, kJ/(kJ · K) | $c_p^{calc}$, kJ/(kJ · K) | $\delta c_p$, % |
|---|---|---|---|---|
| 345.15 | 1.019 | 0.842 | 0.850 | −1.8 |
| 345.35 | 1.393 | 0.862 | 0.873 | −1.3 |
| 345.35 | 1.663 | 0.876 | 0.890 | −1.6 |
| 345.35 | 1.949 | 0.894 | 0.910 | −1.8 |
| 374.35 | 1.969 | 0.907 | 0.917 | −1.1 |
| 374.35 | 1.597 | 0.892 | 0.900 | −0.9 |
| 374.35 | 1.199 | 0.876 | 0.883 | −0.7 |
| 374.45 | 0.773 | 0.861 | 0.866 | −0.5 |
| 374.45 | 0.288 | 0.845 | 0.848 | −0.3 |
| 400.55 | 0.281 | 0.876 | 0.880 | −0.4 |
| 400.55 | 0.717 | 0.889 | 0.893 | −0.5 |
| 400.75 | 1.169 | 0.901 | 0.907 | −0.7 |
| 400.75 | 1.566 | 0.913 | 0.919 | −0.7 |

pressure it is recommended that the generalized equation (0.35) be used in the interval $\omega = 0-1.9$. Where $\omega \geq 1.9$, Eq. (0.41) could be used with the coefficients as shown in Table 4.

Viscosity of Freon-23 at the reference point $\eta_{0.7}(\tau = 0.7$ and $\pi = 0.7)$, obtained from the data in [0.15, 4.7–4.9], is equal to 220 μPa · s.

The coefficients $\bar{b}_{\eta,i}$, $\bar{\alpha}_{\eta,i}$, $\bar{\beta}_{\eta,i}$ [of which only $\bar{\alpha}$ and $\bar{\beta}$ are used in Eq. (0.41)] are obtained on the basis of analyzing concordant experimental data about viscosity of three freons. The initial set of data included ~200 experimental values for viscosity of Freon-23 in the range $T = 203-453$ K and $p = 0.1-59$ MPa, obtained in Refs. [0.15, 2.30, 4.7–4.9]. The results of comparison between calculated and experimental data are shown in Fig. 37.

Thermal conductivity of Freon-23 has been studied only lately (Table 49). Therefore, reference tables are lacking in this regard.

Up to the present time, information about the temperature dependence of

**TABLE 48.** Experimental Studies of Viscosity of Freon-23

| Year | Authors | Tempera-ture, K | Pressure, MPa | Phase[a] | Number of experi-mental points | Meth-od[b] | Reference |
|---|---|---|---|---|---|---|---|
| 1953 | Coughlin | 236—334 | 0.1 | G | 6 | RB | [2.38] |
| 1958 | McCullum | 363—473 | 0.1 | G | 5 | RB | [2.38] |
| 1959 | Kamien and Witzell | 303—363 | 0.1—2 | G | 24 | RB | [2.30] |
| 1959 | Tsui | 363—423 | 0.1 | G | 3 | RB | [2.38] |
| 1961 | Wilbers | 278—288 | 0.1 | G | 2 | RB | [2.38] |
| 1970 | Phillips and Murphy | 190—247 | ~$p_s$ | L | 7 | Ca | [1.50] |
| 1974 | Ivanchenko | 273—432 | 1—59 | G, L | 93 | Ca | [0.15] |
| 1974, 1975 | Raskazov et al. | 203—453 | 0.2—40 | G, L | 162 | Ca | [4.7, 4.8, 4.9] |
| 1977 | Raskazov et al. | 117—188 | 0.1 | L | 10 | Ca | [4.10] |
| 1977 | Sagaedakova | 243—343 | 0.1—3.9 | G, L | 12 | Ca | [4.11] |

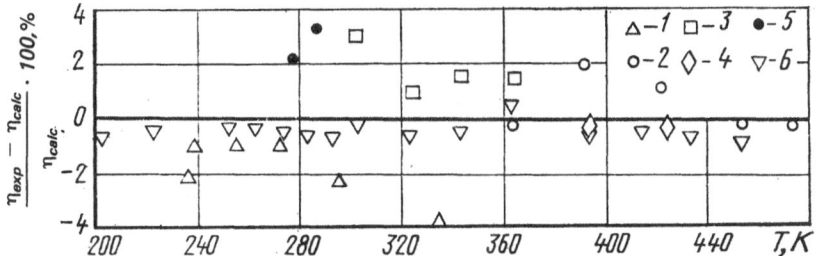

**FIG. 37.** Deviation of experimental values of viscosity of gaseous Freon-23 at atmospheric pressure from the adopted results in the present work: 1, Coughlin [2.38]; 2, McCullum [2.38]; 3, Kamien and Witzell [2.30]; 4, Tsui [2.38]; 5, Wilbers [2.38]; 6, Raskazov et al. [4.7, 4.9].

thermal conductivity for Freon-23 gas at atmospheric pressure is composed of 12 experimental values for $\lambda_T$ in the temperature interval 283–435 K and given in [4.1]. These values served as initial data to derive the coefficients of Eq. (1.20):

$$a_0 = -2.880 \cdot 10^{-3}; \, a_1 = 5.425 \cdot 10^{-5}$$

where $\lambda_T$ is in W/(m · K).

To describe the temperature dependence of thermal conductivity of liquid Freon-23 along the saturation curve, the data of Tauscher [1.63] in the temperature interval 148–268 K (eight experimental points) were used. The data of V. Z. Geller and coworkers [0.13, 4.2] in the interval 118–292 K (32 experimental points) were also used. Since the results of the indicated works agree well with each other (discrepancy does not exceed 2.5%), they were considered to be equally accurate during the analysis. The coefficients of the recommended equation (0.46) are:

$$\lambda_{cr} = 421 \cdot 10^{-4} \text{ W/(m · K)};$$
$$c_1 = 0.04288; \qquad\qquad c_4 = -1.29820;$$
$$c_2 = 1.86268; \qquad\qquad c_5 = 4.16977;$$
$$c_3 = -2.08484; \qquad\qquad c_6 = 1.825558.$$

To calculate the variation of thermal conductivity of gaseous and liquid

**TABLE 49.** Experimental Studies of Thermal Conductivity of Freon-23

| Year | Authors | Temperature, K | Pressure, MPa | Phase[a] | Number of experimental points | Method[b] | Source |
|------|---------|----------------|---------------|----------|-------------------------------|-----------|--------|
| 1967 | Tauscher | 148—268 | $\sim p_s$ | L | 8 | NH | [1.63] |
| 1975 | Geller and Peredri | 193—435 | 0,1—59 | G, L | 147 | H | [4.1] |
| 1975, 1979 | Geller et al. | 118—269 | 0,1—59 | L | 75 | H | [0.13, 4.2] |

[a] G, Gas; L, liquid.
[b] H, Heated filament; NH, nonstationary heated filament.

Freon-23 with pressure, the following system of corresponding equations is recommended: in the interval $\omega = 0-1.9$, Eq. (0.34) with individual coefficients; at $\omega \geq 1.9$, the generalized equation (0.45). The numerical values of the coefficients $b_{\lambda, i}$ for Eq. (0.34) are found using the experimental data in [4.1] in the density interval $10-230$ kg/m³ ($\lambda \cdot 10^4$ is in W/(m · K)):

$$b_1 = 0.051855; \qquad b_3 = 0.047425;$$
$$b_2 = -0.012754; \qquad b_4 = -0.0211362.$$

The coefficients of the generalized equation (0.45) were presented in Table 5; they were obtained from $\lambda$ measurements for three freons. These measurements encompassed 96 experimental values of $\lambda$ for Freon-23 in the range $T = 118-290$ K and $p = p_s - 59$ MPa [0.13]. The numerical values of the coefficients of Eq. (0.34) are obtained from the experimental data [4.1] in the interval $\rho = 0-800$ k/m³. Figure 38 shows the results of comparison of calculated and experimental data.

## 4.4. TABLES OF THERMOPHYSICAL PROPERTIES

Thermodynamic tables for gaseous and liquid Freon-23 are computed from equations presented in Sec. 4.2 and encompass the region $T = 233-453$ K and $p = 0.01-20$ MPa (Tables 50–51).

The tables of transport properties recommended in this book (Tables 52–53), encompass the same region of state as for thermodynamic tables.

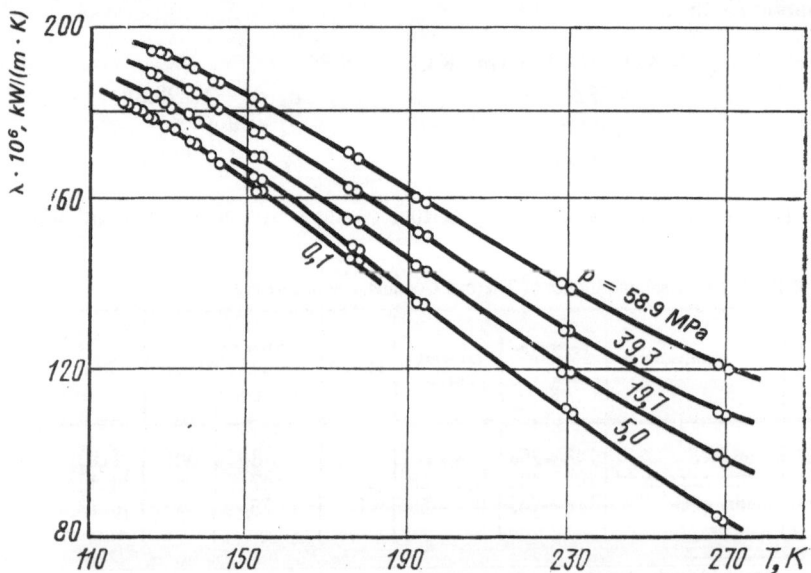

**FIG. 38.** Thermal conductivity of Freon-23 [0.13, 4.2] in the low-temperature region.

**TABLE 50.** Thermodynamic Properties of Freon-23 on the Saturation Line

| $T$ | $p$ | $\rho'$ | $\rho''$ | $h'$ | $h''$ | $s'$ | $s''$ | $c_p'$ | $c_p''$ | $r$ | $\sigma \cdot 10^3$ |
|---|---|---|---|---|---|---|---|---|---|---|---|
| 233.15 | 0.7077 | 1263.7 | 29.96 | 248.4 | 440.0 | 2.424 | 3.247 | 0.828 | 0.856 | 191.6 | 8.60 |
| 235.15 | 0.7615 | 1254.6 | 32.24 | 250.2 | 440.4 | 2.432 | 3.242 | 0.895 | 0.872 | 190.2 | 8.28 |
| 237.15 | 0.8184 | 1245.3 | 34.66 | 252.1 | 440.8 | 2.440 | 3.237 | 0.959 | 0.889 | 188.7 | 7.96 |
| 238.15 | 0.8481 | 1240.6 | 35.93 | 253.0 | 441.0 | 2.444 | 3.234 | 0.989 | 0.897 | 188.0 | 7.80 |
| 239.15 | 0.8785 | 1235.8 | 37.24 | 254.0 | 441.2 | 2.448 | 3.232 | 1.018 | 0.906 | 187.2 | 7.64 |
| 241.15 | 0.9418 | 1226.1 | 39.99 | 256.2 | 441.5 | 2.458 | 3.227 | 1.074 | 0.925 | 185.3 | 7.32 |
| 243.15 | 1.0085 | 1216.2 | 42.90 | 258.4 | 441.8 | 2.467 | 3.222 | 1.127 | 0.944 | 183.4 | 7.00 |
| 245.15 | 1.0787 | 1206.0 | 45.99 | 260.7 | 442.0 | 2.477 | 3.217 | 1.177 | 0.965 | 181.3 | 6.69 |
| 247.15 | 1.1525 | 1195.6 | 49.28 | 263.1 | 442.2 | 2.487 | 3.212 | 1.225 | 0.987 | 179.1 | 6.38 |
| 248.15 | 1.1908 | 1190.3 | 51.01 | 264.4 | 442.3 | 2.492 | 3.210 | 1.248 | 0.999 | 177.9 | 6.23 |
| 249.15 | 1.2300 | 1184.9 | 52.78 | 265.6 | 442.4 | 2.497 | 3.207 | 1.270 | 1.011 | 176.8 | 6.08 |
| 251.15 | 1.3113 | 1173.9 | 56.50 | 268.2 | 442.5 | 2.508 | 3.202 | 1.314 | 1.037 | 174.3 | 5.77 |
| 253.15 | 1.3966 | 1162.7 | 60.45 | 270.9 | 442.6 | 2.519 | 3.197 | 1.356 | 1.064 | 171.7 | 5.47 |
| 255.15 | 1.4860 | 1151.2 | 64.66 | 273.6 | 442.6 | 2.530 | 3.192 | 1.398 | 1.094 | 169.0 | 5.18 |
| 257.15 | 1.5795 | 1139.5 | 69.14 | 276.5 | 442.5 | 2.541 | 3.187 | 1.439 | 1.126 | 166.0 | 4.88 |

**TABLE 50.** Thermodynamic Properties of Freon-23 on the Saturation Line (*Continued*)

| $T$ | $p$ | $\rho'$ | $\rho''$ | $h'$ | $h''$ | $s'$ | $s''$ | $c_p'$ | $c_p''$ | $r$ | $\sigma \cdot 10^3$ |
|---|---|---|---|---|---|---|---|---|---|---|---|
| 258.15 | 1.6279 | 1133.4 | 71.49 | 277.9 | 442.5 | 2.547 | 3.184 | 1.459 | 1.143 | 164.6 | 4.74 |
| 259.15 | 1.6773 | 1127.3 | 73.92 | 279.4 | 442.4 | 2.553 | 3.182 | 1.480 | 1.161 | 163.0 | 4.59 |
| 261.15 | 1.7797 | 1114.9 | 79.01 | 282.4 | 442.3 | 2.564 | 3.177 | 1.521 | 1.199 | 159.9 | 4.31 |
| 263.15 | 1.8866 | 1102.1 | 84.45 | 285.4 | 442.0 | 2.576 | 3.171 | 1.563 | 1.241 | 156.6 | 4.03 |
| 265.15 | 1.9982 | 1088.8 | 90.28 | 288.5 | 441.7 | 2.588 | 3.166 | 1.607 | 1.288 | 153.2 | 3.75 |
| 267.15 | 2.115 | 1075.2 | 96.52 | 291.7 | 441.3 | 2.600 | 3.160 | 1.653 | 1.340 | 149.6 | 3.47 |
| 268.15 | 2.175 | 1068.2 | 99.81 | 293.3 | 441.1 | 2.606 | 3.157 | 1.677 | 1.368 | 147.8 | 3.34 |
| 269.15 | 2.237 | 1061.1 | 103.22 | 295.0 | 440.9 | 2.612 | 3.154 | 1.703 | 1.398 | 145.9 | 3.20 |
| 271.15 | 2.363 | 1046.4 | 110.44 | 298.3 | 440.3 | 2.624 | 3.148 | 1.756 | 1.463 | 142.0 | 2.94 |
| 273.15 | 2.496 | 1031.2 | 118.23 | 301.7 | 439.6 | 2.637 | 3.142 | 1.815 | 1.538 | 137.9 | 2.68 |
| 275.15 | 2.634 | 1015.3 | 126.68 | 305.2 | 438.8 | 2.650 | 3.135 | 1.881 | 1.624 | 133.6 | 2.42 |
| 277.15 | 2.777 | 998.7 | 135.87 | 308.9 | 437.9 | 2.662 | 3.128 | 1.956 | 1.725 | 129.0 | 2.17 |
| 278.15 | 2.851 | 990.1 | 140.77 | 310.7 | 437.3 | 2.669 | 3.124 | 1.998 | 1.782 | 126.6 | 2.05 |
| 279.15 | 2.927 | 981.3 | 145.91 | 312.6 | 436.8 | 2.676 | 3.121 | 2.043 | 1.845 | 124.2 | 1.92 |
| 281.15 | 3.083 | 962.9 | 156.95 | 316.4 | 435.5 | 2.689 | 3.113 | 2.147 | 1.989 | 119.1 | 1.69 |
| 283.15 | 3.245 | 943.4 | 169.17 | 320.4 | 434.0 | 2.703 | 3.104 | 2.273 | 2.167 | 113.6 | 1.45 |
| 285.15 | 3.414 | 922.5 | 182.81 | 324.5 | 432.3 | 2.717 | 3.095 | 2.431 | 2.392 | 107.8 | 1.23 |
| 287.15 | 3.590 | 899.9 | 198.21 | 328.8 | 430.3 | 2.732 | 3.085 | 2.636 | 2.686 | 101.5 | 1.01 |
| 288.15 | 3.681 | 887.9 | 206.69 | 331.1 | 429.2 | 2.739 | 3.080 | 2.764 | 2.869 | 98.1 | 0.91 |

**TABLE 51.** Thermodynamic Properties of Freon-23 in the Single-Phase Region

| | | | | | $T=233.15$ K | | | | |
|---|---|---|---|---|---|---|---|---|---|
| $p$ | $\rho$ | $z$ | $h$ | $s$ | $c_p$ | $w$ | $\mu$ | $\alpha \cdot 10^3$ |
| 0.01 | 0.36 | 0.9982 | 454.7 | 3.800 | 0.643 | 184.2 | 28.62 | 4.318 |
| 0.02 | 0.73 | 0.9963 | 454.5 | 3.717 | 0.645 | 183.9 | 28.61 | 4.346 |
| 0.03 | 1.09 | 0.9945 | 454.3 | 3.668 | 0.647 | 183.7 | 28.60 | 4.376 |
| 0.04 | 1.46 | 0.9927 | 454.1 | 3.633 | 0.649 | 183.5 | 28.60 | 4.405 |
| 0.05 | 1.82 | 0.9908 | 453.9 | 3.606 | 0.651 | 183.3 | 28.59 | 4.435 |
| 0.1 | 3.68 | 0.9815 | 453.0 | 3.521 | 0.663 | 182.3 | 28.57 | 4.588 |
| 0.2 | 7.51 | 0.9623 | 451.1 | 3.433 | 0.687 | 180.1 | 28.53 | 4.920 |
| 0.3 | 11.50 | 0.9426 | 449.1 | 3.378 | 0.713 | 177.8 | 28.51 | 5.291 |
| 0.4 | 15.67 | 0.9221 | 447.0 | 3.338 | 0.742 | 175.5 | 28.52 | 5.711 |
| 0.5 | 20.05 | 0.9007 | 444.8 | 3.304 | 0.774 | 173.0 | 28.54 | 6.189 |
| 0.6 | 24.67 | 0.8785 | 442.6 | 3.275 | 0.811 | 170.4 | 28.60 | 6.742 |
| 0.7 | 29.57 | 0.8551 | 440.2 | 3.249 | 0.852 | 167.6 | 28.68 | 7.388 |
| 0.8 | 1264.17 | 0.0229 | 248.5 | 2.424 | 0.826 | 747.5 | −0.14 | 3.639 |
| 0.9 | 1264.72 | 0.0257 | 248.5 | 2.423 | 0.825 | 749.4 | −0.15 | 3.628 |
| 1.0 | 1265.27 | 0.0285 | 248.5 | 2.423 | 0.824 | 751.2 | −0.15 | 3.617 |
| 1.5 | 1267.98 | 0.0427 | 248.5 | 2.422 | 0.817 | 760.5 | −0.16 | 3.564 |
| 2.0 | 1270.62 | 0.0569 | 248.6 | 2.420 | 0.810 | 769.7 | −0.18 | 3.513 |
| 2.5 | 1273.22 | 0.0709 | 248.7 | 2.419 | 0.803 | 778.8 | −0.19 | 3.465 |
| 3.0 | 1275.76 | 0.0849 | 248.8 | 2.418 | 0.797 | 787.8 | −0.20 | 3.419 |
| 3.5 | 1278.25 | 0.0989 | 248.8 | 2.416 | 0.791 | 796.8 | −0.21 | 3.375 |
| 4.0 | 1280.69 | 0.1128 | 248.9 | 2.415 | 0.785 | 805.6 | −0.22 | 3.332 |
| 4.5 | 1283.09 | 0.1267 | 249.0 | 2.414 | 0.780 | 814.4 | −0.23 | 3.292 |
| 5.0 | 1285.44 | 0.1405 | 249.1 | 2.412 | 0.774 | 823.1 | −0.24 | 3.253 |
| 5.5 | 1287.75 | 0.1543 | 249.2 | 2.411 | 0.769 | 831.7 | −0.25 | 3.215 |
| 6.0 | 1290.02 | 0.1680 | 249.3 | 2.410 | 0.764 | 840.3 | −0.26 | 3.179 |
| 6.5 | 1292.25 | 0.1817 | 249.4 | 2.409 | 0.759 | 848.8 | −0.27 | 3.144 |
| 7.0 | 1294.45 | 0.1953 | 249.5 | 2.407 | 0.754 | 857.2 | −0.28 | 3.111 |
| 7.5 | 1296.61 | 0.2089 | 249.6 | 2.406 | 0.749 | 865.6 | −0.29 | 3.079 |
| 8.0 | 1298.73 | 0.2225 | 249.7 | 2.405 | 0.745 | 873.9 | −0.30 | 3.047 |
| 8.5 | 1300.83 | 0.2360 | 249.8 | 2.404 | 0.740 | 882.2 | −0.31 | 3.017 |
| 9.0 | 1302.89 | 0.2495 | 250.0 | 2.403 | 0.736 | 890.4 | −0.32 | 2.988 |
| 9.5 | 1304.92 | 0.2629 | 250.1 | 2.402 | 0.732 | 898.6 | −0.32 | 2.959 |
| 10.0 | 1306.92 | 0.2764 | 250.2 | 2.400 | 0.727 | 906.8 | −0.33 | 2.932 |
| 11.0 | 1310.83 | 0.3031 | 250.4 | 2.398 | 0.719 | 923.0 | −0.35 | 2.880 |
| 12.0 | 1314.64 | 0.3297 | 250.7 | 2.396 | 0.712 | 939.0 | −0.36 | 2.830 |
| 13.0 | 1318.35 | 0.3562 | 251.0 | 2.394 | 0.705 | 954.9 | −0.38 | 2.783 |
| 14.0 | 1321.96 | 0.3825 | 251.2 | 2.392 | 0.698 | 970.6 | −0.39 | 2.739 |
| 15.0 | 1325.49 | 0.4087 | 251.5 | 2.390 | 0.691 | 986.3 | −0.40 | 2.697 |
| 16.0 | 1328.93 | 0.4349 | 251.8 | 2.388 | 0.684 | 1001.8 | −0.42 | 2.658 |
| 17.0 | 1332.30 | 0.4609 | 252.1 | 2.386 | 0.678 | 1017.3 | −0.43 | 2.620 |
| 18.0 | 1335.58 | 0.4868 | 252.4 | 2.384 | 0.672 | 1032.7 | −0.44 | 2.583 |
| 19.0 | 1338.80 | 0.5126 | 252.7 | 2.382 | 0.667 | 1048.0 | −0.46 | 2.549 |
| 20.0 | 1341.95 | 0.5383 | 253.0 | 2.380 | 0.661 | 1063.2 | −0.47 | 2.516 |

**TABLE 51.** Thermodynamic Properties of Freon-23 in the Single-Phase Region (*Continued*)

$T = 243.15$ K

| $p$ | $\rho$ | $z$ | $h$ | $s$ | $c_p$ | $w$ | $\mu$ | $\alpha \cdot 10^3$ |
|---|---|---|---|---|---|---|---|---|
| 0.01 | 0.35 | 0.9984 | 461.2 | 3.830 | 0.656 | 187.7 | 25.04 | 4.136 |
| 0.02 | 0.69 | 0.9969 | 461.1 | 3.747 | 0.658 | 187.5 | 25.04 | 4.160 |
| 0.03 | 1.04 | 0.9953 | 460.9 | 3.698 | 0.659 | 187.3 | 25.04 | 4.184 |
| 0.04 | 1.39 | 0.9937 | 460.7 | 3.664 | 0.661 | 187.1 | 25.04 | 4.208 |
| 0.05 | 1.75 | 0.9921 | 460.6 | 3.637 | 0.663 | 186.9 | 25.04 | 4.232 |
| 0.1 | 3.52 | 0.9841 | 459.7 | 3.552 | 0.672 | 186.0 | 25.02 | 4.356 |
| 0.2 | 7.16 | 0.9678 | 458.0 | 3.464 | 0.692 | 184.1 | 25.01 | 4.622 |
| 0.3 | 10.92 | 0.9511 | 456.3 | 3.411 | 0.714 | 182.2 | 25.01 | 4.915 |
| 0.4 | 14.83 | 0.9338 | 454.5 | 3.371 | 0.737 | 180.1 | 25.02 | 5.238 |
| 0.5 | 18.90 | 0.9160 | 452.6 | 3.339 | 0.762 | 178.0 | 25.04 | 5.597 |
| 0.6 | 23.15 | 0.8976 | 450.6 | 3.311 | 0.790 | 175.8 | 25.08 | 6.000 |
| 0.7 | 27.60 | 0.8785 | 448.6 | 3.287 | 0.821 | 173.6 | 25.13 | 6.455 |
| 0.8 | 32.27 | 0.8586 | 446.5 | 3.264 | 0.856 | 171.2 | 25.21 | 6.975 |
| 0.9 | 37.20 | 0.8379 | 444.3 | 3.243 | 0.895 | 168.6 | 25.30 | 7.576 |
| 1.0 | 42.44 | 0.8161 | 442.0 | 3.223 | 0.940 | 166.0 | 25.42 | 8.281 |
| 1.5 | 1219.55 | 0.0426 | 258.4 | 2.465 | 1.117 | 593.2 | 0.02 | 4.250 |
| 2.0 | 1222.89 | 0.0566 | 258.4 | 2.464 | 1.108 | 600.6 | 0.01 | 4.168 |
| 2.5 | 1226.14 | 0.0706 | 258.4 | 2.462 | 1.099 | 608.0 | —0.00 | 4.091 |
| 3.0 | 1229.30 | 0.0845 | 258.4 | 2.460 | 1.091 | 615.1 | —0.02 | 4.018 |
| 3.5 | 1232.38 | 0.0984 | 258.4 | 2.459 | 1.083 | 622.1 | —0.03 | 3.949 |
| 4.0 | 1235.39 | 0.1121 | 258.4 | 2.457 | 1.076 | 629.0 | —0.04 | 3.884 |
| 4.5 | 1238.32 | 0.1259 | 258.4 | 2.455 | 1.068 | 635.7 | —0.05 | 3.822 |
| 5.0 | 1241.19 | 0.1395 | 258.4 | 2.454 | 1.061 | 642.4 | —0.06 | 3.764 |
| 5.5 | 1243.99 | 0.1531 | 258.5 | 2.452 | 1.055 | 648.9 | —0.08 | 3.708 |
| 6.0 | 1246.73 | 0.1667 | 258.5 | 2.451 | 1.048 | 655.3 | —0.08 | 3.655 |
| 6.5 | 1249.42 | 0.1802 | 258.6 | 2.449 | 1.042 | 661.5 | —0.10 | 3.604 |
| 7.0 | 1252.05 | 0.1936 | 258.6 | 2.448 | 1.037 | 667.7 | —0.10 | 3.555 |
| 7.5 | 1254.63 | 0.2070 | 258.7 | 2.447 | 1.031 | 673.8 | —0.11 | 3.509 |
| 8.0 | 1257.15 | 0.2204 | 258.7 | 2.445 | 1.025 | 679.8 | —0.12 | 3.464 |
| 8.5 | 1259.64 | 0.2337 | 258.8 | 2.444 | 1.020 | 685.7 | —0.13 | 3.421 |
| 9.0 | 1262.07 | 0.2470 | 258.9 | 2.443 | 1.015 | 691.6 | —0.14 | 3.380 |
| 9.5 | 1264.46 | 0.2602 | 258.9 | 2.441 | 1.010 | 697.3 | —0.15 | 3.340 |
| 10.0 | 1266.81 | 0.2734 | 259.0 | 2.440 | 1.006 | 703.0 | —0.16 | 3.302 |
| 11.0 | 1271.30 | 0.2996 | 259.2 | 2.437 | 0.997 | 714.1 | —0.17 | 3.230 |
| 12.0 | 1275.83 | 0.3257 | 259.4 | 2.435 | 0.988 | 724.9 | —0.18 | 3.162 |
| 13.0 | 1280.12 | 0.3517 | 259.5 | 2.432 | 0.980 | 735.5 | —0.20 | 3.099 |
| 14.0 | 1284.29 | 0.3775 | 259.7 | 2.430 | 0.973 | 745.8 | —0.21 | 3.040 |
| 15.0 | 1288.34 | 0.4032 | 260.0 | 2.428 | 0.965 | 755.9 | —0.22 | 2.984 |
| 16.0 | 1292.28 | 0.4288 | 260.2 | 2.425 | 0.959 | 765.8 | —0.23 | 2.932 |
| 17.0 | 1296.12 | 0.4542 | 260.4 | 2.423 | 0.952 | 775.4 | —0.24 | 2.882 |
| 18.0 | 1299.85 | 0.4796 | 260.6 | 2.421 | 0.946 | 784.9 | —0.25 | 2.835 |
| 19.0 | 1303.50 | 0.5048 | 260.9 | 2.419 | 0.940 | 794.2 | —0.26 | 2.791 |
| 20.0 | 1307.06 | 0.5299 | 261.1 | 2.417 | 0.935 | 803.3 | —0.27 | 2.749 |

**TABLE 51.** Thermodynamic Properties of Freon-23 in the Single-Phase Region (*Continued*)

$T = 253.15$ K

| $p$ | $\rho$ | $z$ | $h$ | $s$ | $c_p$ | $w$ | $\mu$ | $\alpha \cdot 10^3$ |
|---|---|---|---|---|---|---|---|---|
| 0.01 | 0.33 | 0.9986 | 468.0 | 3.860 | 0.669 | 191.1 | 22.06 | 3.970 |
| 0.02 | 0.67 | 0.9973 | 467.8 | 3.777 | 0.670 | 190.9 | 22.06 | 3.989 |
| 0.03 | 1.00 | 0.9959 | 467.7 | 3.728 | 0.672 | 190.8 | 22.06 | 4.009 |
| 0.04 | 1.34 | 0.9946 | 467.5 | 3.694 | 0.673 | 190.6 | 22.06 | 4.029 |
| 0.05 | 1.67 | 0.9932 | 467.4 | 3.667 | 0.675 | 190.4 | 22.06 | 4.049 |
| 0.1 | 3.37 | 0.9863 | 466.6 | 3.582 | 0.683 | 189.6 | 22.05 | 4.151 |
| 0.2 | 6.84 | 0.9723 | 465.1 | 3.496 | 0.700 | 187.9 | 22.05 | 4.367 |
| 0.3 | 10.42 | 0.9580 | 463.5 | 3.443 | 0.718 | 186.2 | 22.06 | 4.601 |
| 0.4 | 14.11 | 0.9433 | 461.9 | 3.404 | 0.737 | 184.5 | 22.07 | 4.856 |
| 0.5 | 17.92 | 0.9282 | 460.3 | 3.373 | 0.757 | 182.7 | 22.08 | 5.134 |
| 0.6 | 21.87 | 0.9128 | 458.6 | 3.346 | 0.779 | 180.8 | 22.11 | 5.438 |
| 0.7 | 25.96 | 0.8968 | 456.8 | 3.323 | 0.803 | 178.9 | 22.15 | 5.774 |
| 0.8 | 30.23 | 0.8804 | 455.0 | 3.301 | 0.829 | 176.9 | 22.19 | 6.147 |
| 0.9 | 34.67 | 0.8635 | 453.2 | 3.282 | 0.858 | 174.8 | 22.25 | 6.564 |
| 1.0 | 39.32 | 0.8459 | 451.2 | 3.263 | 0.890 | 172.6 | 22.32 | 7.035 |
| 1.5 | 1163.68 | 0.0429 | 270.9 | 2.518 | 1.353 | 495.8 | 0.20 | 5.193 |
| 2.0 | 1168.15 | 0.0570 | 270.7 | 2.516 | 1.338 | 504.0 | 0.18 | 5.047 |
| 2.5 | 1172.44 | 0.0709 | 270.6 | 2.514 | 1.324 | 511.9 | 0.16 | 4.913 |
| 3.0 | 1176.58 | 0.0848 | 270.5 | 2.512 | 1.312 | 519.6 | 0.14 | 4.790 |
| 3.5 | 1180.58 | 0.0986 | 270.4 | 2.510 | 1.300 | 527.0 | 0.12 | 4.676 |
| 4.0 | 1184.45 | 0.1123 | 270.4 | 2.508 | 1.288 | 534.2 | 0.10 | 4.570 |
| 4.5 | 1188.19 | 0.1260 | 270.3 | 2.506 | 1.278 | 541.3 | 0.09 | 4.471 |
| 5.0 | 1191.82 | 0.1396 | 270.3 | 2.504 | 1.268 | 548.2 | 0.07 | 4.379 |
| 5.5 | 1195.35 | 0.1531 | 270.2 | 2.502 | 1.259 | 554.9 | 0.06 | 4.292 |
| 6.0 | 1198.78 | 0.1665 | 270.2 | 2.501 | 1.250 | 561.5 | 0.04 | 4.210 |
| 6.5 | 1202.11 | 0.1799 | 270.2 | 2.499 | 1.242 | 567.9 | 0.03 | 4.134 |
| 7.0 | 1205.36 | 0.1932 | 270.1 | 2.497 | 1.234 | 574.2 | 0.02 | 4.061 |
| 7.5 | 1208.53 | 0.2064 | 270.1 | 2.495 | 1.227 | 580.3 | 0.01 | 3.992 |
| 8.0 | 1211.62 | 0.2196 | 270.1 | 2.494 | 1.220 | 586.4 | —0.00 | 3.927 |
| 8.5 | 1214.64 | 0.2328 | 270.1 | 2.492 | 1.213 | 592.3 | —0.02 | 3.865 |
| 9.0 | 1217.59 | 0.2459 | 270.2 | 2.491 | 1.207 | 598.1 | —0.02 | 3.806 |
| 9.5 | 1220.47 | 0.2589 | 270.2 | 2.489 | 1.201 | 603.8 | —0.04 | 3.750 |
| 10.0 | 1223.30 | 0.2719 | 270.2 | 2.488 | 1.195 | 609.4 | —0.04 | 3.696 |
| 11.0 | 1228.77 | 0.2978 | 270.3 | 2.485 | 1.184 | 620.3 | —0.06 | 3.596 |
| 12.0 | 1234.03 | 0.3235 | 270.3 | 2.482 | 1.173 | 630.9 | —0.08 | 3.504 |
| 13.0 | 1239.10 | 0.3490 | 270.4 | 2.479 | 1.164 | 641.1 | —0.09 | 3.418 |
| 14.0 | 1243.99 | 0.3744 | 270.6 | 2.476 | 1.155 | 651.1 | —0.11 | 3.340 |
| 15.0 | 1248.71 | 0.3996 | 270.7 | 2.474 | 1.147 | 660.8 | —0.12 | 3.266 |
| 16.0 | 1253.28 | 0.4247 | 270.8 | 2.471 | 1.139 | 670.2 | —0.13 | 3.198 |
| 17.0 | 1257.71 | 0.4496 | 271.0 | 2.468 | 1.132 | 679.3 | —0.14 | 3.133 |
| 18.0 | 1262.01 | 0.4744 | 271.2 | 2.466 | 1.125 | 688.3 | —0.16 | 3.073 |
| 19.0 | 1266.19 | 0.4992 | 271.3 | 2.464 | 1.118 | 697.0 | —0.17 | 3.016 |
| 20.0 | 1270.25 | 0.5237 | 271.5 | 2.461 | 1.112 | 705.5 | —0.18 | 2.963 |

**TABLE 51.** Thermodynamic Properties of Freon-23 in the Single-Phase Region (*Continued*)

$T = 263.15$ K

| $p$ | $\rho$ | $z$ | $h$ | $s$ | $c_p$ | $w$ | $\mu$ | $\alpha \cdot 10^3$ |
|---|---|---|---|---|---|---|---|---|
| 0.01 | 0.32 | 0.9988 | 474.8 | 3.889 | 0.682 | 194.4 | 19.53 | 3.816 |
| 0.02 | 0.64 | 0.9976 | 474.7 | 3.806 | 0.683 | 194.3 | 19.53 | 3.833 |
| 0.03 | 0.96 | 0.9965 | 474.6 | 3.758 | 0.685 | 194.1 | 19.53 | 3.849 |
| 0.04 | 1.29 | 0.9953 | 474.4 | 3.723 | 0.686 | 194.0 | 19.53 | 3.866 |
| 0.05 | 1.61 | 0.9941 | 474.3 | 3.696 | 0.687 | 193.8 | 19.53 | 3.882 |
| 0.1 | 3.24 | 0.9881 | 473.6 | 3.612 | 0.694 | 193.1 | 19.53 | 3.967 |
| 0.2 | 6.56 | 0.9760 | 472.3 | 3.526 | 0.709 | 191.6 | 19.54 | 4.145 |
| 0.3 | 9.96 | 0.9637 | 470.9 | 3.474 | 0.724 | 190.1 | 19.54 | 4.335 |
| 0.4 | 13.46 | 0.9511 | 469.4 | 3.436 | 0.739 | 188.6 | 19.55 | 4.540 |
| 0.5 | 17.05 | 0.9382 | 468.0 | 3.405 | 0.756 | 187.0 | 19.57 | 4.759 |
| 0.6 | 20.76 | 0.9259 | 466.5 | 3.379 | 0.774 | 185.4 | 19.59 | 4,996 |
| 0.7 | 24.57 | 0.9116 | 464.9 | 3.357 | 0.793 | 183.7 | 19.61 | 5,253 |
| 0.8 | 28.51 | 0.8978 | 463.4 | 3.336 | 0.814 | 182.0 | 19.64 | 5,532 |
| 0.9 | 32.59 | 0.8836 | 461.7 | 3.318 | 0.836 | 180.3 | 19.67 | 5,838 |
| 1.0 | 36.82 | 0.8691 | 460.1 | 3.300 | 0.860 | 178.4 | 19.71 | 6,173 |
| 1.5 | 60.83 | 0.7891 | 450.8 | 3.225 | 1.021 | 168.3 | 20.03 | 8.529 |
| 2.0 | 1103.52 | 0.0580 | 285.3 | 2.575 | 1.557 | 420.8 | 0.41 | 6.478 |
| 2.5 | 1109.70 | 0.0721 | 285.0 | 2.572 | 1.530 | 430.4 | 0.37 | 6.204 |
| 3.0 | 1115.55 | 0.0861 | 284.7 | 2.570 | 1.507 | 439.6 | 0.34 | 5.963 |
| 3.5 | 1121.10 | 0.0999 | 284.5 | 2.567 | 1.486 | 448.4 | 0.31 | 5.748 |
| 4.0 | 1126.38 | 0.1136 | 284.3 | 2.564 | 1.467 | 456.8 | 0.28 | 5.555 |
| 4.5 | 1131.43 | 0.1273 | 284.1 | 2.562 | 1.450 | 465.0 | 0.25 | 5.380 |
| 5.0 | 1136.27 | 0.1408 | 283.9 | 2.560 | 1.434 | 472.8 | 0.23 | 5.222 |
| 5.5 | 1140.91 | 0.1543 | 283.8 | 2.557 | 1.420 | 480.5 | 0.21 | 5.076 |
| 6.0 | 1145.38 | 0.1676 | 283.6 | 2.555 | 1.407 | 487.0 | 0.19 | 4.943 |
| 6.5 | 1149.69 | 0.1809 | 283.5 | 2.553 | 1.394 | 495.0 | 0.17 | 4.820 |
| 7.0 | 1153.85 | 0.1941 | 283.4 | 2.551 | 1.383 | 502.0 | 0.15 | 4.705 |
| 7.5 | 1157.87 | 0.2073 | 283.3 | 2.549 | 1.372 | 508.8 | 0.13 | 4.599 |
| 8.0 | 1161.77 | 0.2204 | 283.2 | 2.547 | 1.362 | 515.4 | 0.12 | 4.500 |
| 8.5 | 1165.55 | 0.2334 | 283.1 | 2.545 | 1.353 | 521.8 | 0.10 | 4.407 |
| 9.0 | 1169.23 | 0.2463 | 283.1 | 2.543 | 1.344 | 528.1 | 0.09 | 4.320 |
| 9.5 | 1172.80 | 0.2592 | 283.0 | 2.541 | 1.336 | 534.3 | 0.07 | 4.237 |
| 10.0 | 1176.27 | 0.2721 | 283.0 | 2.540 | 1.328 | 540.3 | 0.06 | 4.160 |
| 11.0 | 1182.95 | 0.2976 | 282.9 | 2.536 | 1.313 | 551.9 | 0.04 | 4.017 |
| 12.0 | 1189.32 | 0.3229 | 282.9 | 2.533 | 1.300 | 563.2 | 0.02 | 3.889 |
| 13.0 | 1195.39 | 0.3480 | 282.9 | 2.530 | 1.288 | 573.9 | —0.01 | 3.772 |
| 14.0 | 1201.21 | 0.3730 | 282.9 | 2.527 | 1.277 | 584.4 | —0.02 | 3.666 |
| 15.0 | 1206.80 | 0.3978 | 282.9 | 2.524 | 1.267 | 594.4 | —0.04 | 3.568 |
| 16.0 | 1212.17 | 0.4224 | 283.0 | 2.521 | 1.257 | 604.2 | —0.06 | 3.478 |
| 17.0 | 1217.34 | 0.4469 | 283.1 | 2.518 | 1.248 | 613.6 | —0.07 | 3.395 |
| 18.0 | 1222.34 | 0.4712 | 283.2 | 2.515 | 1.240 | 622.8 | —0.08 | 3.317 |
| 19.0 | 1227.17 | 0.4955 | 283.3 | 2.512 | 1.233 | 631.7 | —0.10 | 3.245 |
| 20.0 | 1231.85 | 0.5195 | 283.4 | 2.510 | 1.226 | 640.4 | —0.11 | 3.178 |

**TABLE 51.** Thermodynamic Properties of Freon-23 in the Single-Phase Region (*Continued*)

$T=273.15$ K

| $p$ | $\rho$ | $z$ | $h$ | $s$ | $c_p$ | $w$ | $\mu$ | $\alpha \cdot 10^3$ |
|---|---|---|---|---|---|---|---|---|
| 0.01 | 0.31 | 0.9990 | 481.9 | 3.917 | 0.696 | 197.7 | 17.37 | 3.675 |
| 0.02 | 0.62 | 0.9979 | 481.7 | 3.835 | 0.697 | 197.6 | 17.38 | 3.688 |
| 0.03 | 0.93 | 0.9969 | 481.6 | 3.786 | 0.698 | 197.4 | 17.38 | 3.702 |
| 0.04 | 1.24 | 0.9959 | 481.5 | 3.752 | 0.699 | 197.3 | 17.38 | 3.716 |
| 0.05 | 1.55 | 0.9948 | 481.4 | 3.725 | 0.700 | 197.2 | 17.38 | 3.730 |
| 0.1 | 3.12 | 0.9896 | 480.8 | 3.641 | 0.706 | 196.5 | 17.38 | 3.801 |
| 0.2 | 6.30 | 0.9791 | 479.5 | 3.555 | 0.718 | 195.2 | 17.38 | 3.949 |
| 0.3 | 9.55 | 0.9684 | 478.3 | 3.504 | 0.731 | 193.9 | 17.39 | 4.106 |
| 0.4 | 12.88 | 0.9575 | 477.0 | 3.466 | 0.745 | 192.5 | 17.40 | 4.272 |
| 0.5 | 16.29 | 0.9464 | 475.7 | 3.436 | 0.759 | 191.1 | 17.41 | 4.449 |
| 0.6 | 19.78 | 0.9351 | 474.3 | 3.411 | 0.774 | 189.7 | 17.42 | 4.637 |
| 0.7 | 23.37 | 0.9236 | 473.0 | 3.389 | 0.789 | 188.3 | 17.44 | 4.839 |
| 0.8 | 27.05 | 0.9119 | 471.6 | 3.369 | 0.806 | 186.8 | 17.46 | 5.055 |
| 0.9 | 30.83 | 0.8999 | 470.2 | 3.351 | 0.824 | 185.3 | 17.48 | 5.287 |
| 1.0 | 34.73 | 0.8877 | 468.7 | 3.335 | 0.843 | 183.7 | 17.51 | 5.537 |
| 1.5 | 56.26 | 0.8219 | 460.8 | 3.265 | 0.960 | 175.3 | 17.69 | 7.161 |
| 2.0 | 82.70 | 0.7455 | 451.5 | 3.204 | 1.149 | 165.4 | 17.99 | 9.920 |
| 2.5 | 1031.29 | 0.0747 | 301.7 | 2.637 | 1.814 | 347.8 | 0.76 | 8.837 |
| 3.0 | 1040.82 | 0.0889 | 301.1 | 2.633 | 1.756 | 360.2 | 0.68 | 8.196 |
| 3.5 | 1049.51 | 0.1028 | 300.5 | 2.629 | 1.708 | 371.7 | 0.61 | 7.675 |
| 4.0 | 1057.52 | 0.1166 | 300.0 | 2.625 | 1.668 | 382.5 | 0.56 | 7.241 |
| 4.5 | 1064.97 | 0.1303 | 299.6 | 2.622 | 1.633 | 392.7 | 0.50 | 6.873 |
| 5.0 | 1071.93 | 0.1438 | 299.2 | 2.619 | 1.604 | 402.4 | 0.46 | 6.555 |
| 5.5 | 1078.48 | 0.1572 | 298.9 | 2.616 | 1.578 | 411.6 | 0.42 | 6.277 |
| 6.0 | 1084.67 | 0.1705 | 298.5 | 2.613 | 1.554 | 420.4 | 0.38 | 6.032 |
| 6.5 | 1090.55 | 0.1837 | 298.3 | 2.610 | 1.534 | 428.8 | 0.35 | 5.814 |
| 7.0 | 1096.14 | 0.1969 | 298.0 | 2.608 | 1.515 | 437.0 | 0.32 | 5.617 |
| 7.5 | 1101.48 | 0.2099 | 297.8 | 2.605 | 1.498 | 444.8 | 0.29 | 5.439 |
| 8.0 | 1106.60 | 0.2229 | 297.6 | 2.603 | 1.482 | 452.4 | 0.27 | 5.277 |
| 8.5 | 1111.50 | 0.2358 | 297.4 | 2.600 | 1.468 | 459.7 | 0.25 | 5.129 |
| 9.0 | 1116.23 | 0.2486 | 297.2 | 2.598 | 1.455 | 466.9 | 0.22 | 4.993 |
| 9.5 | 1120.78 | 0.2613 | 297.0 | 2.596 | 1.443 | 473.8 | 0.20 | 4.867 |
| 10.0 | 1125.17 | 0.2740 | 296.9 | 2.594 | 1.431 | 480.5 | 0.18 | 4.750 |
| 11.0 | 1133.53 | 0.2992 | 296.7 | 2.590 | 1.411 | 493.4 | 0.15 | 4.539 |
| 12.0 | 1141.39 | 0.3241 | 296.5 | 2.586 | 1.393 | 505.7 | 0.12 | 4.355 |
| 13.0 | 1148.81 | 0.3489 | 296.3 | 2.582 | 1.377 | 517.5 | 0.09 | 4.192 |
| 14.0 | 1155.84 | 0.3734 | 296.2 | 2.578 | 1.362 | 528.7 | 0.07 | 4.046 |
| 15.0 | 1162.54 | 0.3978 | 296.2 | 2.575 | 1.349 | 539.6 | 0.04 | 3.914 |
| 16.0 | 1168.93 | 0.4220 | 296.1 | 2.572 | 1.337 | 550.0 | 0.02 | 3.794 |
| 17.0 | 1175.04 | 0.4460 | 296.1 | 2.569 | 1.326 | 560.0 | 0.00 | 3.685 |
| 18.0 | 1180.91 | 0.4699 | 296.1 | 2.565 | 1.316 | 569.8 | −0.01 | 3.585 |
| 19.0 | 1186.54 | 0.4937 | 296.1 | 2.562 | 1.307 | 579.2 | −0.03 | 3.493 |
| 20.0 | 1191.97 | 0.5173 | 296.2 | 2.560 | 1.299 | 588.3 | −0.04 | 3.408 |

**TABLE 51.** Thermodynamic Properties of Freon-23 in the Single-Phase Region (*Continued*)

$T = 283{,}15$ K

| $p$ | $\rho$ | $z$ | $h$ | $s$ | $c_p$ | $w$ | $\mu$ | $\alpha \cdot 10^3$ |
|---|---|---|---|---|---|---|---|---|
| 0.01 | 0.30 | 0.9991 | 489.0 | 3.945 | 0.709 | 200.9 | 15.52 | 3.543 |
| 0.02 | 0.60 | 0.9982 | 488.9 | 3.862 | 0.710 | 200.8 | 15.52 | 3.555 |
| 0.03 | 0.89 | 0.9973 | 488.8 | 3.814 | 0.711 | 200.6 | 15.52 | 3.567 |
| 0.04 | 1.19 | 0.9964 | 488.7 | 3.779 | 0.712 | 200.5 | 15.52 | 3.578 |
| 0.05 | 1.49 | 0.9955 | 488.5 | 3.752 | 0.713 | 200.4 | 15.52 | 3.590 |
| 0.1 | 3.00 | 0.9909 | 488.0 | 3.669 | 0.719 | 199.8 | 15.52 | 3.650 |
| 0.2 | 6.06 | 0.9817 | 486.9 | 3.583 | 0.729 | 198.7 | 15.53 | 3.774 |
| 0.3 | 9.18 | 0.9723 | 485.7 | 3.532 | 0.740 | 197.5 | 15.53 | 3.904 |
| 0.4 | 12.36 | 0.9628 | 484.6 | 3.495 | 0.752 | 196.3 | 15.54 | 4.042 |
| 0.5 | 15.60 | 0.9532 | 483.4 | 3.466 | 0.764 | 195.0 | 15.55 | 4.186 |
| 0.6 | 18.91 | 0.9434 | 482.2 | 3.441 | 0.776 | 193.8 | 15.56 | 4.339 |
| 0.7 | 22.30 | 0.9335 | 481.0 | 3.419 | 0.789 | 192.5 | 15.57 | 4.500 |
| 0.8 | 25.77 | 0.9234 | 479.7 | 3.400 | 0.803 | 191.2 | 15.59 | 4.671 |
| 0.9 | 29.31 | 0.9132 | 478.5 | 3.383 | 0.818 | 189.9 | 15.60 | 4.852 |
| 1.0 | 32.94 | 0.9027 | 477.2 | 3.367 | 0.833 | 188.5 | 15.62 | 5.045 |
| 1.5 | 52.63 | 0.8476 | 470.3 | 3.301 | 0.924 | 181.4 | 15.73 | 6.232 |
| 2.0 | 75.68 | 0.7859 | 462.5 | 3.245 | 1.052 | 173.4 | 15.89 | 8.000 |
| 2.5 | 104.03 | 0.7147 | 453.4 | 3.193 | 1.259 | 164.1 | 16.11 | 10.979 |
| 3.0 | 142.43 | 0.6264 | 441.7 | 3.137 | 1.684 | 152.6 | 16.40 | 17.419 |
| 3.5 | 952.39 | 0.1093 | 319.6 | 2.699 | 2.175 | 286.2 | 1.27 | 12.826 |
| 4.0 | 967.80 | 0.1229 | 318.4 | 2.693 | 2.033 | 302.4 | 1.10 | 11.186 |
| 4.5 | 981.01 | 0.1364 | 317.4 | 2.688 | 1.932 | 316.8 | 0.97 | 10.028 |
| 5.0 | 992.65 | 0.1498 | 316.5 | 2.683 | 1.856 | 329.9 | 0.86 | 9.158 |
| 5.5 | 1003.10 | 0.1631 | 315.7 | 2.678 | 1.795 | 342.0 | 0.78 | 8.474 |
| 6.0 | 1012.60 | 0.1762 | 315.1 | 2.674 | 1.745 | 353.3 | 0.70 | 7.920 |
| 6.5 | 1021.34 | 0.1893 | 314.5 | 2.670 | 1.704 | 363.9 | 0.64 | 7.459 |
| 7.0 | 1029.43 | 0.2022 | 314.0 | 2.667 | 1.668 | 373.9 | 0.58 | 7.067 |
| 7.5 | 1037.00 | 0.2151 | 313.5 | 2.664 | 1.638 | 383.4 | 0.53 | 6.731 |
| 8.0 | 1044.09 | 0.2279 | 313.1 | 2.660 | 1.611 | 392.4 | 0.49 | 6.437 |
| 8.5 | 1050.79 | 0.2406 | 312.7 | 2.657 | 1.587 | 401.1 | 0.45 | 6.178 |
| 9.0 | 1057.14 | 0.2532 | 312.4 | 2.655 | 1.565 | 409.4 | 0.41 | 5.947 |
| 9.5 | 1063.17 | 0.2657 | 312.1 | 2.652 | 1.546 | 417.4 | 0.38 | 5.740 |
| 10.0 | 1068.92 | 0.2782 | 311.8 | 2.649 | 1.529 | 425.1 | 0.35 | 5.552 |
| 11.0 | 1079.69 | 0.3030 | 311.3 | 2.644 | 1.498 | 439.8 | 0.30 | 5.226 |
| 12.0 | 1089.64 | 0.3275 | 310.9 | 2.639 | 1.472 | 453.6 | 0.25 | 4.951 |
| 13.0 | 1098.89 | 0.3518 | 310.6 | 2.635 | 1.450 | 466.6 | 0.21 | 4.715 |
| 14.0 | 1107.54 | 0.3759 | 310.3 | 2.631 | 1.430 | 479.0 | 0.18 | 4.509 |
| 15.0 | 1115.68 | 0.3998 | 310.1 | 2.627 | 1.413 | 490.8 | 0.14 | 4.328 |
| 16.0 | 1123.37 | 0.4236 | 309.9 | 2.623 | 1.397 | 502.1 | 0.12 | 4.167 |
| 17.0 | 1130.67 | 0.4472 | 309.8 | 2.619 | 1.383 | 513.0 | 0.09 | 4.023 |
| 18.0 | 1137.61 | 0.4706 | 309.6 | 2.616 | 1.371 | 523.4 | 0.07 | 3.893 |
| 19.0 | 1144.24 | 0.4938 | 309.6 | 2.613 | 1.359 | 533.5 | 0.04 | 3.774 |
| 20.0 | 1150.58 | 0.5170 | 309.5 | 2.609 | 1.349 | 543.2 | 0.02 | 3.666 |

**TABLE 51.** Thermodynamic Properties of Freon-23 in the Single-Phase Region (*Continued*)

$T = 293.15$ K

| $p$ | $\rho$ | $z$ | $h$ | $s$ | $c_p$ | $w$ | $\mu$ | $\alpha \cdot 10^3$ |
|---|---|---|---|---|---|---|---|---|
| 0.01 | 0.29 | 0.9992 | 496.2 | 3.971 | 0.723 | 204.0 | 13.91 | 3.421 |
| 0.02 | 0.58 | 0.9984 | 496.1 | 3.889 | 0.724 | 203.9 | 13.91 | 3.431 |
| 0.03 | 0.86 | 0.9976 | 496.0 | 3.840 | 0.725 | 203.8 | 13.92 | 3.441 |
| 0.04 | 1.15 | 0.9968 | 495.9 | 3.806 | 0.726 | 203.7 | 13.92 | 3.451 |
| 0.05 | 1.44 | 0.9960 | 495.8 | 3.779 | 0.727 | 203.6 | 13.92 | 3.461 |
| 0.1 | 2.90 | 0.9920 | 495.3 | 3.695 | 0.731 | 203.1 | 13.92 | 3.512 |
| 0.2 | 5.84 | 0.9839 | 494.3 | 3.611 | 0.741 | 202.0 | 13.92 | 3.617 |
| 0.3 | 8.83 | 0.9757 | 493.3 | 3.560 | 0.750 | 200.9 | 13.93 | 3.726 |
| 0.4 | 11.88 | 0.9673 | 492.2 | 3.523 | 0.760 | 199.9 | 13.94 | 3.840 |
| 0.5 | 14.98 | 0.9589 | 491.2 | 3.494 | 0.770 | 198.8 | 13.94 | 3.960 |
| 0.6 | 18.13 | 0.9504 | 490.1 | 3.469 | 0.781 | 197.7 | 13.95 | 4.085 |
| 0.7 | 21.35 | 0.9418 | 489.0 | 3.448 | 0.792 | 196.5 | 13.96 | 4.217 |
| 0.8 | 24.63 | 0.9331 | 487.9 | 3.430 | 0.804 | 195.4 | 13.97 | 4.355 |
| 0.9 | 27.97 | 0.9242 | 486.7 | 3.413 | 0.816 | 194.2 | 13.98 | 4.500 |
| 1.0 | 31.39 | 0.9152 | 485.6 | 3.397 | 0.828 | 193.1 | 13.99 | 4.652 |
| 1.5 | 49.63 | 0.8682 | 479.5 | 3.334 | 0.901 | 186.9 | 14.06 | 5.556 |
| 2.0 | 70.31 | 0.8171 | 472.9 | 3.282 | 0.996 | 180.2 | 14.15 | 6.790 |
| 2.5 | 94.44 | 0.7604 | 465.3 | 3.236 | 1.129 | 172.7 | 14.25 | 8.594 |
| 3.0 | 123.85 | 0.6958 | 456.6 | 3.190 | 1.337 | 164.3 | 14.36 | 11.522 |
| 3.5 | 162.63 | 0.6182 | 445.7 | 3.141 | 1.730 | 154.4 | 14.44 | 17.274 |
| 4.0 | 224.63 | 0.5115 | 429.9 | 3.078 | 2.958 | 141.7 | 14.30 | 35.834 |
| 4.5 | 847.47 | 0.1525 | 340.6 | 2.770 | 3.067 | 222.8 | 2.31 | 23.880 |
| 5.0 | 878.15 | 0.1636 | 337.8 | 2.758 | 2.544 | 246.1 | 1.84 | 17.402 |
| 5.5 | 900.75 | 0.1754 | 335.8 | 2.749 | 2.279 | 265.1 | 1.54 | 14.178 |
| 6.0 | 918.96 | 0.1876 | 334.2 | 2.742 | 2.113 | 281.4 | 1.32 | 12.189 |
| 6.5 | 934.35 | 0.1998 | 332.9 | 2.736 | 1.998 | 295.9 | 1.16 | 10.816 |
| 7.0 | 947.75 | 0.2122 | 331.9 | 2.730 | 1.911 | 309.1 | 1.03 | 9.801 |
| 7.5 | 959.68 | 0.2245 | 330.9 | 2.725 | 1.843 | 321.3 | 0.93 | 9.013 |
| 8.0 | 970.45 | 0.2368 | 330.1 | 2.721 | 1.788 | 332.6 | 0.84 | 8.380 |
| 8.5 | 980.29 | 0.2491 | 329.4 | 2.717 | 1.743 | 343.2 | 0.76 | 7.859 |
| 9.0 | 989.37 | 0.2613 | 328.8 | 2.713 | 1.704 | 353.2 | 0.70 | 7.419 |
| 9.5 | 997.81 | 0.2735 | 328.2 | 2.709 | 1.670 | 362.8 | 0.64 | 7.043 |
| 10.0 | 1005.71 | 0.2856 | 327.7 | 2.706 | 1.641 | 371.8 | 0.59 | 6.717 |
| 11.0 | 1020.14 | 0.3097 | 326.9 | 2.699 | 1.593 | 388.8 | 0.50 | 6.177 |
| 12.0 | 1033.10 | 0.3337 | 326.1 | 2.694 | 1.553 | 404.5 | 0.43 | 5.745 |
| 13.0 | 1044.90 | 0.3574 | 325.5 | 2.688 | 1.521 | 419.2 | 0.36 | 5.391 |
| 14.0 | 1055.74 | 0.3809 | 325.0 | 2.683 | 1.493 | 433.0 | 0.31 | 5.093 |
| 15.0 | 1065.79 | 0.4043 | 324.6 | 2.679 | 1.470 | 446.1 | 0.27 | 4.839 |
| 16.0 | 1075.17 | 0.4275 | 324.2 | 2.674 | 1.449 | 458.4 | 0.23 | 4.618 |
| 17.0 | 1083.97 | 0.4505 | 323.9 | 2.670 | 1.431 | 470.2 | 0.19 | 4.425 |
| 18.0 | 1092.26 | 0.4734 | 323.7 | 2.666 | 1.415 | 481.5 | 0.16 | 4.253 |
| 19.0 | 1100.11 | 0.4961 | 323.5 | 2.662 | 1.400 | 492.4 | 0.13 | 4.100 |
| 20.0 | 1107.56 | 0.5187 | 323.3 | 2.659 | 1.387 | 502.8 | 0.10 | 3.962 |

**TABLE 51.** Thermodynamic Properties of Freon-23 in the Single-Phase Region (*Continued*)

$T = 303,15$ K

| $p$ | $\rho$ | $z$ | $h$ | $s$ | $c_p$ | $w$ | $\mu$ | $\alpha \cdot 10^3$ |
|------|--------|--------|-------|-------|-------|-------|-------|-----------|
| 0.01 | 0.28 | 0.9993 | 503.6 | 3.997 | 0.737 | 207.1 | 12.52 | 3.307 |
| 0.02 | 0.56 | 0.9986 | 503.5 | 3.914 | 0.738 | 207.0 | 12.52 | 3.316 |
| 0.03 | 0.84 | 0.9979 | 503.4 | 3.866 | 0.739 | 206.9 | 12.52 | 3.324 |
| 0.04 | 1.11 | 0.9972 | 503.3 | 3.831 | 0.740 | 206.8 | 12.52 | 3.333 |
| 0.05 | 1.39 | 0.9965 | 503.2 | 3.805 | 0.741 | 206.7 | 12.52 | 3.341 |
| 0.1 | 2.80 | 0.9929 | 502.8 | 3.721 | 0.745 | 206.2 | 12.52 | 3.385 |
| 0.2 | 5.64 | 0.9857 | 501.8 | 3.637 | 0.753 | 205.3 | 12.53 | 3.474 |
| 0.3 | 8.52 | 0.9785 | 500.9 | 3.586 | 0.761 | 204.3 | 12.54 | 3.567 |
| 0.4 | 11.44 | 0.9712 | 499.9 | 3.550 | 0.770 | 203.3 | 12.54 | 3.663 |
| 0.5 | 14.41 | 0.9638 | 499.0 | 3.521 | 0.779 | 202.4 | 12.55 | 3.763 |
| 0.6 | 17.43 | 0.9563 | 498.0 | 3.497 | 0.788 | 201.4 | 12.55 | 3.867 |
| 0.7 | 20.49 | 0.9488 | 497.0 | 3.476 | 0.797 | 200.4 | 12.56 | 3.976 |
| 0.8 | 23.61 | 0.9411 | 496.0 | 3.458 | 0.807 | 199.4 | 12.57 | 4.089 |
| 0.9 | 26.78 | 0.9334 | 495.0 | 3.441 | 0.817 | 198.3 | 12.57 | 4.207 |
| 1.0 | 30.01 | 0.9256 | 493.9 | 3.426 | 0.828 | 197.3 | 12.58 | 4.330 |
| 1.5 | 47.07 | 0.8851 | 488.5 | 3.365 | 0.887 | 191.9 | 12.62 | 5.038 |
| 2.0 | 65.99 | 0.8419 | 482.7 | 3.316 | 0.961 | 186.2 | 12.67 | 5.949 |
| 2.5 | 87.33 | 0.7952 | 476.3 | 3.273 | 1.056 | 180.0 | 12.72 | 7.169 |
| 3.0 | 111.98 | 0.7442 | 469.2 | 3.233 | 1.186 | 173.3 | 12.76 | 8.889 |
| 3.5 | 141.48 | 0.6872 | 461.0 | 3.193 | 1.379 | 166.0 | 12.77 | 11.514 |
| 4.0 | 178.79 | 0.6215 | 451.3 | 3.151 | 1.702 | 157.9 | 12.69 | 16.039 |
| 4.5 | 231.10 | 0.5409 | 438.8 | 3.101 | 2.387 | 148.8 | 12.39 | 25.844 |
| 5.0 | 326.48 | 0.4254 | 419.0 | 3.030 | 5.044 | 138.3 | 11.25 | 64.434 |
| 5.5 | 667.34 | 0.2289 | 370.7 | 2.867 | 7.851 | 159.4 | 5.02 | 90.076 |
| 6.0 | 760.77 | 0.2191 | 360.7 | 2.832 | 3.663 | 195.1 | 3.17 | 32.422 |
| 6.5 | 804.90 | 0.2243 | 356.2 | 2.815 | 2.843 | 219.5 | 2.43 | 21.614 |
| 7.0 | 835.10 | 0.2328 | 353.3 | 2.803 | 2.470 | 239.2 | 1.99 | 16.829 |
| 7.5 | 858.41 | 0.2427 | 351.2 | 2.794 | 2.251 | 256.2 | 1.69 | 14.065 |
| 8.0 | 877.57 | 0.2532 | 349.5 | 2.787 | 2.104 | 271.2 | 1.47 | 12.239 |
| 8.5 | 893.93 | 0.2641 | 348.1 | 2.780 | 1.997 | 284.8 | 1.30 | 10.930 |
| 9.0 | 908.25 | 0.2753 | 346.9 | 2.774 | 1.914 | 297.3 | 1.16 | 9.939 |
| 9.5 | 921.02 | 0.2865 | 345.8 | 2.769 | 1.849 | 308.9 | 1.04 | 9.158 |
| 10.0 | 932.57 | 0.2979 | 344.9 | 2.764 | 1.796 | 319.7 | 0.95 | 8.525 |
| 11.0 | 952.88 | 0.3207 | 343.4 | 2.756 | 1.712 | 339.7 | 0.79 | 7.553 |
| 12.0 | 970.40 | 0.3435 | 342.2 | 2.748 | 1.650 | 357.8 | 0.67 | 6.836 |
| 13.0 | 985.86 | 0.3663 | 341.2 | 2.742 | 1.601 | 374.4 | 0.57 | 6.283 |
| 14.0 | 999.74 | 0.3890 | 340.3 | 2.736 | 1.562 | 389.9 | 0.49 | 5.839 |
| 15.0 | 1012.35 | 0.4116 | 339.6 | 2.730 | 1.529 | 404.3 | 0.43 | 5.474 |
| 16.0 | 1023.92 | 0.4341 | 339.0 | 2.725 | 1.501 | 417.9 | 0.37 | 5.167 |
| 17.0 | 1034.64 | 0.4564 | 338.5 | 2.720 | 1.477 | 430.8 | 0.32 | 4.905 |
| 18.0 | 1044.62 | 0.4786 | 338.1 | 2.715 | 1.456 | 443.0 | 0.28 | 4.678 |
| 19.0 | 1053.97 | 0.5008 | 337.7 | 2.711 | 1.437 | 454.7 | 0.24 | 4.478 |
| 20.0 | 1062.78 | 0.5227 | 337.4 | 2.707 | 1.421 | 465.8 | 0.20 | 4.301 |

**TABLE 51.** Thermodynamic Properties of Freon-23 in the Single-Phase Region (*Continued*)

$T = 313.15$ K

| $p$ | $\rho$ | $z$ | $h$ | $s$ | $c_p$ | $w$ | $\mu$ | $\alpha \cdot 10^3$ |
|------|--------|--------|--------|--------|--------|--------|--------|---------|
| 0.01 | 0.27 | 0.9994 | 511.1 | 4.022 | 0.752 | 210.1 | 11.31 | 3.201 |
| 0.02 | 0.54 | 0.9987 | 511.0 | 3.939 | 0.752 | 210.0 | 11.31 | 3.208 |
| 0.03 | 0.81 | 0.9981 | 510.9 | 3.891 | 0.753 | 209.9 | 11.31 | 3.215 |
| 0.04 | 1.08 | 0.9975 | 510.9 | 3.856 | 0.754 | 209.8 | 11.31 | 3.223 |
| 0.05 | 1.35 | 0.9969 | 510.8 | 3.830 | 0.754 | 209.7 | 11.31 | 3.230 |
| 0.1 | 2.71 | 0.9937 | 510.3 | 3.746 | 0.758 | 209.3 | 11.31 | 3.267 |
| 0.2 | 5.45 | 0.9873 | 509.5 | 3.662 | 0.765 | 208.5 | 11.32 | 3.344 |
| 0.3 | 8.22 | 0.9809 | 508.6 | 3.612 | 0.772 | 207.6 | 11.32 | 3.423 |
| 0.4 | 11.04 | 0.9745 | 507.7 | 3.576 | 0.780 | 206.7 | 11.32 | 3.505 |
| 0.5 | 13.89 | 0.9679 | 506.8 | 3.547 | 0.788 | 205.8 | 11.33 | 3.589 |
| 0.6 | 16.78 | 0.9614 | 505.9 | 3.523 | 0.796 | 205.0 | 11.33 | 3.677 |
| 0.7 | 19.72 | 0.9547 | 505.0 | 3.503 | 0.804 | 204.1 | 11.34 | 3.767 |
| 0.8 | 22.69 | 0.9480 | 504.1 | 3.485 | 0.812 | 203.1 | 11.34 | 3.861 |
| 0.9 | 25.71 | 0.9412 | 503.2 | 3.469 | 0.821 | 202.2 | 11.35 | 3.959 |
| 1.0 | 28.78 | 0.9344 | 502.3 | 3.454 | 0.830 | 201.3 | 11.36 | 4.060 |
| 1.5 | 44.86 | 0.8992 | 497.4 | 3.394 | 0.880 | 196.6 | 11.38 | 4.627 |
| 2.0 | 62.38 | 0.8622 | 492.2 | 3.348 | 0.939 | 191.6 | 11.41 | 5.327 |
| 2.5 | 81.70 | 0.8228 | 486.7 | 3.308 | 1.011 | 186.3 | 11.43 | 6.208 |
| 3.0 | 103.31 | 0.7809 | 480.6 | 3.271 | 1.103 | 180.8 | 11.43 | 7.352 |
| 3.5 | 127.93 | 0.7357 | 474.0 | 3.236 | 1.223 | 175.0 | 11.41 | 8.895 |
| 4.0 | 156.68 | 0.6865 | 466.6 | 3.201 | 1.390 | 168.9 | 11.34 | 11.080 |
| 4.5 | 191.45 | 0.6321 | 458.1 | 3.165 | 1.637 | 162.5 | 11.18 | 14.385 |
| 5.0 | 235.67 | 0.5705 | 448.1 | 3.125 | 2.041 | 155.9 | 10.84 | 19.853 |
| 5.5 | 296.28 | 0.4992 | 435.6 | 3.079 | 2.785 | 149.6 | 10.16 | 29.952 |
| 6.0 | 387.62 | 0.4162 | 419.3 | 3.022 | 4.222 | 145.5 | 8.80 | 49.169 |
| 6.5 | 518.99 | 0.3368 | 400.0 | 2.957 | 5.477 | 150.2 | 6.62 | 63.309 |
| 7.0 | 632.71 | 0.2975 | 385.9 | 2.909 | 4.346 | 168.6 | 4.66 | 44.075 |
| 7.5 | 701.29 | 0.2876 | 378.3 | 2.883 | 3.332 | 190.0 | 3.48 | 29.170 |
| 8.0 | 746.15 | 0.2883 | 373.6 | 2.865 | 2.795 | 209.4 | 2.77 | 21.665 |
| 8.5 | 779.05 | 0.2934 | 370.2 | 2.853 | 2.402 | 226.7 | 2.30 | 17.420 |
| 9.0 | 805.00 | 0.3006 | 367.7 | 2.843 | 2.277 | 242.2 | 1.97 | 14.721 |
| 9.5 | 826.47 | 0.3091 | 365.7 | 2.834 | 2.133 | 256.3 | 1.72 | 12.856 |
| 10.0 | 844.82 | 0.3183 | 364.0 | 2.827 | 2.026 | 269.3 | 1.52 | 11.486 |
| 11.0 | 875.12 | 0.3380 | 361.4 | 2.815 | 1.875 | 292.7 | 1.22 | 9.601 |
| 12.0 | 899.79 | 0.3586 | 359.3 | 2.805 | 1.773 | 313.5 | 1.01 | 8.356 |
| 13.0 | 920.63 | 0.3797 | 357.7 | 2.796 | 1.698 | 332.2 | 0.86 | 7.464 |
| 14.0 | 938.73 | 0.4010 | 356.4 | 2.788 | 1.641 | 349.4 | 0.73 | 6.791 |
| 15.0 | 954.79 | 0.4225 | 355.3 | 2.781 | 1.595 | 365.3 | 0.63 | 6.261 |
| 16.0 | 969.23 | 0.4439 | 354.4 | 2.775 | 1.557 | 380.2 | 0.55 | 5.831 |
| 17.0 | 982.39 | 0.4653 | 353.6 | 2.769 | 1.526 | 394.1 | 0.48 | 5.474 |
| 18.0 | 994.48 | 0.4867 | 352.9 | 2.764 | 1.499 | 407.3 | 0.42 | 5.172 |
| 19.0 | 1005.67 | 0.5080 | 352.3 | 2.759 | 1.475 | 419.9 | 0.36 | 4.913 |
| 20.0 | 1016.11 | 0.5293 | 351.8 | 2.754 | 1.455 | 431.8 | 0.32 | 4.688 |

**TABLE 51.** Thermodynamic Properties of Freon-23 in the Single-Phase Region (*Continued*)

$T=323,15$ K

| $p$ | $\rho$ | $z$ | $h$ | $s$ | $c_p$ | $w$ | $\mu$ | $a \cdot 10^9$ |
|---|---|---|---|---|---|---|---|---|
| 0.01 | 0.26 | 0.9994 | 518.7 | 4.046 | 0,766 | 213,0 | 10.24 | 3.101 |
| 0.02 | 0.52 | 0.9989 | 518,7 | 3,964 | 0,767 | 213,0 | 10.24 | 3.107 |
| 0.03 | 0.78 | 0.9983 | 518.6 | 3.915 | 0,767 | 212,9 | 10.24 | 3.114 |
| 0.04 | 1.04 | 0.9978 | 518.5 | 3.881 | 0.768 | 212.8 | 10.25 | 3.120 |
| 0.05 | 1.31 | 0.9972 | 518.4 | 3.854 | 0.768 | 212.7 | 10.25 | 3.126 |
| 0.1 | 2,62 | 0.9944 | 518.0 | 3.771 | 0,772 | 212.4 | 10.25 | 3.159 |
| 0.2 | 5.27 | 0,9887 | 517.2 | 3.687 | 0.778 | 211,6 | 10.25 | 3.225 |
| 0.3 | 7.95 | 0,9830 | 516.4 | 3.637 | 0.784 | 210,8 | 10.26 | 3.292 |
| 0,4 | 10,67 | 0.9773 | 515.6 | 3,601 | 0.791 | 210.0 | 10.26 | 3.362 |
| 0,5 | 13.41 | 0,9715 | 514.8 | 3.573 | 0.798 | 209.2 | 10.26 | 3,434 |
| 0.6 | 16,19 | 0,9657 | 514.0 | 3.549 | 0.805 | 208,4 | 10.27 | 3.509 |
| 0.7 | 19.01 | 0,9598 | 513,2 | 3,529 | 0.812 | 207.6 | 10.27 | 3.585 |
| 0,8 | 21.86 | 0.9539 | 512.3 | 3.511 | 0.819 | 206.8 | 10.28 | 3.664 |
| 0.9 | 24.74 | 0.9479 | 511.5 | 3.495 | 0.827 | 206.0 | 10.28 | 3.745 |
| 1.0 | 27.67 | 0.9419 | 510.6 | 3.481 | 0.834 | 205.1 | 10.28 | 3.829 |
| 1.5 | 42.90 | 0.9111 | 506.2 | 3.422 | 0.876 | 200.9 | 10.30 | 4.293 |
| 2.0 | 59.29 | 0.8790 | 501.6 | 3.377 | 0.925 | 196,6 | 10.31 | 4.845 |
| 2.5 | 77.06 | 0.8454 | 496.7 | 3.339 | 0.982 | 192.1 | 10.32 | 5.511 |
| 3.0 | 96.51 | 0.8101 | 491.4 | 3.305 | 1.051 | 187.4 | 10.31 | 6.330 |
| 3.5 | 118.01 | 0.7729 | 485.8 | 3.273 | 1.136 | 182,6 | 10.28 | 7.356 |
| 4.0 | 142.11 | 0.7335 | 479.7 | 3.243 | 1.242 | 177.6 | 10.21 | 8.671 |
| 4,5 | 169,56 | 0,6916 | 473,1 | 3.212 | 1.380 | 172.6 | 10.09 | 10.402 |
| 5,0 | 201.43 | 0.6468 | 465.7 | 3,181 | 1.564 | 167.7 | 9.89 | 12.742 |
| 5.5 | 239.30 | 0.5989 | 457.5 | 3.149 | 1,818 | 162.9 | 9.57 | 15.982 |
| 6.0 | 285.36 | 0,5479 | 448.3 | 3.114 | 2.172 | 158.7 | 9,07 | 20.487 |
| 6.5 | 342,28 | 0.4949 | 437.8 | 3,077 | 2.644 | 155.9 | 8.31 | 26.374 |
| 7.0 | 411,33 | 0.4435 | 426.5 | 3,037 | 3.156 | 155.8 | 7.30 | 32.403 |
| 7.5 | 488,07 | 0.4004 | 415.3 | 2.999 | 3.462 | 159,9 | 6,12 | 35.091 |
| 8.0 | 561.17 | 0.3715 | 405.7 | 2.967 | 3.382 | 169,2 | 4.99 | 32.395 |
| 8,5 | 621,80 | 0.3562 | 398.4 | 2.942 | 3.080 | 182.1 | 4.06 | 27.128 |
| 9.0 | 669,33 | 0.3504 | 393.1 | 2.923 | 2.771 | 196.4 | 3.35 | 22.296 |
| 9.5 | 706.80 | 0.3503 | 389,0 | 2.908 | 2.524 | 210.8 | 2.82 | 18.645 |
| 10.0 | 737,16 | 0.3535 | 385.8 | 2.896 | 2.337 | 224.7 | 2.42 | 15.979 |
| 11,0 | 784,09 | 0.3656 | 381,2 | 2.877 | 2.083 | 250.2 | 1.86 | 12,501 |
| 12.0 | 819.63 | 0,3815 | 377.8 | 2.863 | 1.923 | 273.9 | 1.50 | 10.385 |
| 13.0 | 848,20 | 0.3994 | 375.3 | 2.852 | 1.812 | 293,5 | 1.24 | 8.973 |
| 14.0 | 872.13 | 0.4183 | 373.3 | 2.842 | 1.731 | 312.2 | 1,04 | 7.963 |
| 15.0 | 892.74 | 0.4378 | 371.6 | 2.833 | 1.669 | 329.5 | 0.89 | 7.203 |
| 16.0 | 910.87 | 0.4577 | 370.3 | 2.826 | 1.619 | 345.5 | 0.77 | 6.608 |
| 17,0 | 927.08 | 0.4778 | 369.1 | 2.819 | 1.578 | 360.5 | 0.67 | 6.128 |
| 18.0 | 941.75 | 0.4981 | 368,2 | 2.812 | 1.544 | 374.6 | 0.59 | 5.733 |
| 19.0 | 955.16 | 0.5184 | 367.3 | 2.806 | 1.515 | 387.9 | 0.52 | 5.399 |
| 20.0 | 967.54 | 0.5387 | 366.6 | 2,801 | 1,490 | 400.6 | 0.45 | 5.115 |

**TABLE 51.** Thermodynamic Properties of Freon-23 in the Single-Phase Region (*Continued*)

$T = 333.15$ K

| $p$ | $\rho$ | $z$ | $h$ | $s$ | $c_p$ | $w$ | $\mu$ | $\alpha \cdot 10^3$ |
|---|---|---|---|---|---|---|---|---|
| 0.01 | 0.25 | 0.9995 | 526.5 | 4.070 | 0.780 | 216.0 | 9.31 | 3.007 |
| 0.02 | 0.51 | 0.9990 | 526.4 | 3.987 | 0.781 | 215.9 | 9.31 | 3.013 |
| 0.03 | 0.76 | 0.9985 | 526.3 | 3.939 | 0.781 | 215.8 | 9.31 | 3.018 |
| 0.04 | 1.01 | 0.9980 | 526.3 | 3.905 | 0.782 | 215.8 | 9.31 | 3.024 |
| 0.05 | 1.27 | 0.9975 | 526.2 | 3.878 | 0.782 | 215.7 | 9.31 | 3.029 |
| 0.1 | 2.54 | 0.9950 | 525.8 | 3.795 | 0.785 | 215.3 | 9.32 | 3.057 |
| 0.2 | 5.11 | 0.9899 | 525.1 | 3.711 | 0.791 | 214.6 | 9.32 | 3.115 |
| 0.3 | 7.70 | 0.9848 | 524.4 | 3.661 | 0.797 | 213.9 | 9.32 | 3.173 |
| 0.4 | 10.32 | 0.9797 | 523.6 | 3.625 | 0.802 | 213.2 | 9.32 | 3.233 |
| 0.5 | 12.97 | 0.9746 | 522.9 | 3.597 | 0.808 | 212.5 | 9.33 | 3.295 |
| 0.6 | 15.64 | 0.9694 | 522.1 | 3.574 | 0.815 | 211.8 | 9.33 | 3.359 |
| 0.7 | 18.35 | 0.9642 | 521.3 | 3.554 | 0.821 | 211.0 | 9.34 | 3.424 |
| 0.8 | 21.09 | 0.9589 | 520.6 | 3.536 | 0.827 | 210.3 | 9.34 | 3.491 |
| 0.9 | 23.85 | 0.9536 | 519.8 | 3.521 | 0.834 | 209.6 | 9.34 | 3.559 |
| 1.0 | 26.65 | 0.9483 | 519.0 | 3.506 | 0.840 | 208.8 | 9.34 | 3.630 |
| 1.5 | 41.16 | 0.9212 | 515.0 | 3.449 | 0.876 | 205.1 | 9.35 | 4.014 |
| 2.0 | 56.60 | 0.8931 | 510.8 | 3.406 | 0.917 | 201.2 | 9.36 | 4.460 |
| 2.5 | 73.13 | 0.8641 | 506.4 | 3.369 | 0.964 | 197.3 | 9.35 | 4.981 |
| 3.0 | 90.94 | 0.8339 | 501.8 | 3.337 | 1.018 | 193.3 | 9.34 | 5.596 |
| 3.5 | 110.24 | 0.8025 | 496.9 | 3.307 | 1.081 | 189.2 | 9.30 | 6.330 |
| 4.0 | 131.33 | 0.7699 | 491.7 | 3.279 | 1.157 | 185.1 | 9.24 | 7.217 |
| 4.5 | 154.57 | 0.7359 | 486.2 | 3.252 | 1.248 | 181.0 | 9.15 | 8.299 |
| 5.0 | 180.41 | 0.7005 | 480.3 | 3.225 | 1.359 | 177.0 | 9.01 | 9.631 |
| 5.5 | 209.44 | 0.6638 | 473.9 | 3.199 | 1.496 | 173.1 | 8.80 | 11.280 |
| 6.0 | 242.35 | 0.6258 | 467.1 | 3.171 | 1.665 | 169.7 | 8.51 | 13.311 |
| 6.5 | 279.87 | 0.5871 | 459.7 | 3.144 | 1.870 | 166.9 | 8.12 | 15.751 |
| 7.0 | 322.61 | 0.5485 | 451.9 | 3.115 | 2.107 | 165.0 | 7.60 | 18.504 |
| 7.5 | 370.60 | 0.5115 | 443.8 | 3.086 | 2.352 | 164.6 | 6.97 | 21.221 |
| 8.0 | 422.67 | 0.4784 | 435.7 | 3.058 | 2.560 | 166.2 | 6.24 | 23.272 |
| 8.5 | 476.19 | 0.4512 | 428.0 | 3.032 | 2.680 | 170.2 | 5.49 | 24.015 |
| 9.0 | 527.76 | 0.4310 | 421.1 | 3.008 | 2.691 | 176.8 | 4.76 | 23.285 |
| 9.5 | 574.65 | 0.4179 | 415.2 | 2.988 | 2.616 | 185.4 | 4.10 | 21.520 |
| 10.0 | 615.68 | 0.4105 | 410.3 | 2.971 | 2.499 | 195.6 | 3.54 | 19.362 |
| 11.0 | 681.30 | 0.4081 | 403.0 | 2.944 | 2.255 | 217.6 | 2.70 | 15.428 |
| 12.0 | 730.49 | 0.4152 | 397.8 | 2.924 | 2.063 | 239.7 | 2.12 | 12.601 |
| 13.0 | 768.92 | 0.4274 | 394.0 | 2.909 | 1.924 | 260.5 | 1.72 | 10.650 |
| 14.0 | 800.17 | 0.4422 | 391.1 | 2.896 | 1.821 | 279.8 | 1.43 | 9.264 |
| 15.0 | 826.42 | 0.4588 | 388.7 | 2.885 | 1.742 | 297.8 | 1.21 | 8.240 |
| 16.0 | 849.04 | 0.4763 | 386.8 | 2.876 | 1.681 | 314.6 | 1.04 | 7.455 |
| 17.0 | 868.91 | 0.4945 | 385.2 | 2.868 | 1.631 | 330.3 | 0.90 | 6.835 |
| 18.0 | 886.63 | 0.5132 | 383.8 | 2.860 | 1.589 | 345.1 | 0.79 | 6.332 |
| 19.0 | 902.64 | 0.5321 | 382.7 | 2.853 | 1.555 | 359.0 | 0.69 | 5.916 |
| 20.0 | 917.23 | 0.5511 | 381.7 | 2.847 | 1.525 | 372.3 | 0.61 | 5.565 |

**TABLE 51.** Thermodynamic Properties of Freon-23 in the Single-Phase Region (*Continued*)

$$T=343.15 \ K$$

| $p$ | $\rho$ | $z$ | $h$ | $s$ | $c_p$ | $w$ | $\mu$ | $\alpha \cdot 10^3$ |
|------|--------|--------|-------|-------|-------|-------|------|----------|
| 0.01 | 0.25 | 0.9996 | 534.4 | 4.093 | 0.794 | 218.8 | 8.49 | 2.919 |
| 0.02 | 0.49 | 0.9991 | 534.3 | 4.011 | 0.795 | 218.8 | 8.49 | 2.924 |
| 0.03 | 0.74 | 0.9987 | 534.2 | 3.962 | 0.795 | 218.7 | 8.49 | 2.929 |
| 0.04 | 0.98 | 0.9982 | 534.2 | 3.928 | 0.796 | 218.6 | 8.49 | 2.934 |
| 0.05 | 1.23 | 0.9978 | 534.1 | 3.901 | 0.796 | 218.6 | 8.49 | 2.938 |
| 0.1 | 2.47 | 0.9955 | 533.8 | 3.818 | 0.799 | 218.3 | 8.50 | 2.963 |
| 0.2 | 4.95 | 0.9910 | 533.1 | 3.735 | 0.804 | 217.6 | 8.50 | 3.013 |
| 0.3 | 7.46 | 0.9864 | 532.4 | 3.685 | 0.809 | 217.0 | 8.50 | 3.064 |
| 0.4 | 10.00 | 0.9818 | 531.7 | 3.649 | 0.814 | 216.3 | 8.50 | 3.116 |
| 0.5 | 12.56 | 0.9772 | 531.0 | 3.621 | 0.820 | 215.7 | 8.50 | 3.169 |
| 0.6 | 15.14 | 0.9726 | 530.3 | 3.598 | 0.825 | 215.0 | 8.51 | 3.224 |
| 0.7 | 17.75 | 0.9680 | 529.6 | 3.578 | 0.830 | 214.4 | 8.51 | 3.280 |
| 0.8 | 20.38 | 0.9633 | 528.9 | 3.561 | 0.836 | 213.7 | 8.51 | 3.337 |
| 0.9 | 23.04 | 0.9586 | 528.2 | 3.545 | 0.842 | 213.0 | 8.51 | 3.395 |
| 1.0 | 25.73 | 0.9539 | 527.5 | 3.531 | 0.848 | 212.4 | 8.52 | 3.455 |
| 1.5 | 39.59 | 0.9299 | 523.8 | 3.475 | 0.879 | 209.0 | 8.52 | 3.778 |
| 2.0 | 54.22 | 0.9052 | 520.0 | 3.433 | 0.914 | 205.6 | 8.52 | 4.144 |
| 2.5 | 69.73 | 0.8798 | 516.0 | 3.398 | 0.952 | 202.1 | 8.51 | 4.562 |
| 3.0 | 86.24 | 0.8537 | 511.8 | 3.367 | 0.996 | 198.6 | 8.49 | 5.040 |
| 3.5 | 103.88 | 0.8268 | 507.5 | 3.339 | 1.046 | 195.1 | 8.46 | 5.593 |
| 4.0 | 122.82 | 0.7992 | 503.0 | 3.313 | 1.103 | 191.6 | 8.40 | 6.233 |
| 4.5 | 143.26 | 0.7708 | 498.2 | 3.288 | 1.169 | 188.2 | 8.33 | 6.978 |
| 5.0 | 165.43 | 0.7417 | 493.3 | 3.264 | 1.246 | 184.9 | 8.22 | 7.848 |
| 5.5 | 189.59 | 0.7119 | 488.0 | 3.240 | 1.334 | 181.7 | 8.07 | 8.861 |
| 6.0 | 216.01 | 0.6816 | 482.5 | 3.217 | 1.437 | 178.8 | 7.87 | 10.034 |
| 6.5 | 245.00 | 0.6511 | 476.7 | 3.194 | 1.555 | 176.3 | 7.62 | 11.370 |
| 7.0 | 276.77 | 0.6207 | 470.7 | 3.171 | 1.688 | 174.4 | 7.30 | 12.846 |
| 7.5 | 311.44 | 0.5910 | 464.4 | 3.147 | 1.830 | 173.2 | 6.91 | 14.390 |
| 8.0 | 348.84 | 0.5628 | 458.0 | 3.124 | 1.973 | 173.0 | 6.46 | 15.873 |
| 8.5 | 388.43 | 0.5370 | 451.7 | 3.102 | 2.103 | 174.0 | 5.96 | 17.108 |
| 9.0 | 429.24 | 0.5145 | 445.6 | 3.081 | 2.204 | 176.4 | 5.44 | 17.905 |
| 9.5 | 470.01 | 0.4960 | 439.8 | 3.060 | 2.266 | 180.3 | 4.91 | 18.144 |
| 10.0 | 509.44 | 0.4817 | 434.5 | 3.042 | 2.285 | 185.7 | 4.40 | 17.832 |
| 11.0 | 580.50 | 0.4650 | 425.6 | 3.011 | 2.225 | 200.1 | 3.50 | 16.081 |
| 12.0 | 639.05 | 0.4608 | 418.8 | 2.986 | 2.106 | 217.5 | 2.80 | 13.879 |
| 13.0 | 686.37 | 0.4648 | 413.6 | 2.967 | 1.986 | 235.9 | 2.27 | 11.932 |
| 14.0 | 725.05 | 0.4738 | 409.6 | 2.951 | 1.883 | 254.1 | 1.88 | 10.387 |
| 15.0 | 757.31 | 0.4861 | 406.5 | 2.938 | 1.799 | 271.7 | 1.58 | 9.190 |
| 16.0 | 784.81 | 0.5003 | 403.9 | 2.927 | 1.731 | 288.4 | 1.35 | 8.256 |
| 17.0 | 808.70 | 0.5159 | 401.7 | 2.917 | 1.676 | 304.3 | 1.16 | 7.515 |
| 18.0 | 829.77 | 0.5323 | 400.0 | 2.908 | 1.629 | 319.3 | 1.02 | 6.915 |
| 19.0 | 848.61 | 0.5494 | 398.4 | 2.900 | 1.590 | 333.6 | 0.89 | 6.420 |
| 20.0 | 865.65 | 0.5670 | 397.1 | 2.893 | 1.557 | 347.2 | 0.79 | 6.006 |

**TABLE 51.** Thermodynamic Properties of Freon-23 in the Single-Phase Region (*Continued*)

$T = 353.15$ K

| $p$ | $\rho$ | $z$ | $h$ | $s$ | $c_p$ | $w$ | $\mu$ | $\alpha \cdot 10^3$ |
|------|--------|--------|-------|-------|-------|-------|------|---------|
| 0.01 | 0.24 | 0.9996 | 542.4 | 4.116 | 0.808 | 221.7 | 7.77 | 2.836 |
| 0.02 | 0.48 | 0.9992 | 542.3 | 4.033 | 0.809 | 221.6 | 7.77 | 2.840 |
| 0.03 | 0.72 | 0.9988 | 542.2 | 3.985 | 0.809 | 221.6 | 7.77 | 2.844 |
| 0.04 | 0.96 | 0.9984 | 542.2 | 3.951 | 0.810 | 221.5 | 7.77 | 2.849 |
| 0.05 | 1.19 | 0.9980 | 542.1 | 3.924 | 0.810 | 221.4 | 7.77 | 2.853 |
| 0.1 | 2.39 | 0.9960 | 541.8 | 3.841 | 0.812 | 221.1 | 7.77 | 2.874 |
| 0.2 | 4.81 | 0.9919 | 541.2 | 3.758 | 0.817 | 220.6 | 7.77 | 2.918 |
| 0.3 | 7.24 | 0.9878 | 540.5 | 3.708 | 0.822 | 220.0 | 7.77 | 2.963 |
| 0.4 | 9.70 | 0.9837 | 539.9 | 3.673 | 0.826 | 219.4 | 7.78 | 3.008 |
| 0.5 | 12.17 | 0.9796 | 539.2 | 3.645 | 0.831 | 218.8 | 7.78 | 3.054 |
| 0.6 | 14.67 | 0.9754 | 538.6 | 3.622 | 0.836 | 218.2 | 7.78 | 3.102 |
| 0.7 | 17.19 | 0.9713 | 537.9 | 3.602 | 0.841 | 217.6 | 7.78 | 3.150 |
| 0.8 | 19.73 | 0.9673 | 537.3 | 3.585 | 0.846 | 217.0 | 7.78 | 3.199 |
| 0.9 | 22.29 | 0.9629 | 536.6 | 3.570 | 0.851 | 216.4 | 7.78 | 3.250 |
| 1.0 | 24.87 | 0.9587 | 536.0 | 3.556 | 0.856 | 215.8 | 7.79 | 3.301 |
| 1.5 | 38.16 | 0.9373 | 532.6 | 3.500 | 0.883 | 212.8 | 7.79 | 3.575 |
| 2.0 | 52.09 | 0.9155 | 529.1 | 3.459 | 0.913 | 209.7 | 7.79 | 3.880 |
| 2.5 | 66.74 | 0.8932 | 525.5 | 3.425 | 0.946 | 206.7 | 7.77 | 4.221 |
| 3.0 | 82.18 | 0.8704 | 521.7 | 3.395 | 0.982 | 203.6 | 7.75 | 4.604 |
| 3.5 | 98.51 | 0.8472 | 517.9 | 3.368 | 1.023 | 200.6 | 7.72 | 5.034 |
| 4.0 | 115.83 | 0.8234 | 513.8 | 3.344 | 1.068 | 197.5 | 7.67 | 5.518 |
| 4.5 | 134.25 | 0.7993 | 509.7 | 3.320 | 1.118 | 194.6 | 7.60 | 6.065 |
| 5.0 | 153.90 | 0.7747 | 505.3 | 3.298 | 1.175 | 191.8 | 7.52 | 6.680 |
| 5.5 | 174.90 | 0.7498 | 500.8 | 3.277 | 1.238 | 189.1 | 7.40 | 7.370 |
| 6.0 | 197.40 | 0.7248 | 496.2 | 3.256 | 1.309 | 186.7 | 7.25 | 8.140 |
| 6.5 | 221.51 | 0.6997 | 491.3 | 3.236 | 1.388 | 184.5 | 7.07 | 8.988 |
| 7.0 | 247.35 | 0.6748 | 486.4 | 3.216 | 1.474 | 182.7 | 6.85 | 9.903 |
| 7.5 | 274.96 | 0.6504 | 481.2 | 3.196 | 1.565 | 181.5 | 6.59 | 10.862 |
| 8.0 | 304.31 | 0.6269 | 476.1 | 3.176 | 1.660 | 180.8 | 6.28 | 11.824 |
| 8.5 | 335.22 | 0.6046 | 470.8 | 3.157 | 1.754 | 180.8 | 5.95 | 12.731 |
| 9.0 | 367.39 | 0.5841 | 465.7 | 3.138 | 1.841 | 181.7 | 5.58 | 13.513 |
| 9.5 | 400.35 | 0.5658 | 460.6 | 3.120 | 1.916 | 183.5 | 5.19 | 14.100 |
| 10.0 | 433.51 | 0.5500 | 455.7 | 3.103 | 1.973 | 186.3 | 4.79 | 14.441 |
| 11.0 | 498.01 | 0.5267 | 446.9 | 3.072 | 2.028 | 194.9 | 4.02 | 14.339 |
| 12.0 | 556.75 | 0.5139 | 439.5 | 3.045 | 2.012 | 206.8 | 3.34 | 13.431 |
| 13.0 | 607.85 | 0.5100 | 433.4 | 3.023 | 1.956 | 221.1 | 2.77 | 12.170 |
| 14.0 | 651.42 | 0.5125 | 428.5 | 3.005 | 1.886 | 236.5 | 2.32 | 10.901 |
| 15.0 | 688.51 | 0.5195 | 424.6 | 2.990 | 1.818 | 252.3 | 1.96 | 9.776 |
| 16.0 | 720.36 | 0.5296 | 421.3 | 2.977 | 1.756 | 268.0 | 1.68 | 8.829 |
| 17.0 | 748.03 | 0.5419 | 418.7 | 2.965 | 1.702 | 283.2 | 1.45 | 8.044 |
| 18.0 | 772.39 | 0.5557 | 416.4 | 2.955 | 1.656 | 297.9 | 1.26 | 7.393 |
| 19.0 | 794.07 | 0.5706 | 414.5 | 2.946 | 1.616 | 312.1 | 1.11 | 6.849 |
| 20.0 | 813.57 | 0.5862 | 412.8 | 2.938 | 1.581 | 325.7 | 0.98 | 6.390 |

**TABLE 51.** Thermodynamic Properties of Freon-23 in the Single-Phase Region (*Continued*)

$T=363.15$ K

| $p$ | $\rho$ | $z$ | $h$ | $s$ | $c_p$ | $w$ | $\mu$ | $\alpha \cdot 10^3$ |
|---|---|---|---|---|---|---|---|---|
| 0.01 | 0.23 | 0.9996 | 550.5 | 4.138 | 0.822 | 224.5 | 7.13 | 2.757 |
| 0.02 | 0.46 | 0.9993 | 550.5 | 4.056 | 0.823 | 224.4 | 7.13 | 2.761 |
| 0.03 | 0.70 | 0.9989 | 550.4 | 4.008 | 0.823 | 224.4 | 7.13 | 2.765 |
| 0.04 | 0.93 | 0.9985 | 550.3 | 3.973 | 0.824 | 224.3 | 7.13 | 2.769 |
| 0.05 | 1.16 | 0.9982 | 550.3 | 3.947 | 0.824 | 224.3 | 7.13 | 2.772 |
| 0.1 | 2.33 | 0.9964 | 550.0 | 3.864 | 0.826 | 224.0 | 7.13 | 2.791 |
| 0.2 | 4.67 | 0.9927 | 549.4 | 3.780 | 0.830 | 223.4 | 7.13 | 2.830 |
| 0.3 | 7.03 | 0.9890 | 548.8 | 3.731 | 0.834 | 222.9 | 7.13 | 2.869 |
| 0.4 | 9.41 | 0.9853 | 548.2 | 3.696 | 0.838 | 222.4 | 7.13 | 2.909 |
| 0.5 | 11.81 | 0.9816 | 547.6 | 3.668 | 0.843 | 221.8 | 7.14 | 2.949 |
| 0.6 | 14.23 | 0.9779 | 547.0 | 3.645 | 0.847 | 221.3 | 7.14 | 2.990 |
| 0.7 | 16.66 | 0.9742 | 546.4 | 3.626 | 0.851 | 220.8 | 7.14 | 3.032 |
| 0.8 | 19.12 | 0.9704 | 545.8 | 3.609 | 0.856 | 220.2 | 7.14 | 3.075 |
| 0.9 | 21.59 | 0.9667 | 545.2 | 3.593 | 0.860 | 219.7 | 7.14 | 3.119 |
| 1.0 | 24.08 | 0.9629 | 544.6 | 3.580 | 0.865 | 219.1 | 7.14 | 3.163 |
| 1.5 | 36.85 | 0.9439 | 541.4 | 3.525 | 0.889 | 216.4 | 7.14 | 3.398 |
| 2.0 | 50.17 | 0.9245 | 538.2 | 3.484 | 0.915 | 213.7 | 7.14 | 3.655 |
| 2.5 | 64.07 | 0.9048 | 534.9 | 3.451 | 0.943 | 211.0 | 7.12 | 3.939 |
| 3.0 | 78.63 | 0.8847 | 531.5 | 3.422 | 0.974 | 208.3 | 7.10 | 4.251 |
| 3.5 | 93.89 | 0.8644 | 528.0 | 3.396 | 1.007 | 205.6 | 7.07 | 4.595 |
| 4.0 | 109.93 | 0.8438 | 524.4 | 3.373 | 1.044 | 203.0 | 7.03 | 4.974 |
| 4.5 | 126.80 | 0.8230 | 520.6 | 3.351 | 1.084 | 200.4 | 6.97 | 5.392 |
| 5.0 | 144.58 | 0.8019 | 516.8 | 3.330 | 1.128 | 198.0 | 6.90 | 5.851 |
| 5.5 | 163.35 | 0.7808 | 512.9 | 3.310 | 1.176 | 195.7 | 6.80 | 6.353 |
| 6.0 | 183.17 | 0.7596 | 508.8 | 3.291 | 1.229 | 193.6 | 6.69 | 6.899 |
| 6.5 | 204.10 | 0.7385 | 504.7 | 3.273 | 1.286 | 191.8 | 6.55 | 7.487 |
| 7.0 | 226.20 | 0.7176 | 500.4 | 3.255 | 1.347 | 190.2 | 6.38 | 8.110 |
| 7.5 | 249.47 | 0.6971 | 496.1 | 3.237 | 1.412 | 189.0 | 6.19 | 8.758 |
| 8.0 | 273.90 | 0.6773 | 491.7 | 3.219 | 1.479 | 188.2 | 5.97 | 9.414 |
| 8.5 | 299.43 | 0.6583 | 487.2 | 3.202 | 1.546 | 188.0 | 5.73 | 10.056 |
| 9.0 | 325.90 | 0.6404 | 482.8 | 3.186 | 1.612 | 188.2 | 5.46 | 10.654 |
| 9.5 | 353.14 | 0.6238 | 478.5 | 3.170 | 1.674 | 189.1 | 5.17 | 11.178 |
| 10.0 | 380.86 | 0.6088 | 474.2 | 3.154 | 1.730 | 190.7 | 4.88 | 11.598 |
| 11.0 | 436.50 | 0.5844 | 466.1 | 3.125 | 1.813 | 196.0 | 4.26 | 12.044 |
| 12.0 | 490.22 | 0.5676 | 458.8 | 3.099 | 1.852 | 204.0 | 3.68 | 11.940 |
| 13.0 | 539.92 | 0.5583 | 452.5 | 3.076 | 1.852 | 214.4 | 3.14 | 11.407 |
| 14.0 | 584.55 | 0.5554 | 447.1 | 3.057 | 1.827 | 226.6 | 2.68 | 10.642 |
| 15.0 | 623.98 | 0.5574 | 442.6 | 3.040 | 1.788 | 239.9 | 2.30 | 9.813 |
| 16.0 | 658.63 | 0.5633 | 438.9 | 3.025 | 1.745 | 253.6 | 1.98 | 9.018 |
| 17.0 | 689.15 | 0.5720 | 435.7 | 3.012 | 1.702 | 267.5 | 1.72 | 8.302 |
| 18.0 | 716.19 | 0.5828 | 433.0 | 3.001 | 1.662 | 281.3 | 1.50 | 7.675 |
| 19.0 | 740.34 | 0.5951 | 430.7 | 2.991 | 1.626 | 294.8 | 1.32 | 7.133 |
| 20.0 | 762.08 | 0.6086 | 428.7 | 2.982 | 1.593 | 308.0 | 1.17 | 6.664 |

**TABLE 51.** Thermodynamic Properties of Freon-23 in the Single-Phase Region (*Continued*)

$T = 373.15$ K

| $p$ | $\rho$ | $z$ | $h$ | $s$ | $c_p$ | $w$ | $\mu$ | $\alpha \cdot 10^3$ |
|---|---|---|---|---|---|---|---|---|
| 0.01 | 0.23 | 0.9997 | 558.8 | 4.161 | 0.836 | 227.2 | 6.56 | 2.683 |
| 0.02 | 0.45 | 0.9993 | 558.7 | 4.078 | 0.836 | 227.2 | 6.56 | 2.687 |
| 0.03 | 0.68 | 0.9990 | 558.7 | 4.030 | 0.837 | 227.1 | 6.56 | 2.690 |
| 0.04 | 0.90 | 0.9987 | 558.6 | 3.996 | 0.837 | 227.1 | 6.56 | 2.693 |
| 0.05 | 1.13 | 0.9984 | 558.6 | 3.969 | 0.838 | 227.0 | 6.56 | 2.697 |
| 0.1 | 2.26 | 0.9967 | 558.3 | 3.886 | 0.839 | 226.8 | 6.56 | 2.713 |
| 0.2 | 4.54 | 0.9934 | 557.7 | 3.803 | 0.843 | 226.3 | 6.56 | 2.747 |
| 0.3 | 6.84 | 0.9901 | 557.2 | 3.754 | 0.847 | 225.8 | 6.56 | 2.782 |
| 0.4 | 9.15 | 0.9868 | 556.6 | 3.718 | 0.850 | 225.3 | 6.56 | 2.817 |
| 0.5 | 11.47 | 0.9834 | 556.1 | 3.691 | 0.854 | 224.8 | 6.56 | 2.852 |
| 0.6 | 13.82 | 0.9801 | 555.5 | 3.668 | 0.858 | 224.3 | 6.56 | 2.888 |
| 0.7 | 16.17 | 0.9767 | 554.9 | 3.649 | 0.862 | 223.8 | 6.56 | 2.925 |
| 0.8 | 18.55 | 0.9734 | 554.4 | 3.632 | 0.866 | 223.4 | 6.56 | 2.963 |
| 0.9 | 20.94 | 0.9700 | 553.8 | 3.617 | 0.870 | 222.9 | 6.57 | 3.000 |
| 1.0 | 23.35 | 0.9666 | 553.2 | 3.603 | 0.874 | 222.4 | 6.57 | 3.039 |
| 1.5 | 35.65 | 0.9495 | 550.3 | 3.549, | 0.895 | 219.9 | 6.56 | 3.241 |
| 2.0 | 48.41 | 0.9323 | 547.4 | 3.509 | 0.918 | 217.5 | 6.56 | 3.461 |
| 2.5 | 61.68 | 0.9148 | 544.3 | 3.476 | 0.943 | 215.0 | 6.54 | 3.700 |
| 3.0 | 75.47 | 0.8971 | 541.2 | 3.448 | 0.969 | 212.6 | 6.52 | 3.958 |
| 3.5 | 89.84 | 0.8792 | 538.0 | 3.423 | 0.997 | 210.3 | 6.50 | 4.240 |
| 4.0 | 104.82 | 0.8611 | 534.7 | 3.401 | 1.028 | 208.0 | 6.46 | 4.544 |
| 4.5 | 120.47 | 0.8430 | 531.3 | 3.380 | 1.061 | 205.8 | 6.41 | 4.874 |
| 5.0 | 136.81 | 0.8248 | 527.9 | 3.360 | 1.096 | 203.7 | 6.34 | 5.230 |
| 5.5 | 153.89 | 0.8065 | 524.4 | 3.341 | 1.135 | 201.7 | 6.27 | 5.612 |
| 6.0 | 171.75 | 0.7884 | 520.8 | 3.324 | 1.176 | 199.9 | 6.17 | 6.020 |
| 6.5 | 190.42 | 0.7703 | 517.1 | 3.306 | 1.219 | 198.3 | 6.06 | 6.452 |
| 7.0 | 209.91 | 0.7525 | 513.4 | 3.290 | 1.265 | 197.0 | 5.93 | 6.904 |
| 7.5 | 230.24 | 0.7351 | 509.6 | 3.273 | 1.314 | 195.9 | 5.79 | 7.370 |
| 8.0 | 251.39 | 0.7182 | 505.8 | 3.258 | 1.364 | 195.2 | 5.62 | 7.842 |
| 8.5 | 273.31 | 0.7018 | 502.0 | 3.242 | 1.414 | 194.8 | 5.43 | 8.309 |
| 9.0 | 295.94 | 0.6863 | 498.1 | 3.227 | 1.465 | 194.8 | 5.23 | 8.758 |
| 9.5 | 319.16 | 0.6717 | 494.3 | 3.213 | 1.514 | 195.3 | 5.02 | 9.174 |
| 10.0 | 342.85 | 0.6582 | 490.6 | 3.199 | 1.560 | 196.2 | 4.79 | 9.542 |
| 11.0 | 390.94 | 0.6350 | 483.3 | 3.172 | 1.640 | 199.7 | 4.30 | 10.077 |
| 12.0 | 438.68 | 0.6173 | 476.5 | 3.147 | 1.696 | 205.3 | 3.82 | 10.295 |
| 13.0 | 484.56 | 0.6554 | 470.4 | 3.125 | 1.726 | 233.0 | 3.36 | 10.199 |
| 14.0 | 527.46 | 0.5990 | 464.9 | 3.105 | 1.732 | 222.4 | 2.93 | 9.857 |
| 15.0 | 566.74 | 0.5973 | 460.2 | 3.087 | 1.722 | 233.1 | 2.56 | 9.364 |
| 16.0 | 602.29 | 0.5995 | 456.1 | 3.072 | 1.701 | 244.7 | 2.23 | 8.807 |
| 17.0 | 634.27 | 0.6049 | 452.6 | 3.058 | 1.674 | 256.9 | 1.96 | 8.245 |
| 18.0 | 663.02 | 0.6127 | 449.5 | 3.046 | 1.646 | 269.3 | 1.72 | 7.713 |
| 19.0 | 688.94 | 0.6224 | 446.9 | 3.035 | 1.617 | 281.8 | 1.52 | 7.225 |
| 20.0 | 712.42 | 0.6335 | 444.6 | 3.025 | 1.590 | 294.1 | 1.35 | 6.787 |

**TABLE 51.** Thermodynamic Properties of Freon-23 in the Single-Phase Region (*Continued*)

$T = 383.15$ K

| $p$ | $\rho$ | $z$ | $h$ | $s$ | $c_p$ | $w$ | $\mu$ | $\alpha \cdot 10^3$ |
|---|---|---|---|---|---|---|---|---|
| 0.01 | 0.22 | 0.9997 | 567.2 | 4.183 | 0.850 | 229.9 | 6.05 | 2.613 |
| 0.02 | 0.44 | 0.9994 | 567.1 | 4.100 | 0.850 | 229.9 | 6.05 | 2.616 |
| 0.03 | 0.66 | 0.9991 | 567.1 | 4.052 | 0.850 | 229.9 | 6.05 | 2.619 |
| 0.04 | 0.88 | 0.9988 | 567.0 | 4.018 | 0.851 | 229.8 | 6.05 | 2.622 |
| 0.05 | 1.10 | 0.9985 | 567.0 | 3.991 | 0.851 | 229.8 | 6.05 | 2.625 |
| 0.1 | 2.20 | 0.9970 | 566.7 | 3.908 | 0.853 | 229.5 | 6.05 | 2.640 |
| 0.2 | 4.42 | 0.9940 | 566.2 | 3.825 | 0.856 | 229.1 | 6.05 | 2.670 |
| 0.3 | 6.65 | 0.9910 | 565.7 | 3.776 | 0.859 | 228.7 | 6.05 | 2.700 |
| 0.4 | 8.90 | 0.9880 | 565.2 | 3.741 | 0.863 | 228.2 | 6.05 | 2.731 |
| 0.5 | 11.16 | 0.9850 | 564.7 | 3.713 | 0.866 | 227.8 | 6.05 | 2.763 |
| 0.6 | 13.43 | 0.9820 | 564.1 | 3.691 | 0.870 | 227.3 | 6.05 | 2.794 |
| 0.7 | 15.71 | 0.9790 | 563.6 | 3.671 | 0.873 | 226.9 | 6.05 | 2.827 |
| 0.8 | 18.02 | 0.9760 | 563.1 | 3.654 | 0.877 | 226.4 | 6.06 | 2.860 |
| 0.9 | 20.33 | 0.9729 | 562.5 | 3.639 | 0.880 | 226.0 | 6.06 | 2.893 |
| 1.0 | 22.66 | 0.9699 | 562.0 | 3.626 | 0.884 | 225.5 | 6.06 | 2.926 |
| 1.5 | 34.54 | 0.9546 | 559.3 | 3.572 | 0.903 | 223.3 | 6.05 | 3.102 |
| 2.0 | 46.81 | 0.9391 | 556.5 | 3.533 | 0.923 | 221.1 | 6.04 | 3.291 |
| 2.5 | 59.50 | 0.9235 | 553.7 | 3.501 | 0.944 | 218.9 | 6.03 | 3.494 |
| 3.0 | 72.63 | 0.9078 | 550.8 | 3.473 | 0.967 | 216.8 | 6.01 | 3.712 |
| 3.5 | 86.24 | 0.8926 | 547.9 | 3.449 | 0.991 | 214.7 | 5.99 | 3.946 |
| 4.0 | 100.35 | 0.8761 | 544.9 | 3.427 | 1.017 | 212.7 | 5.95 | 4.195 |
| 4.5 | 114.98 | 0.8602 | 541.9 | 3.407 | 1.045 | 210.8 | 5.91 | 4.462 |
| 5.0 | 130.16 | 0.8442 | 538.7 | 3.389 | 1.074 | 209.0 | 5.85 | 4.746 |
| 5.5 | 145.92 | 0.8284 | 535.6 | 3.371 | 1.106 | 207.3 | 5.79 | 5.046 |
| 6.0 | 162.27 | 0.8126 | 532.3 | 3.354 | 1.139 | 205.7 | 5.71 | 5.363 |
| 6.5 | 179.23 | 0.7971 | 529.1 | 3.338 | 1.173 | 204.3 | 5.62 | 5.693 |
| 7.0 | 196.81 | 0.7817 | 525.8 | 3.322 | 1.210 | 203.2 | 5.51 | 6.036 |
| 7.5 | 214.99 | 0.7667 | 522.4 | 3.307 | 1.248 | 202.2 | 5.40 | 6.387 |
| 8.0 | 233.78 | 0.7521 | 519.0 | 3.292 | 1.286 | 201.6 | 5.26 | 6.742 |
| 8.5 | 253.14 | 0.7380 | 515.6 | 3.278 | 1.326 | 201.2 | 5.12 | 7.094 |
| 9.0 | 273.02 | 0.7245 | 512.2 | 3.264 | 1.366 | 201.1 | 4.96 | 7.437 |
| 9.5 | 293.36 | 0.7117 | 508.9 | 3.251 | 1.405 | 201.4 | 4.79 | 7.763 |
| 10.0 | 314.09 | 0.6997 | 505.5 | 3.238 | 1.443 | 202.0 | 4.61 | 8.063 |
| 11.0 | 356.29 | 0.6786 | 499.0 | 3.213 | 1.512 | 204.5 | 4.23 | 8.556 |
| 12.0 | 398.69 | 0.6615 | 492.8 | 3.190 | 1.568 | 208.6 | 3.83 | 8.865 |
| 13.0 | 440.31 | 0.6489 | 487.0 | 3.169 | 1.608 | 214.4 | 3.44 | 8.970 |
| 14.0 | 480.26 | 0.6407 | 481.7 | 3.149 | 1.632 | 221.8 | 3.07 | 8.883 |
| 15.0 | 517.87 | 0.6366 | 477.0 | 3.132 | 1.641 | 230.4 | 2.72 | 8.647 |
| 16.0 | 552.78 | 0.6361 | 472.8 | 3.116 | 1.638 | 240.1 | 2.41 | 8.312 |
| 17.0 | 584.90 | 0.6388 | 469.1 | 3.101 | 1.627 | 250.5 | 2.14 | 7.925 |
| 18.0 | 614.29 | 0.6440 | 465.8 | 3.089 | 1.611 | 261.4 | 1.90 | 7.522 |
| 19.0 | 641.14 | 0.6513 | 462.9 | 3.077 | 1.593 | 272.7 | 1.69 | 7.125 |
| 20.0 | 665.71 | 0.6603 | 460.4 | 3.066 | 1.573 | 284.0 | 1.52 | 6.750 |

**TABLE 51.** Thermodynamic Properties of Freon-23 in the Single-Phase Region (*Continued*)

$T = 393.15$ K

| $p$ | $\rho$ | $z$ | $h$ | $s$ | $c_p$ | $w$ | $\mu$ | $\alpha \cdot 10^3$ |
|------|--------|--------|-------|-------|--------|-------|-------|----------|
| 0.01 | 0.21 | 0.9997 | 575.7 | 4.204 | 0.863 | 232.6 | 5.59 | 2.546 |
| 0.02 | 0.43 | 0.9995 | 575.7 | 4.122 | 0.863 | 232.6 | 5.59 | 2.549 |
| 0.03 | 0.64 | 0.9992 | 575.6 | 4.074 | 0.864 | 232.6 | 5.59 | 2.551 |
| 0.04 | 0.86 | 0.9989 | 575.6 | 4.040 | 0.864 | 232.5 | 5.59 | 2.554 |
| 0.05 | 1.07 | 0.9986 | 575.5 | 4.013 | 0.864 | 232.5 | 5.59 | 2.557 |
| 0.1 | 2.15 | 0.9973 | 575.3 | 3.930 | 0.866 | 232.3 | 5.60 | 2.570 |
| 0.2 | 4.31 | 0.9946 | 574.8 | 3.847 | 0.869 | 231.9 | 5.60 | 2.597 |
| 0.3 | 6.48 | 0.9919 | 574.3 | 3.798 | 0.872 | 231.5 | 5.60 | 2.624 |
| 0.4 | 8.66 | 0.9892 | 573.8 | 3.763 | 0.875 | 231.0 | 5.60 | 2.651 |
| 0.5 | 10.86 | 0.9864 | 573.3 | 3.735 | 0.878 | 230.6 | 5.60 | 2.679 |
| 0.6 | 13.06 | 0.9837 | 572.9 | 3.713 | 0.881 | 230.2 | 5.60 | 2.707 |
| 0.7 | 15.28 | 0.9810 | 572.4 | 3.694 | 0.884 | 229.8 | 5.60 | 2.736 |
| 0.8 | 17.52 | 0.9782 | 571.9 | 3.677 | 0.887 | 229.4 | 5.60 | 2.765 |
| 0.9 | 19.76 | 0.9755 | 571.4 | 3.662 | 0.891 | 229.0 | 5.60 | 2.794 |
| 1.0 | 22.02 | 0.9728 | 570.9 | 3.648 | 0.894 | 228.6 | 5.60 | 2.824 |
| 1.5 | 33.50 | 0.9590 | 568.3 | 3.595 | 0.911 | 226.6 | 5.60 | 2.978 |
| 2.0 | 45.33 | 0.9451 | 565.8 | 3.556 | 0.929 | 224.6 | 5.59 | 3.142 |
| 2.5 | 57.51 | 0.9312 | 563.2 | 3.525 | 0.948 | 222.7 | 5.57 | 3.316 |
| 3.0 | 70.06 | 0.9171 | 560.5 | 3.498 | 0.968 | 220.8 | 5.56 | 3.501 |
| 3.5 | 83.01 | 0.9031 | 557.8 | 3.475 | 0.989 | 218.9 | 5.53 | 3.698 |
| 4.0 | 96.37 | 0.8890 | 555.0 | 3.453 | 1.011 | 217.2 | 5.50 | 3.906 |
| 4.5 | 110.15 | 0.8750 | 552.2 | 3.434 | 1.034 | 215.5 | 5.46 | 4.126 |
| 5.0 | 124.38 | 0.8610 | 549.4 | 3.416 | 1.059 | 213.9 | 5.41 | 4.357 |
| 5.5 | 139.06 | 0.8471 | 546.5 | 3.399 | 1.085 | 212.4 | 5.36 | 4.599 |
| 6.0 | 154.21 | 0.8334 | 543.6 | 3.383 | 1.113 | 211.1 | 5.29 | 4.852 |
| 6.5 | 169.82 | 0.8198 | 540.6 | 3.367 | 1.141 | 209.9 | 5.21 | 5.113 |
| 7.0 | 185.91 | 0.8065 | 537.6 | 3.352 | 1.171 | 208.9 | 5.13 | 5.382 |
| 7.5 | 202.46 | 0.7934 | 534.6 | 3.338 | 1.201 | 208.1 | 5.03 | 5.655 |
| 8.0 | 219.47 | 0.7808 | 531.6 | 3.324 | 1.232 | 207.5 | 4.92 | 5.931 |
| 8.5 | 236.90 | 0.7685 | 528.5 | 3.311 | 1.264 | 207.1 | 4.81 | 6.205 |
| 9.0 | 254.73 | 0.7568 | 525.5 | 3.298 | 1.296 | 207.0 | 4.68 | 6.473 |
| 9.5 | 272.92 | 0.7456 | 522.5 | 3.286 | 1.328 | 207.2 | 4.54 | 6.731 |
| 10.0 | 291.42 | 0.7350 | 519.5 | 3.274 | 1.359 | 207.7 | 4.40 | 6.974 |
| 11.0 | 329.06 | 0.7160 | 513.6 | 3.250 | 1.418 | 209.6 | 4.09 | 7.397 |
| 12.0 | 367.07 | 0.7002 | 507.9 | 3.229 | 1.469 | 212.8 | 3.76 | 7.708 |
| 13.0 | 404.78 | 0.6879 | 502.5 | 3.208 | 1.511 | 217.4 | 3.44 | 7.888 |
| 14.0 | 441.56 | 0.6791 | 497.5 | 3.190 | 1.541 | 223.3 | 3.11 | 7.932 |
| 15.0 | 476.85 | 0.6738 | 493.0 | 3.172 | 1.560 | 230.3 | 2.81 | 7.856 |
| 16.0 | 510.26 | 0.6716 | 488.8 | 3.157 | 1.569 | 238.4 | 2.52 | 7.685 |
| 17.0 | 541.58 | 0.6723 | 485.0 | 3.142 | 1.570 | 247.3 | 2.27 | 7.446 |
| 18.0 | 570.73 | 0.6755 | 481.7 | 3.129 | 1.565 | 256.8 | 2.04 | 7.168 |
| 19.0 | 597.74 | 0.6808 | 478.6 | 3.117 | 1.556 | 266.8 | 1.83 | 6.872 |
| 20.0 | 622.74 | 0.6879 | 476.0 | 3.106 | 1.545 | 277.0 | 1.65 | 6.574 |

**TABLE 51.** Thermodynamic Properties of Freon-23 in the Single-Phase Region (*Continued*)

$T = 403.15$ K

| $p$ | $\rho$ | $z$ | $h$ | $s$ | $c_p$ | $w$ | $\mu$ | $\alpha \cdot 10^3$ |
|---|---|---|---|---|---|---|---|---|
| 0.01 | 0.21 | 0.9998 | 584.4 | 4.226 | 0.876 | 235.3 | 5.19 | 2.483 |
| 0.02 | 0.42 | 0.9995 | 584.4 | 4.144 | 0.876 | 235.3 | 5.19 | 2.485 |
| 0.03 | 0.63 | 0.9993 | 584.3 | 4.095 | 0.877 | 235.2 | 5.19 | 2.488 |
| 0.04 | 0.84 | 0.9990 | 584.3 | 4.061 | 0.877 | 235.2 | 5.19 | 2.490 |
| 0.05 | 1.05 | 0.9988 | 584.2 | 4.035 | 0.877 | 235.2 | 5.19 | 2.492 |
| 0.1 | 2.09 | 0.9975 | 584.0 | 3.952 | 0.879 | 235.0 | 5.19 | 2.504 |
| 0.2 | 4.20 | 0.9951 | 583.5 | 3.869 | 0.881 | 234.6 | 5.19 | 2.528 |
| 0.3 | 6.31 | 0.9926 | 583.1 | 3.820 | 0.884 | 234.2 | 5.19 | 2.552 |
| 0.4 | 8.44 | 0.9902 | 582.6 | 3.785 | 0.887 | 233.8 | 5.19 | 2.577 |
| 0.5 | 10.57 | 0.9877 | 582.2 | 3.757 | 0.890 | 233.5 | 5.19 | 2.602 |
| 0.6 | 12.72 | 0.9852 | 581.7 | 3.735 | 0.893 | 233.1 | 5.19 | 2.627 |
| 0.7 | 14.88 | 0.9828 | 581.2 | 3.716 | 0.895 | 232.7 | 5.19 | 2.652 |
| 0.8 | 17.05 | 0.9803 | 580.8 | 3.699 | 0.898 | 232.4 | 5.19 | 2.678 |
| 0.9 | 19.23 | 0.9778 | 580.3 | 3.684 | 0.901 | 232.0 | 5.19 | 2.704 |
| 1.0 | 21.42 | 0.9753 | 579.8 | 3.671 | 0.904 | 231.6 | 5.19 | 2.730 |
| 1.5 | 32.54 | 0.9629 | 577.5 | 3.618 | 0.919 | 229.8 | 5.18 | 2.865 |
| 2.0 | 43.95 | 0.9504 | 575.1 | 3.580 | 0.935 | 228.0 | 5.18 | 3.008 |
| 2.5 | 55.68 | 0.9379 | 572.6 | 3.548 | 0.952 | 226.3 | 5.16 | 3.159 |
| 3.0 | 67.71 | 0.9254 | 570.2 | 3.522 | 0.970 | 224.6 | 5.15 | 3.319 |
| 3.5 | 80.08 | 0.9129 | 567.6 | 3.499 | 0.988 | 223.0 | 5.12 | 3.486 |
| 4.0 | 92.79 | 0.9004 | 565.1 | 3.478 | 1.008 | 221.4 | 5.09 | 3.662 |
| 4.5 | 105.85 | 0.8880 | 562.5 | 3.459 | 1.028 | 219.9 | 5.06 | 3.846 |
| 5.0 | 119.27 | 0.8756 | 559.9 | 3.442 | 1.049 | 218.5 | 5.02 | 4.037 |
| 5.5 | 133.06 | 0.8634 | 557.2 | 3.425 | 1.071 | 217.3 | 4.97 | 4.237 |
| 6.0 | 147.22 | 0.8513 | 554.6 | 3.410 | 1.094 | 216.1 | 4.91 | 4.443 |
| 6.5 | 161.74 | 0.8394 | 551.9 | 3.395 | 1.118 | 215.1 | 4.85 | 4.654 |
| 7.0 | 176.64 | 0.8278 | 549.2 | 3.381 | 1.143 | 214.2 | 4.77 | 4.870 |
| 7.5 | 191.90 | 0.8164 | 546.4 | 3.368 | 1.168 | 213.6 | 4.69 | 5.089 |
| 8.0 | 207.50 | 0.8053 | 543.7 | 3.355 | 1.194 | 213.1 | 4.60 | 5.309 |
| 8.5 | 223.44 | 0.7946 | 540.9 | 3.342 | 1.220 | 212.8 | 4.51 | 5.527 |
| 9.0 | 239.68 | 0.7843 | 538.2 | 3.330 | 1.246 | 212.7 | 4.40 | 5.742 |
| 9.5 | 256.20 | 0.7745 | 535.4 | 3.318 | 1.272 | 212.8 | 4.29 | 5.949 |
| 10.0 | 272.96 | 0.7652 | 532.7 | 3.307 | 1.298 | 213.2 | 4.17 | 6.147 |
| 11.0 | 307.02 | 0.7484 | 527.4 | 3.285 | 1.348 | 214.8 | 3.92 | 6.503 |
| 12.0 | 341.46 | 0.7341 | 522.2 | 3.264 | 1.393 | 217.4 | 3.65 | 6.787 |
| 13.0 | 375.83 | 0.7225 | 517.2 | 3.245 | 1.433 | 221.2 | 3.37 | 6.985 |
| 14.0 | 409.65 | 0.7138 | 512.5 | 3.227 | 1.464 | 226.0 | 3.10 | 7.088 |
| 15.0 | 442.51 | 0.7080 | 508.2 | 3.210 | 1.488 | 231.9 | 2.83 | 7.101 |
| 16.0 | 474.07 | 0.7050 | 504.1 | 3.195 | 1.503 | 238.7 | 2.58 | 7.034 |
| 17.0 | 504.07 | 0.7044 | 500.4 | 3.181 | 1.512 | 246.4 | 2.34 | 6.904 |
| 18.0 | 532.40 | 0.7062 | 497.0 | 3.168 | 1.516 | 254.6 | 2.12 | 6.727 |
| 19.0 | 558.99 | 0.7100 | 494.0 | 3.155 | 1.514 | 263.4 | 1.92 | 6.521 |
| 20.0 | 583.88 | 0.7155 | 491.2 | 3.144 | 1.510 | 272.6 | 1.75 | 6.299 |

**TABLE 51.** Thermodynamic Properties of Freon-23 in the Single-Phase Region (*Continued*)

$T=413.15$ K

| $p$ | $\rho$ | $z$ | $h$ | $s$ | $c_p$ | $w$ | $\mu$ | $\alpha \cdot 10^3$ |
|------|--------|--------|-------|-------|-------|-------|------|--------|
| 0.01 | 0.20 | 0.9998 | 593.2 | 4.247 | 0.889 | 237.9 | 4.82 | 2.423 |
| 0.02 | 0.41 | 0.9996 | 593.2 | 4.165 | 0.889 | 237.9 | 4.82 | 2.425 |
| 0.03 | 0.61 | 0.9993 | 593.1 | 4.117 | 0.890 | 237.9 | 4.82 | 2.427 |
| 0.04 | 0.82 | 0.9991 | 593.1 | 4.083 | 0.890 | 237.8 | 4.82 | 2.429 |
| 0.05 | 1.02 | 0.9989 | 593.0 | 4.056 | 0.890 | 237.8 | 4.82 | 2.431 |
| 0.1 | 2.04 | 0.9978 | 592.8 | 3.973 | 0.891 | 237.6 | 4.82 | 2.442 |
| 0.2 | 4.09 | 0.9955 | 592.4 | 3.890 | 0.894 | 237.3 | 4.82 | 2.463 |
| 0.3 | 6.16 | 0.9933 | 592.0 | 3.841 | 0.896 | 236.9 | 4.82 | 2.485 |
| 0.4 | 8.23 | 0.9911 | 591.5 | 3.806 | 0.899 | 236.6 | 4.82 | 2.507 |
| 0.5 | 10.31 | 0.9888 | 591.1 | 3.779 | 0.901 | 236.3 | 4.82 | 2.529 |
| 0.6 | 12.40 | 0.9866 | 590.7 | 3.757 | 0.904 | 235.9 | 4.82 | 2.551 |
| 0.7 | 14.49 | 0.9844 | 590.2 | 3.737 | 0.907 | 235.6 | 4.82 | 2.574 |
| 0.8 | 16.60 | 0.9821 | 589.8 | 3.721 | 0.909 | 235.3 | 4.82 | 2.597 |
| 0.9 | 18.72 | 0.9799 | 589.3 | 3.706 | 0.912 | 234.9 | 4.82 | 2.620 |
| 1.0 | 20.85 | 0.9776 | 588.9 | 3.693 | 0.915 | 234.6 | 4.82 | 2.643 |
| 1.5 | 31.64 | 0.9664 | 586.7 | 3.641 | 0.928 | 232.9 | 4.82 | 2.763 |
| 2.0 | 42.68 | 0.9552 | 584.4 | 3.602 | 0.943 | 231.3 | 4.81 | 2.889 |
| 2.5 | 53.98 | 0.9439 | 582.2 | 3.572 | 0.958 | 229.8 | 4.80 | 3.020 |
| 3.0 | 65.56 | 0.9327 | 579.8 | 3.546 | 0.973 | 228.3 | 4.78 | 3.158 |
| 3.5 | 77.41 | 0.9215 | 577.5 | 3.523 | 0.990 | 226.8 | 4.76 | 3.302 |
| 4.0 | 89.55 | 0.9104 | 575.1 | 3.503 | 1.007 | 225.4 | 4.73 | 3.453 |
| 4.5 | 101.98 | 0.8994 | 572.7 | 3.484 | 1.024 | 224.1 | 4.70 | 3.608 |
| 5.0 | 114.71 | 0.8884 | 570.3 | 3.467 | 1.043 | 222.9 | 4.66 | 3.770 |
| 5.5 | 127.74 | 0.8776 | 567.9 | 3.451 | 1.062 | 221.8 | 4.62 | 3.936 |
| 6.0 | 141.06 | 0.8669 | 565.4 | 3.436 | 1.082 | 220.8 | 4.57 | 4.107 |
| 6.5 | 154.69 | 0.8565 | 562.9 | 3.422 | 1.102 | 220.0 | 4.51 | 4.282 |
| 7.0 | 168.60 | 0.8462 | 560.5 | 3.409 | 1.123 | 219.3 | 4.45 | 4.459 |
| 7.5 | 182.80 | 0.8362 | 558.0 | 3.396 | 1.144 | 218.7 | 4.38 | 4.638 |
| 8.0 | 197.27 | 0.8266 | 555.4 | 3.383 | 1.166 | 218.3 | 4.30 | 4.817 |
| 8.5 | 212.01 | 0.8172 | 552.9 | 3.371 | 1.188 | 218.0 | 4.22 | 4.994 |
| 9.0 | 226.98 | 0.8082 | 550.4 | 3.360 | 1.210 | 218.0 | 4.14 | 5.169 |
| 9.5 | 242.17 | 0.7996 | 547.9 | 3.349 | 1.232 | 218.1 | 4.04 | 5.339 |
| 10.0 | 257.55 | 0.7914 | 545.5 | 3.338 | 1.254 | 218.5 | 3.94 | 5.502 |
| 11.0 | 288.74 | 0.7765 | 540.6 | 3.317 | 1.296 | 219.9 | 3.73 | 5.801 |
| 12.0 | 320.27 | 0.7637 | 535.8 | 3.297 | 1.336 | 222.1 | 3.51 | 6.052 |
| 13.0 | 351.82 | 0.7531 | 531.2 | 3.279 | 1.371 | 225.3 | 3.27 | 6.242 |
| 14.0 | 383.04 | 0.7450 | 526.8 | 3.262 | 1.402 | 229.5 | 3.04 | 6.367 |
| 15.0 | 413.60 | 0.7392 | 522.7 | 3.246 | 1.426 | 234.5 | 2.80 | 6.425 |
| 16.0 | 443.24 | 0.7358 | 518.8 | 3.231 | 1.445 | 240.4 | 2.58 | 6.421 |
| 17.0 | 471.73 | 0.7345 | 515.2 | 3.217 | 1.459 | 247.0 | 2.37 | 6.362 |
| 18.0 | 498.93 | 0.7353 | 511.9 | 3.204 | 1.467 | 254.2 | 2.17 | 6.259 |
| 19.0 | 524.74 | 0.7380 | 508.9 | 3.192 | 1.471 | 262.0 | 1.98 | 6.124 |
| 20.0 | 549.13 | 0.7423 | 506.1 | 3.181 | 1.472 | 270.2 | 1.81 | 5.967 |

**TABLE 51.** Thermodynamic Properties of Freon-23 in the Single-Phase Region (*Continued*)

$T = 423.15$ K

| $p$ | $\rho$ | $z$ | $h$ | $s$ | $c_p$ | $w$ | $\mu$ | $\alpha \cdot 10^3$ |
|------|--------|--------|-------|-------|-------|-------|------|---------|
| 0.01 | 0.20 | 0.9998 | 602.1 | 4.269 | 0.902 | 240.5 | 4.49 | 2.365 |
| 0.02 | 0.40 | 0.9996 | 602.1 | 4.186 | 0.902 | 240.5 | 4.49 | 2.367 |
| 0.03 | 0.60 | 0.9994 | 602.1 | 4.138 | 0.902 | 240.5 | 4.49 | 2.369 |
| 0.04 | 0.80 | 0.9992 | 602.0 | 4.104 | 0.902 | 240.4 | 4.49 | 2.371 |
| 0.05 | 1.00 | 0.9990 | 602.0 | 4.077 | 0.903 | 240.4 | 4.49 | 2.373 |
| 0.1 | 1.99 | 0.9980 | 601.8 | 3.995 | 0.904 | 240.3 | 4.49 | 2.382 |
| 0.2 | 4.00 | 0.9959 | 601.4 | 3.911 | 0.906 | 239.9 | 4.49 | 2.402 |
| 0.3 | 6.01 | 0.9939 | 601.0 | 3.863 | 0.908 | 239.6 | 4.49 | 2.421 |
| 0.4 | 8.03 | 0.9919 | 600.6 | 3.828 | 0.911 | 239.3 | 4.49 | 2.441 |
| 0.5 | 10.05 | 0.9899 | 600.1 | 3.800 | 0.913 | 239.0 | 4.49 | 2.461 |
| 0.6 | 12.09 | 0.9878 | 599.7 | 3.778 | 0.915 | 238.7 | 4.49 | 2.481 |
| 0.7 | 14.13 | 0.9858 | 599.3 | 3.759 | 0.918 | 238.4 | 4.49 | 2.501 |
| 0.8 | 16.18 | 0.9838 | 598.9 | 3.742 | 0.920 | 238.1 | 4.49 | 2.521 |
| 0.9 | 18.24 | 0.9817 | 598.5 | 3.728 | 0.923 | 237.8 | 4.49 | 2.542 |
| 1.0 | 20.31 | 0.9797 | 598.1 | 3.715 | 0.925 | 237.5 | 4.49 | 2.563 |
| 1.5 | 30.79 | 0.9695 | 596.0 | 3.663 | 0.937 | 236.0 | 4.48 | 2.669 |
| 2.0 | 41.49 | 0.9594 | 593.9 | 3.625 | 0.950 | 234.5 | 4.48 | 2.780 |
| 2.5 | 52.41 | 0.9493 | 591.7 | 3.594 | 0.964 | 233.1 | 4.46 | 2.896 |
| 3.0 | 63.57 | 0.9392 | 589.6 | 3.569 | 0.978 | 231.8 | 4.45 | 3.016 |
| 3.5 | 74.96 | 0.9292 | 587.4 | 3.547 | 0.992 | 230.5 | 4.43 | 3.141 |
| 4.0 | 86.59 | 0.9193 | 585.2 | 3.527 | 1.007 | 229.3 | 4.40 | 3.271 |
| 4.5 | 98.47 | 0.9094 | 583.0 | 3.509 | 1.023 | 228.2 | 4.37 | 3.404 |
| 5.0 | 110.60 | 0.8997 | 580.7 | 3.492 | 1.039 | 227.1 | 4.34 | 3.542 |
| 5.5 | 122.97 | 0.8901 | 578.4 | 3.476 | 1.056 | 226.2 | 4.30 | 3.683 |
| 6.0 | 135.58 | 0.8807 | 576.2 | 3.462 | 1.073 | 225.3 | 4.26 | 3.827 |
| 6.5 | 148.44 | 0.8714 | 573.9 | 3.448 | 1.090 | 224.6 | 4.21 | 3.973 |
| 7.0 | 161.53 | 0.8624 | 571.6 | 3.435 | 1.108 | 224.0 | 4.15 | 4.121 |
| 7.5 | 174.85 | 0.8536 | 569.3 | 3.423 | 1.127 | 223.5 | 4.09 | 4.269 |
| 8.0 | 188.39 | 0.8451 | 567.0 | 3.411 | 1.145 | 223.2 | 4.03 | 4.418 |
| 8.5 | 202.13 | 0.8369 | 564.7 | 3.399 | 1.164 | 223.0 | 3.96 | 4.565 |
| 9.0 | 216.06 | 0.8290 | 562.4 | 3.388 | 1.183 | 223.0 | 3.88 | 4.710 |
| 9.5 | 230.16 | 0.8214 | 560.1 | 3.377 | 1.202 | 223.2 | 3.81 | 4.851 |
| 10.0 | 244.41 | 0.8142 | 557.8 | 3.367 | 1.220 | 223.5 | 3.72 | 4.987 |
| 11.0 | 273.26 | 0.8011 | 553.3 | 3.347 | 1.257 | 224.8 | 3.54 | 5.239 |
| 12.0 | 302.40 | 0.7897 | 548.9 | 3.329 | 1.291 | 226.8 | 3.35 | 5.457 |
| 13.0 | 331.57 | 0.7802 | 544.6 | 3.311 | 1.323 | 229.6 | 3.15 | 5.632 |
| 14.0 | 360.53 | 0.7728 | 540.6 | 3.295 | 1.352 | 233.2 | 2.95 | 5.760 |
| 15.0 | 389.03 | 0.7673 | 536.7 | 3.279 | 1.376 | 237.6 | 2.75 | 5.839 |
| 16.0 | 416.85 | 0.7638 | 533.0 | 3.265 | 1.396 | 242.8 | 2.55 | 5.869 |
| 17.0 | 443.80 | 0.7623 | 529.6 | 3.251 | 1.411 | 248.6 | 2.36 | 5.855 |
| 18.0 | 469.73 | 0.7626 | 526.3 | 3.238 | 1.423 | 255.0 | 2.18 | 5.803 |
| 19.0 | 494.56 | 0.7645 | 523.4 | 3.226 | 1.430 | 261.9 | 2.01 | 5.720 |
| 20.0 | 518.24 | 0.7680 | 520.6 | 3.215 | 1.435 | 269.3 | 1.85 | 5.614 |

**TABLE 51.** Thermodynamic Properties of Freon-23 in the Single-Phase Region (*Continued*)

$T=433.15$ K

| $p$ | $\rho$ | $z$ | $h$ | $s$ | $c_p$ | $w$ | $\mu$ | $\alpha \cdot 10^3$ |
|---|---|---|---|---|---|---|---|---|
| 0.01 | 0.19 | 0.9998 | 611.2 | 4.290 | 0.914 | 243.1 | 4.19 | 2.310 |
| 0.02 | 0.39 | 0.9996 | 611.2 | 4.207 | 0.914 | 243.1 | 4.19 | 2.312 |
| 0.03 | 0.58 | 0.9994 | 611.1 | 4.159 | 0.915 | 243.1 | 4.19 | 2.314 |
| 0.04 | 0.78 | 0.9993 | 611.1 | 4.125 | 0.915 | 243.0 | 4.19 | 2.316 |
| 0.05 | 0.97 | 0.9991 | 611.0 | 4.098 | 0.915 | 243.0 | 4.19 | 2.317 |
| 0.1 | 1.95 | 0.9982 | 610.9 | 4.016 | 0.916 | 242.9 | 4.19 | 2.326 |
| 0.2 | 3.90 | 0.9963 | 610.5 | 3.933 | 0.918 | 242.6 | 4.19 | 2.343 |
| 0.3 | 5.86 | 0.9945 | 610.1 | 3.884 | 0.920 | 242.3 | 4.19 | 2.361 |
| 0.4 | 7.83 | 0.9926 | 609.7 | 3.849 | 0.922 | 242.0 | 4.19 | 2.379 |
| 0.5 | 9.81 | 0.9908 | 609.3 | 3.822 | 0.924 | 241.7 | 4.19 | 2.396 |
| 0.6 | 11.80 | 0.9889 | 608.9 | 3.799 | 0.927 | 241.4 | 4.19 | 2.414 |
| 0.7 | 13.79 | 0.9871 | 608.5 | 3.780 | 0.929 | 241.2 | 4.19 | 2.433 |
| 0.8 | 15.79 | 0.9852 | 608.1 | 3.764 | 0.931 | 240.9 | 4.19 | 2.451 |
| 0.9 | 17.79 | 0.9834 | 607.8 | 3.749 | 0.933 | 240.6 | 4.19 | 2.469 |
| 1.0 | 19.81 | 0.9816 | 607.4 | 3.736 | 0.935 | 240.3 | 4.19 | 2.488 |
| 1.5 | 29.99 | 0.9724 | 605.4 | 3.684 | 0.947 | 239.0 | 4.18 | 2.583 |
| 2.0 | 40.37 | 0.9632 | 603.4 | 3.647 | 0.959 | 237.7 | 4.18 | 2.682 |
| 2.5 | 50.94 | 0.9541 | 601.4 | 3.617 | 0.971 | 236.4 | 4.16 | 2.784 |
| 3.0 | 61.72 | 0.9450 | 599.4 | 3.592 | 0.983 | 235.2 | 4.15 | 2.890 |
| 3.5 | 72.69 | 0.9360 | 597.3 | 3.570 | 0.996 | 234.1 | 4.13 | 2.999 |
| 4.0 | 83.88 | 0.9271 | 595.2 | 3.550 | 1.010 | 233.0 | 4.11 | 3.112 |
| 4.5 | 95.26 | 0.9184 | 593.2 | 3.532 | 1.024 | 232.0 | 4.08 | 3.227 |
| 5.0 | 106.86 | 0.9097 | 591.1 | 3.516 | 1.038 | 231.1 | 4.05 | 3.345 |
| 5.5 | 118.65 | 0.9012 | 589.0 | 3.501 | 1.052 | 230.3 | 4.01 | 3.466 |
| 6.0 | 130.65 | 0.8928 | 586.9 | 3.487 | 1.067 | 229.6 | 3.98 | 3.589 |
| 6.5 | 142.85 | 0.8846 | 584.7 | 3.473 | 1.083 | 229.0 | 3.93 | 3.713 |
| 7.0 | 155.23 | 0.8767 | 582.6 | 3.461 | 1.098 | 228.5 | 3.88 | 3.838 |
| 7.5 | 167.81 | 0.8689 | 580.5 | 3.449 | 1.114 | 228.1 | 3.83 | 3.963 |
| 8.0 | 180.55 | 0.8614 | 578.3 | 3.437 | 1.130 | 227.9 | 3.78 | 4.088 |
| 8.5 | 193.46 | 0.8542 | 576.2 | 3.426 | 1.146 | 227.8 | 3.72 | 4.211 |
| 9.0 | 206.52 | 0.8472 | 574.1 | 3.415 | 1.163 | 227.8 | 3.65 | 4.333 |
| 9.5 | 219.71 | 0.8406 | 571.9 | 3.405 | 1.179 | 228.0 | 3.58 | 4.451 |
| 10.0 | 233.03 | 0.8343 | 569.8 | 3.395 | 1.195 | 228.4 | 3.51 | 4.566 |
| 11.0 | 259.93 | 0.8227 | 565.7 | 3.376 | 1.227 | 229.5 | 3.37 | 4.780 |
| 12.0 | 287.06 | 0.8127 | 561.6 | 3.358 | 1.257 | 231.4 | 3.19 | 4.969 |
| 13.0 | 314.23 | 0.8043 | 557.7 | 3.341 | 1.286 | 234.0 | 3.02 | 5.127 |
| 14.0 | 341.24 | 0.7976 | 553.9 | 3.326 | 1.312 | 237.2 | 2.85 | 5.250 |
| 15.0 | 367.92 | 0.7926 | 550.2 | 3.311 | 1.335 | 241.1 | 2.67 | 5.335 |
| 16.0 | 394.06 | 0.7894 | 546.7 | 3.297 | 1.354 | 245.7 | 2.50 | 5.383 |
| 17.0 | 419.54 | 0.7878 | 543.4 | 3.283 | 1.371 | 250.9 | 2.33 | 5.396 |
| 18.0 | 444.21 | 0.7878 | 540.4 | 3.271 | 1.384 | 256.6 | 2.16 | 5.377 |
| 19.0 | 467.99 | 0.7893 | 537.5 | 3.259 | 1.394 | 262.9 | 2.01 | 5.331 |
| 20.0 | 490.81 | 0.7922 | 534.8 | 3.248 | 1.401 | 269.5 | 1.86 | 5.263 |

**TABLE 51.** Thermodynamic Properties of Freon-23 in the Single-Phase Region (*Continued*)

$T=443.15$ K

| $p$ | $\rho$ | $z$ | $h$ | $s$ | $c_p$ | $w$ | $\mu$ | $\alpha \cdot 10^3$ |
|---|---|---|---|---|---|---|---|---|
| 0.01 | 0.19 | 0.9998 | 620.4 | 4.311 | 0.926 | 245.7 | 3.92 | 2.258 |
| 0.02 | 0.38 | 0.9997 | 620.4 | 4.228 | 0.926 | 245.6 | 3.92 | 2.260 |
| 0.03 | 0.57 | 0.9995 | 620.3 | 4.180 | 0.927 | 245.6 | 3.92 | 2.261 |
| 0.04 | 0.76 | 0.9993 | 620.3 | 4.146 | 0.927 | 245.6 | 3.92 | 2.263 |
| 0.05 | 0.95 | 0.9992 | 620.2 | 4.119 | 0.927 | 245.6 | 3.92 | 2.264 |
| 0.1 | 1.90 | 0.9983 | 620.1 | 4.037 | 0.928 | 245.4 | 3.92 | 2.272 |
| 0.2 | 3.81 | 0.9966 | 619.7 | 3.954 | 0.930 | 245.2 | 3.92 | 2.288 |
| 0.3 | 5.73 | 0.9950 | 619.3 | 3.905 | 0.932 | 244.9 | 3.92 | 2.304 |
| 0.4 | 7.65 | 0.9933 | 619.0 | 3.870 | 0.934 | 244.7 | 3.92 | 2.320 |
| 0.5 | 9.58 | 0.9916 | 618.6 | 3.843 | 0.936 | 244.4 | 3.92 | 2.336 |
| 0.6 | 11.52 | 0.9899 | 618.2 | 3.821 | 0.938 | 244.1 | 3.92 | 2.352 |
| 0.7 | 13.46 | 0.9882 | 617.9 | 3.802 | 0.940 | 243.9 | 3.92 | 2.369 |
| 0.8 | 15.41 | 0.9866 | 617.5 | 3.785 | 0.942 | 243.6 | 3.92 | 2.385 |
| 0.9 | 17.36 | 0.9849 | 617.1 | 3.771 | 0.944 | 243.4 | 3.92 | 2.402 |
| 1.0 | 19.33 | 0.9832 | 616.8 | 3.757 | 0.946 | 243.1 | 3.92 | 2.418 |
| 1.5 | 29.24 | 0.9749 | 614.9 | 3.706 | 0.956 | 241.9 | 3.91 | 2.503 |
| 2.0 | 39.32 | 0.9666 | 613.0 | 3.669 | 0.967 | 240.8 | 3.90 | 2.591 |
| 2.5 | 49.57 | 0.9584 | 611.1 | 3.639 | 0.978 | 239.6 | 3.89 | 2.682 |
| 3.0 | 59.99 | 0.9502 | 609.2 | 3.614 | 0.989 | 238.6 | 3.88 | 2.776 |
| 3.5 | 70.59 | 0.9422 | 607.3 | 3.592 | 1.001 | 237.6 | 3.86 | 2.872 |
| 4.0 | 81.36 | 0.9342 | 605.3 | 3.573 | 1.013 | 236.6 | 3.84 | 2.970 |
| 4.5 | 92.31 | 0.9263 | 603.4 | 3.556 | 1.025 | 235.8 | 3.81 | 3.071 |
| 5.0 | 103.43 | 0.9186 | 601.4 | 3.540 | 1.038 | 235.0 | 3.79 | 3.174 |
| 5.5 | 114.72 | 0.9110 | 599.5 | 3.525 | 1.051 | 234.3 | 3.75 | 3.278 |
| 6.0 | 126.18 | 0.9036 | 597.5 | 3.511 | 1.064 | 233.7 | 3.72 | 3.384 |
| 6.5 | 137.80 | 0.8964 | 595.5 | 3.498 | 1.078 | 233.2 | 3.68 | 3.490 |
| 7.0 | 149.57 | 0.8893 | 593.5 | 3.486 | 1.092 | 232.8 | 3.64 | 3.597 |
| 7.5 | 161.50 | 0.8825 | 591.5 | 3.474 | 1.106 | 232.5 | 3.59 | 3.704 |
| 8.0 | 173.37 | 0.8759 | 589.6 | 3.463 | 1.120 | 232.3 | 3.54 | 3.810 |
| 8.5 | 185.76 | 0.8695 | 587.6 | 3.452 | 1.134 | 232.0 | 3.49 | 3.915 |
| 9.0 | 198.08 | 0.8634 | 585.6 | 3.442 | 1.148 | 232.4 | 3.43 | 4.018 |
| 9.5 | 210.51 | 0.8576 | 583.6 | 3.432 | 1.162 | 232.6 | 3.37 | 4.119 |
| 10.0 | 223.03 | 0.8520 | 581.7 | 3.422 | 1.176 | 233.0 | 3.31 | 4.217 |
| 11.0 | 248.28 | 0.8419 | 577.8 | 3.404 | 1.204 | 234.1 | 3.18 | 4.400 |
| 12.0 | 273.71 | 0.8331 | 574.0 | 3.386 | 1.231 | 235.9 | 3.04 | 4.565 |
| 13.0 | 299.17 | 0.8257 | 570.4 | 3.370 | 1.256 | 238.3 | 2.89 | 4.705 |
| 14.0 | 324.51 | 0.8198 | 566.8 | 3.355 | 1.280 | 241.2 | 2.73 | 4.819 |
| 15.0 | 349.57 | 0.8154 | 563.4 | 3.341 | 1.301 | 244.8 | 2.58 | 4.904 |
| 16.0 | 374.22 | 0.8125 | 560.1 | 3.327 | 1.320 | 248.9 | 2.42 | 4.960 |
| 17.0 | 398.33 | 0.8110 | 557.0 | 3.314 | 1.336 | 253.6 | 2.27 | 4.988 |
| 18.0 | 421.78 | 0.8109 | 554.0 | 3.302 | 1.350 | 258.8 | 2.13 | 4.990 |
| 19.0 | 444.51 | 0.8122 | 551.2 | 3.290 | 1.361 | 264.5 | 1.99 | 4.969 |
| 20.0 | 466.43 | 0.8148 | 548.6 | 3.280 | 1.370 | 270.6 | 1.85 | 4.928 |

**TABLE 51.** Thermodynamic Properties of Freon-23 in the Single-Phase Region (*Continued*)

$T = 453.15$ K

| $p$ | $\rho$ | $z$ | $h$ | $s$ | $c_p$ | $w$ | $\mu$ | $\alpha \cdot 10^3$ |
|---|---|---|---|---|---|---|---|---|
| 0.01 | 0.19 | 0.9998 | 629.7 | 4.331 | 0.938 | 248.2 | 3.68 | 2.208 |
| 0.02 | 0.37 | 0.9997 | 629.7 | 4.249 | 0.938 | 248.2 | 3.68 | 2.210 |
| 0.03 | 0.56 | 0.9995 | 629.6 | 4.201 | 0.938 | 248.1 | 3.68 | 2.211 |
| 0.04 | 0.74 | 0.9994 | 629.6 | 4.166 | 0.939 | 248.1 | 3.68 | 2.212 |
| 0.05 | 0.93 | 0.9992 | 629.6 | 4.140 | 0.939 | 248.1 | 3.68 | 2.214 |
| 0.1 | 1.86 | 0.9985 | 629.4 | 4.057 | 0.940 | 248.0 | 3.68 | 2.221 |
| 0.2 | 3.73 | 0.9969 | 629.0 | 3.974 | 0.942 | 247.7 | 3.68 | 2.235 |
| 0.3 | 5.60 | 0.9954 | 628.7 | 3.926 | 0.943 | 247.5 | 3.68 | 2.250 |
| 0.4 | 7.48 | 0.9939 | 628.3 | 3.891 | 0.945 | 247.3 | 3.68 | 2.264 |
| 0.5 | 9.36 | 0.9924 | 628.0 | 3.864 | 0.947 | 247.0 | 3.68 | 2.279 |
| 0.6 | 11.25 | 0.9908 | 627.6 | 3.842 | 0.949 | 246.8 | 3.68 | 2.293 |
| 0.7 | 13.15 | 0.9893 | 627.3 | 3.823 | 0.951 | 246.6 | 3.68 | 2.308 |
| 0.8 | 15.05 | 0.9878 | 627.0 | 3.806 | 0.952 | 246.4 | 3.67 | 2.323 |
| 0.9 | 16.96 | 0.9863 | 626.6 | 3.792 | 0.954 | 246.1 | 3.67 | 2.338 |
| 1.0 | 18.87 | 0.9847 | 626.2 | 3.779 | 0.956 | 245.9 | 3.67 | 2.353 |
| 1.5 | 28.53 | 0.9772 | 624.5 | 3.727 | 0.966 | 244.8 | 3.67 | 2.430 |
| 2.0 | 38.33 | 0.9697 | 622.7 | 3.690 | 0.975 | 243.8 | 3.66 | 2.508 |
| 2.5 | 48.28 | 0.9623 | 620.9 | 3.661 | 0.985 | 242.8 | 3.64 | 2.589 |
| 3.0 | 58.38 | 0.9549 | 619.1 | 3.636 | 0.996 | 241.8 | 3.63 | 2.673 |
| 3.5 | 68.63 | 0.9477 | 617.3 | 3.615 | 1.006 | 240.9 | 3.61 | 2.758 |
| 4.0 | 79.03 | 0.9405 | 615.5 | 3.596 | 1.017 | 240.1 | 3.59 | 2.844 |
| 4.5 | 89.58 | 0.9335 | 613.6 | 3.578 | 1.028 | 239.4 | 3.57 | 2.933 |
| 5.0 | 100.28 | 0.9266 | 611.8 | 3.563 | 1.040 | 238.7 | 3.55 | 3.023 |
| 5.5 | 111.11 | 0.9198 | 610.0 | 3.548 | 1.051 | 238.1 | 3.52 | 3.113 |
| 6.0 | 122.09 | 0.9132 | 608.1 | 3.535 | 1.063 | 237.6 | 3.48 | 3.205 |
| 6.5 | 133.20 | 0.9068 | 606.3 | 3.522 | 1.075 | 237.2 | 3.45 | 3.297 |
| 7.0 | 144.44 | 0.9006 | 604.4 | 3.510 | 1.087 | 236.9 | 3.41 | 3.389 |
| 7.5 | 155.81 | 0.8945 | 602.5 | 3.498 | 1.100 | 236.7 | 3.37 | 3.481 |
| 8.0 | 167.28 | 0.8887 | 600.7 | 3.487 | 1.112 | 236.6 | 3.33 | 3.573 |
| 8.5 | 178.86 | 0.8831 | 598.8 | 3.477 | 1.125 | 236.6 | 3.28 | 3.663 |
| 9.0 | 190.54 | 0.8777 | 597.0 | 3.467 | 1.137 | 236.8 | 3.23 | 3.751 |
| 9.5 | 202.31 | 0.8726 | 595.2 | 3.457 | 1.150 | 237.0 | 3.18 | 3.838 |
| 10.0 | 214.14 | 0.8678 | 593.4 | 3.448 | 1.162 | 237.4 | 3.12 | 3.922 |
| 11.0 | 237.98 | 0.8589 | 589.8 | 3.430 | 1.187 | 238.6 | 3.01 | 4.081 |
| 12.0 | 261.96 | 0.8513 | 586.2 | 3.414 | 1.211 | 240.3 | 2.88 | 4.224 |
| 13.0 | 285.95 | 0.8448 | 582.8 | 3.398 | 1.234 | 242.5 | 2.75 | 4.348 |
| 14.0 | 309.82 | 0.8397 | 579.4 | 3.383 | 1.255 | 245.2 | 2.62 | 4.452 |
| 15.0 | 333.48 | 0.8359 | 576.2 | 3.369 | 1.274 | 248.5 | 2.48 | 4.534 |
| 16.0 | 356.78 | 0.8334 | 573.1 | 3.356 | 1.292 | 252.4 | 2.34 | 4.592 |
| 17.0 | 379.64 | 0.8321 | 570.2 | 3.343 | 1.308 | 256.7 | 2.21 | 4.629 |
| 18.0 | 401.96 | 0.8322 | 567.3 | 3.332 | 1.322 | 261.4 | 2.08 | 4.643 |
| 19.0 | 423.67 | 0.8334 | 564.7 | 3.320 | 1.333 | 266.6 | 1.95 | 4.639 |
| 20.0 | 444.71 | 0.8357 | 562.1 | 3.310 | 1.343 | 272.2 | 1.83 | 4.617 |

**TABLE 52.** Transport Properties of Freon-23 on the Saturation Line

| $T$ | $p$ | $\eta' \cdot 10^6$ | $\eta'' \cdot 10^6$ | $\nu' \cdot 10^6$ | $\nu'' \cdot 10^4$ | $\lambda' \cdot 10^6$ | $\lambda'' \cdot 10^6$ | $a' \cdot 10^6$ | $a'' \cdot 10^6$ | $Pr'$ | $Pr''$ |
|---|---|---|---|---|---|---|---|---|---|---|---|
| 233.15 | 0.7077 | 164.3 | 11.68 | 0.1300 | 0.390 | 103.2 | 11.31 | 0.0987 | 0.441 | 1.318 | 0.884 |
| 235.15 | 0.7615 | 160.2 | 11.80 | 0.1277 | 0.366 | 101.8 | 11.54 | 0.0907 | 0.410 | 1.391 | 0.892 |
| 237.15 | 0.8184 | 156.1 | 11.92 | 0.1253 | 0.344 | 100.4 | 11.77 | 0.0841 | 0.382 | 1.490 | 0.901 |
| 238.15 | 0.8481 | 154.0 | 11.98 | 0.1241 | 0.334 | 99.3 | 11.89 | 0.0810 | 0.369 | 1.533 | 0.904 |
| 239.15 | 0.8785 | 152.0 | 12.05 | 0.1230 | 0.323 | 98.6 | 12.01 | 0.0784 | 0.356 | 1.569 | 0.909 |
| 241.15 | 0.9418 | 147.9 | 12.17 | 0.1206 | 0.304 | 97.2 | 12.26 | 0.0739 | 0.331 | 1.633 | 0.918 |
| 243.15 | 1.0085 | 143.9 | 12.30 | 0.1183 | 0.287 | 95.8 | 12.52 | 0.0700 | 0.309 | 1.691 | 0.928 |
| 245.15 | 1.0787 | 139.8 | 12.44 | 0.1159 | 0.270 | 94.5 | 12.78 | 0.0665 | 0.288 | 1.741 | 0.939 |
| 247.15 | 1.1525 | 135.7 | 12.57 | 0.1135 | 0.255 | 93.2 | 13.06 | 0.0636 | 0.268 | 1.785 | 0.950 |
| 248.15 | 1.1908 | 133.7 | 12.64 | 0.1123 | 0.248 | 92.5 | 13.20 | 0.0623 | 0.261 | 1.804 | 0.948 |
| 249.15 | 1.2300 | 131.8 | 12.72 | 0.1112 | 0.241 | 91.8 | 13.34 | 0.0610 | 0.250 | 1.823 | 0.963 |
| 251.15 | 1.3113 | 128.9 | 12.86 | 0.1098 | 0.228 | 90.5 | 13.64 | 0.0586 | 0.233 | 1.873 | 0.978 |
| 253.15 | 1.3966 | 125.8 | 13.01 | 0.1082 | 0.215 | 89.1 | 13.95 | 0.0565 | 0.217 | 1.915 | 0.992 |
| 255.15 | 1.4860 | 122.6 | 13.17 | 0.1065 | 0.204 | 87.8 | 14.27 | 0.0545 | 0.202 | 1.954 | 1.209 |
| 257.15 | 1.5795 | 119.3 | 13.34 | 0.1047 | 0.1929 | 86.1 | 14.61 | 0.0526 | 0.1877 | 1.994 | 1.028 |
| 258.15 | 1.6279 | 117.6 | 13.42 | 0.1038 | 0.1877 | 85.5 | 14.78 | 0.0517 | 0.1808 | 2.01 | 1.038 |

**TABLE 52.** Transport Properties of Freon-23 on the Saturation Line (Continued)

| $T$ | $p$ | $\eta'\cdot10^6$ | $\eta''\cdot10^6$ | $\nu'\cdot10^6$ | $\nu''\cdot10^6$ | $\lambda'\cdot10^6$ | $\lambda''\cdot10^6$ | $a'\cdot10^6$ | $a''\cdot10^6$ | $Pr'$ | $Pr''$ |
|---|---|---|---|---|---|---|---|---|---|---|---|
| 259.15 | 1.6773 | 115.9 | 13.51 | 0.1028 | 0.1827 | 84.8 | 14.96 | 0.0508 | 0.1743 | 2.02 | 1.048 |
| 261.15 | 1.7797 | 112.4 | 13.68 | 0.1008 | 0.1732 | 83.5 | 15.33 | 0.0492 | 0.1618 | 2.05 | 1.070 |
| 263.15 | 1.8866 | 108.8 | 13.87 | 0.0987 | 0.1643 | 82.2 | 15.71 | 0.0477 | 0.1499 | 2.07 | 1.096 |
| 265.15 | 1.9982 | 105.2 | 14.07 | 0.0966 | 0.1558 | 80.9 | 16.12 | 0.0462 | 0.1386 | 2.09 | 1.124 |
| 267.15 | 2.115 | 101.6 | 14.28 | 0.0945 | 0.1479 | 79.6 | 16.54 | 0.0448 | 0.1279 | 2.11 | 1.157 |
| 268.15 | 2.175 | 99.8 | 14.39 | 0.0934 | 0.1441 | 79.0 | 16.76 | 0.0441 | 0.1228 | 2.12 | 1.174 |
| 269.15 | 2.237 | 97.9 | 14.50 | 0.0923 | 0.1404 | 78.3 | 16.99 | 0.0433 | 0.1177 | 2.13 | 1.193 |
| 271.15 | 2.363 | 94.3 | 14.73 | 0.0901 | 0.1334 | 77.0 | 17.46 | 0.0419 | 0.1081 | 2.15 | 1.234 |
| 273.15 | 2.496 | 90.7 | 14.98 | 0.0879 | 0.1267 | 75.3 | 17.97 | 0.0403 | 0.0988 | 2.18 | 1.282 |
| 275.15 | 2.634 | 87.1 | 15.25 | 0.0858 | 0.1204 | 74.0 | 18.50 | 0.0388 | 0.0899 | 2.21 | 1.338 |
| 277.15 | 2.777 | 83.5 | 15.53 | 0.0836 | 0.1143 | 72.7 | 19.08 | 0.0372 | 0.0814 | 2.25 | 1.405 |
| 278.15 | 2.851 | 81.7 | 15.69 | 0.0826 | 0.1114 | 72.0 | 19.38 | 0.0363 | 0.0773 | 2.27 | 1.442 |
| 279.15 | 2.927 | 80.0 | 15.84 | 0.0815 | 0.1086 | 71.3 | 19.69 | 0.0355 | 0.0732 | 2.29 | 1.484 |
| 281.15 | 3.083 | 76.5 | 16.19 | 0.0795 | 0.1031 | 69.8 | 20.4 | 0.0338 | 0.0652 | 2.35 | 1.581 |
| 283.15 | 3.245 | 73.1 | 16.56 | 0.0775 | 0.0979 | 68.3 | 21.1 | 0.0319 | 0.0576 | 2.43 | 1.701 |
| 285.15 | 3.414 | 69.7 | 16.98 | 0.0756 | 0.0929 | 66.8 | 21.9 | 0.0298 | 0.0501 | 2.54 | 1.854 |
| 287.15 | 3.590 | 66.4 | 17.45 | 0.0737 | 0.0881 | 65.1 | 22.8 | 0.0274 | 0.0427 | 2.69 | 2.06 |
| 288.15 | 3.681 | 64.7 | 17.72 | 0.0728 | 0.0857 | 64.2 | 23.3 | 0.0261 | 0.0393 | 2.78 | 2.18 |

**TABLE 53.** Transport Properties of Freon-23 in the Single-Phase Region

| | $T=233,15$ K | | | | $T=243,15$ K | | | |
|---|---|---|---|---|---|---|---|---|
| $p$ | $\eta \cdot 10^6$ | $\lambda \cdot 10^6$ | $a \cdot 10^6$ | Pr | $\eta \cdot 10^6$ | $\lambda \cdot 10^6$ | $a \cdot 10^6$ | Pr |
| 0.1 | 11,54 | 9,76 | 4.0000 | 0.785 | 12.04 | 10.30 | 4.350 | 0.786 |
| 0.5 | 11.61 | 10.79 | 0.6950 | 0.833 | 12.10 | 11.27 | 0.782 | 0.818 |
| 1.0 | 165.0 | 103.3 | 0.0991 | 1.316 | 12.30 | 12.48 | 0.313 | 0,926 |
| 2.0 | 167.3 | 104,4 | 0.1014 | 1.298 | 146.5 | 96.9 | 0.0715 | 1.675 |
| 3.0 | 169.6 | 105.5 | 0.1038 | 1.281 | 149.0 | 98.1 | 0.0731 | 1.658 |
| 4.0 | 171.8 | 106.6 | 0.1060 | 1.265 | 151.4 | 99.3 | 0.0747 | 1.641 |
| 5.0 | 174.1 | 107.7 | 0.1082 | 1.251 | 153.7 | 100.5 | 0.0763 | 1.622 |
| 6.0 | 176.2 | 108.7 | 0.1103 | 1.238 | 156.0 | 101.7 | 0.0778 | 1.608 |
| 8.0 | 180,5 | 110,6 | 0.1143 | 1.216 | 160.5 | 103.8 | 0.0805 | 1.586 |
| 10.0 | 184,7 | 112.3 | 0.1182 | 1.196 | 164.7 | 105.6 | 0.0829 | 1.568 |
| 12.0 | 188.8 | 113.6 | 0.1215 | 1.182 | 168.8 | 107,1 | 0.0850 | 1.556 |
| 14.0 | 192.8 | 114.7 | 0.1243 | 1.173 | 172.8 | 108.2 | 0.0866 | 1.553 |
| 16.0 | 196.8 | 115.5 | 0.1271 | 1.165 | 176.8 | 109.2 | 0.0881 | 1.553 |
| 18.0 | 201.0 | 116.1 | 0.1294 | 1.163 | 180.6 | 109.9 | 0.0894 | 1.554 |
| 20.0 | 205.0 | 116.6 | 0.1314 | 1.162 | 184,4 | 110.5 | 0.0904 | 1.561 |

**TABLE 53.** Transport Properties of Freon-23 in the Single-Phase Region (*Continued*)

| | $T=253,15$ K | | | | $T=263,15$ K | | | |
|---|---|---|---|---|---|---|---|---|
| $p$ | $\eta \cdot 10^6$ | $\lambda \cdot 10^6$ | $a \cdot 10^6$ | Pr | $\eta \cdot 10^6$ | $\lambda \cdot 10^6$ | $a \cdot 10^6$ | Pr |
| 0.1 | 12.54 | 10.84 | 4.7100 | 0,790 | 13.03 | 11.38 | 5.0600 | 0.795 |
| 0.5 | 12.59 | 11.77 | 0.8680 | 0.810 | 13.08 | 12.26 | 0.9510 | 0.807 |
| 1.0 | 12.76 | 12.86 | 0.3670 | 0,886 | 13.23 | 13.28 | 0.4190 | 0.858 |
| 2.0 | 127,4 | 89,6 | 0.0573 | 1.904 | 109.2 | 82.4 | 0.0480 | 2.06 |
| 3.0 | 129,7 | 90,9 | 0.0589 | 1.871 | 112.7 | 83.8 | 0.0498 | 2.03 |
| 4.0 | 131,5 | 92,2 | 0.0604 | 1.838 | 115.8 | 85.2 | 0.0516 | 1.992 |
| 5.0 | 134.2 | 93,5 | 0.0619 | 1.819 | 118.7 | 86.6 | 0.0531 | 1.967 |
| 6.0 | 136.8 | 94.8 | 0.0632 | 1.805 | 121.3 | 88.0 | 0.0546 | 1.940 |
| 8.0 | 141.7 | 97,0 | 0.0656 | 1.782 | 126.0 | 90.5 | 0.0572 | 1.896 |
| 10.0 | 146.2 | 99.0 | 0.0677 | 1.765 | 130.1 | 92.6 | 0.0593 | 1,865 |
| 12.0 | 150.6 | 100,7 | 0.0696 | 1.753 | 133.6 | 94.5 | 0.0611 | 1.838 |
| 14.0 | 154.7 | 102.0 | 0.0710 | 1.752 | 137.9 | 96.0 | 0.0626 | 1.834 |
| 16.0 | 158.7 | 103,1 | 0.0722 | 1.753 | 142.1 | 97.2 | 0.0638 | 1.837 |
| 18.0 | 162,6 | 103,9 | 0.0732 | 1.760 | 146.1 | 98.2 | 0.0648 | 1.845 |
| 20.0 | 166,4 | 104.6 | 0.0740 | 1.770 | 149.9 | 99.0 | 0.0655 | 1.858 |

**TABLE 53.** Transport Properties of Freon-23 in the Single-Phase Region (*Continued*)

| | $T=273.15$ K | | | | $T=283.15$ K | | | |
|---|---|---|---|---|---|---|---|---|
| $p$ | $\eta\cdot10^6$ | $\lambda\cdot10^6$ | $a\cdot10^6$ | Pr | $\eta\cdot10^6$ | $\lambda\cdot10^6$ | $a\cdot10^6$ | Pr |
| 0.1 | 13.53 | 11.92 | 5.4100 | 0.802 | 14.02 | 12.47 | 5.7800 | 0.809 |
| 0.5 | 13.57 | 12.77 | 1.0330 | 0.806 | 14.06 | 13.27 | 1.1130 | 0.810 |
| 1.0 | 13.71 | 13.71 | 0.4680 | 0.844 | 14.18 | 14.16 | 0.5160 | 0.834 |
| 2.0 | 14.34 | 16.15 | 0.1700 | 1.020 | 14.71 | 16.34 | 0.2050 | 0.948 |
| 3.0 | 93.5 | 76.7 | 0.0420 | 2.14 | 15.97 | 19.72 | 0.0822 | 1.364 |
| 4.0 | 97.2 | 78.2 | 0.0443 | 2.07 | 77.6 | 70.6 | 0.0359 | 2.23 |
| 5.0 | 101.0 | 79.7 | 0.0463 | 2.04 | 82.5 | 72.2 | 0.0392 | 2.12 |
| 6.0 | 104.4 | 81.1 | 0.0481 | 2.00 | 86.8 | 73.7 | 0.0417 | 2.06 |
| 8.0 | 110.6 | 83.8 | 0.0511 | 1.956 | 94.3 | 76.6 | 0.0455 | 1.985 |
| 10.0 | 115.9 | 86.2 | 0.0535 | 1.925 | 100.7 | 79.1 | 0.0484 | 1.946 |
| 12.0 | 120.7 | 88.2 | 0.0555 | 1.905 | 106.3 | 81.3 | 0.0507 | 1.924 |
| 14.0 | 124.8 | 89.9 | 0.0571 | 1.891 | 111.4 | 83.2 | 0.0525 | 1.916 |
| 16.0 | 128.6 | 91.3 | 0.0584 | 1.884 | 115.9 | 84.7 | 0.0540 | 1.911 |
| 18.0 | 131.9 | 92.4 | 0.0595 | 1.877 | 119.1 | 86.0 | 0.0551 | 1.900 |
| 20.0 | 134.8 | 93.4 | 0.0603 | 1.875 | 123.8 | 87.0 | 0.0560 | 1.921 |

**TABLE 53.** Transport Properties of Freon-23 in the Single-Phase Region (*Continued*)

| | $T=293.15$ K | | | | $T=303.15$ K | | | |
|---|---|---|---|---|---|---|---|---|
| $p$ | $\eta\cdot10^6$ | $\lambda\cdot10^6$ | $a\cdot10^6$ | Pr | $\eta\cdot10^6$ | $\lambda\cdot10^6$ | $a\cdot10^6$ | Pr |
| 0.1 | 14.50 | 13.01 | 6.1400 | 0.814 | 14.99 | 13.55 | 6.5000 | 0.824 |
| 0.5 | 14.55 | 13.78 | 1.1950 | 0.813 | 15.03 | 14.30 | 1.2740 | 0.819 |
| 1.0 | 14.66 | 14.63 | 0.5630 | 0.830 | 15.13 | 15.10 | 0.6080 | 0.829 |
| 2.0 | 15.12 | 16.61 | 0.2370 | 0.907 | 15.54 | 16.93 | 0.2670 | 0.882 |
| 3.0 | 16.07 | 19.32 | 0.1167 | 1.112 | 16.32 | 19.26 | 0.1450 | 1.005 |
| 4.0 | 18.43 | 24.5 | 0.0369 | 2.22 | 17.76 | 22.7 | 0.0746 | 1.332 |
| 5.0 | 63.6 | 68.1 | 0.0305 | 2.37 | 22.1 | 30.5 | 0.01852 | 3.66 |
| 6.0 | 69.6 | 71.5 | 0.0368 | 2.06 | 50.8 | 59.3 | 0.0213 | 3.13 |
| 8.0 | 78.6 | 75.7 | 0.0436 | 1.858 | 64.0 | 68.6 | 0.0371 | 1.966 |
| 10.0 | 85.8 | 78.6 | 0.0476 | 1.792 | 72.2 | 73.1 | 0.0436 | 1.776 |
| 12.0 | 92.1 | 80.9 | 0.0504 | 1.769 | 79.0 | 76.3 | 0.0476 | 1.710 |
| 14.0 | 97.7 | 82.8 | 0.0525 | 1.763 | 85.0 | 78.7 | 0.0504 | 1.687 |
| 16.0 | 102.8 | 84.4 | 0.0542 | 1.764 | 90.4 | 80.7 | 0.0525 | 1.682 |
| 18.0 | 107.5 | 85.8 | 0.0555 | 1.773 | 95.4 | 82.4 | 0.0542 | 1.684 |
| 20.0 | 111.9 | 87.1 | 0.0567 | 1.782 | 100.0 | 83.9 | 0.0555 | 1.695 |

**TABLE 53.** Transport Properties of Freon-23 in the Single-Phase Region (*Continued*)

| | $T=313.15$ K | | | | $T=323.15$ K | | | |
|---|---|---|---|---|---|---|---|---|
| $p$ | $\eta \cdot 10^6$ | $\lambda \cdot 10^6$ | $a \cdot 10^6$ | Pr | $\eta \cdot 10^6$ | $\lambda \cdot 10^6$ | $a \cdot 10^6$ | Pr |
| 0.1 | 15.47 | 14.09 | 6.8600 | 0.832 | 15.95 | 14.63 | 7.2300 | 0.842 |
| 0.5 | 15.51 | 14.81 | 1.3530 | 0.825 | 15.99 | 15.33 | 1.4320 | 0.832 |
| 1.0 | 15.60 | 15.58 | 0.6520 | 0.831 | 16.08 | 16.06 | 0.6960 | 0.835 |
| 2.0 | 15.98 | 17.29 | 0.2950 | 0.868 | 16.42 | 17.67 | 0.3220 | 0.860 |
| 3.0 | 16.64 | 19.36 | 0.1699 | 0.948 | 17.00 | 19.56 | 0.1928 | 0.913 |
| 4.0 | 17.74 | 22.1 | 0.1015 | 1.115 | 17.90 | 21.9 | 0.1241 | 1.015 |
| 5.0 | 19.70 | 26.2 | 0.0545 | 1.534 | 19.28 | 24.9 | 0.0790 | 1.211 |
| 6.0 | 25.0 | 34.6 | 0.0211 | 3.06 | 21.6 | 29.3 | 0.0473 | 1.600 |
| 8.0 | 49.9 | 58.8 | 0.0282 | 2.37 | 35.5 | 46.0 | 0.0242 | 2.62 |
| 10.0 | 60.3 | 66.5 | 0.0388 | 1.840 | 49.5 | 58.6 | 0.0340 | 1.974 |
| 12.0 | 67.6 | 71.0 | 0.0445 | 1.690 | 57.8 | 65.0 | 0.0412 | 1.711 |
| 14.0 | 72.4 | 73.5 | 0.0481 | 1.617 | 64.2 | 69.3 | 0.0459 | 1.603 |
| 16.0 | 79.3 | 76.7 | 0.0508 | 1.611 | 69.7 | 72.4 | 0.0491 | 1.558 |
| 18.0 | 84.4 | 78.8 | 0.0529 | 1.604 | 74.8 | 75.0 | 0.0516 | 1.539 |
| 20.0 | 89.1 | 80.6 | 0.0545 | 1.609 | 79.4 | 77.1 | 0.0535 | 1.535 |

**TABLE 53.** Transport Properties of Freon-23 in the Single-Phase Region (*Continued*)

| | $T=333.15$ K | | | | $T=343.15$ K | | | |
|---|---|---|---|---|---|---|---|---|
| $p$ | $\eta \cdot 10^6$ | $\lambda \cdot 10^3$ | $a \cdot 10^6$ | Pr | $\eta \cdot 10^6$ | $\lambda \cdot 10^6$ | $a \cdot 10^6$ | Pr |
| 0.1 | 16.43 | 15.18 | 7.6100 | 0.850 | 16.91 | 15.72 | 7.9600 | 0.860 |
| 0.5 | 16.46 | 15.85 | 1.5120 | 0.839 | 16.94 | 16.37 | 1.5890 | 0.849 |
| 1.0 | 16.55 | 16.55 | 0.7390 | 0.844 | 17.01 | 17.05 | 0.7810 | 0.846 |
| 2.0 | 16.86 | 18.08 | 0.3480 | 0.856 | 17.30 | 18.50 | 0.3730 | 0.855 |
| 3.0 | 17.38 | 19.82 | 0.2140 | 0.893 | 17.78 | 20.1 | 0.2340 | 0.880 |
| 4.0 | 18.15 | 21.9 | 0.1441 | 0.959 | 18.45 | 22.0 | 0.1624 | 0.925 |
| 5.0 | 19.24 | 24.4 | 0.0995 | 1.072 | 19.37 | 24.1 | 0.1169 | 1.002 |
| 6.0 | 20.8 | 27.6 | 0.0684 | 1.255 | 20.6 | 26.8 | 0.0863 | 1.105 |
| 8.0 | 27.7 | 37.7 | 0.0348 | 1.883 | 24.8 | 33.9 | 0.0492 | 1.445 |
| 10.0 | 39.9 | 50.3 | 0.0327 | 1.982 | 33.0 | 43.7 | 0.0375 | 1.727 |
| 12.0 | 49.4 | 58.6 | 0.0389 | 1.738 | 42.2 | 52.5 | 0.0390 | 1.693 |
| 14.0 | 56.2 | 64.0 | 0.0439 | 1.600 | 49.4 | 58.8 | 0.0431 | 1.581 |
| 16.0 | 61.8 | 67.9 | 0.0476 | 1.529 | 55.1 | 63.4 | 0.0467 | 1.503 |
| 18.0 | 66.7 | 71.0 | 0.0504 | 1.493 | 59.9 | 66.9 | 0.0495 | 1.458 |
| 20.0 | 71.2 | 73.5 | 0.0525 | 1.479 | 64.3 | 69.8 | 0.0518 | 1.434 |

**TABLE 53.** Transport Properties of Freon-23 in the Single-Phase Region (*Continued*)

| | $T=353.15$ K | | | | $T=363.15$ K | | | |
|---|---|---|---|---|---|---|---|---|
| $p$ | $\eta \cdot 10^6$ | $\lambda \cdot 10^6$ | $a \cdot 10^6$ | Pr | $\eta \cdot 10^6$ | $\lambda \cdot 10^6$ | $a \cdot 10^6$ | Pr |
| 0.1 | 17.38 | 16.26 | 8.3800 | 0.868 | 17.85 | 16.80 | 8.7300 | 0.878 |
| 0.5 | 17.41 | 16.89 | 1.6700 | 0.857 | 17.87 | 17.41 | 1.7490 | 0.865 |
| 1.0 | 17.48 | 17.54 | 0.8240 | 0.853 | 17.94 | 18.04 | 0.8660 | 0.817 |
| 2.0 | 17.75 | 18.93 | 0.3980 | 0.856 | 18.19 | 19.38 | 0.4220 | 0.859 |
| 3.0 | 18.18 | 20.5 | 0.2540 | 0.871 | 18.59 | 20.8 | 0.2720 | 0.869 |
| 4.0 | 18.79 | 22.2 | 0.1795 | 0.904 | 19.14 | 22.4 | 0.1952 | 0.892 |
| 5.0 | 19.58 | 24.1 | 0.1333 | 0.958 | 19.85 | 24.2 | 0.1484 | 0.925 |
| 6.0 | 20.6 | 26.3 | 0.1018 | 1.025 | 20.7 | 26.1 | 0.1159 | 0.975 |
| 8.0 | 23.7 | 32.0 | 0.0633 | 1.230 | 23.2 | 30.9 | 0.0763 | 1.110 |
| 10.0 | 29.1 | 39.4 | 0.0461 | 1.456 | 27.1 | 36.9 | 0.0560 | 1.271 |
| 12.0 | 36.6 | 47.3 | 0.0422 | 1.558 | 32.8 | 43.5 | 0.0479 | 1.397 |
| 14.0 | 43.6 | 53.9 | 0.0439 | 1.525 | 39.0 | 49.7 | 0.0465 | 1.435 |
| 16.0 | 49.4 | 59.0 | 0.0466 | 1.472 | 44.7 | 54.9 | 0.0478 | 1.420 |
| 18.0 | 54.3 | 63.0 | 0.0492 | 1.429 | 49.5 | 59.2 | 0.0497 | 1.391 |
| 20.0 | 58.6 | 66.2 | 0.0515 | 1.399 | 53.8 | 62.7 | 0.0516 | 1.368 |

**TABLE 53.** Transport Properties of Freon-23 in the Single-Phase Region (*Continued*)

| | $T=373.15$ K | | | | $T=383.15$ K | | | |
|---|---|---|---|---|---|---|---|---|
| $p$ | $\eta \cdot 10^6$ | $\lambda \cdot 10^6$ | $a \cdot 10^6$ | Pr | $\eta \cdot 10^6$ | $\lambda \cdot 10^6$ | $a \cdot 10^6$ | Pr |
| 0.1 | 18.31 | 17.34 | 9.1400 | 0.886 | 18.77 | 17.89 | 9.5300 | 0.895 |
| 0.5 | 18.34 | 17.94 | 1.8310 | 0.873 | 18.80 | 18.46 | 1.9100 | 0.882 |
| 1.0 | 18.40 | 18.55 | 0.9090 | 0.867 | 18.86 | 19.06 | 0.9510 | 0.875 |
| 2.0 | 18.64 | 19.83 | 0.4460 | 0.863 | 19.08 | 20.3 | 0.4700 | 0.867 |
| 3.0 | 19.01 | 21.2 | 0.2900 | 0.869 | 19.43 | 21.6 | 0.3070 | 0.871 |
| 4.0 | 19.51 | 22.7 | 0.2110 | 0.882 | 19.89 | 23.0 | 0.2250 | 0.881 |
| 5.0 | 20.2 | 24.3 | 0.1621 | 0.911 | 20.5 | 24.5 | 0.1753 | 0.898 |
| 6.0 | 20.9 | 26.1 | 0.1292 | 0.942 | 21.2 | 26.1 | 0.1412 | 0.925 |
| 8.0 | 23.0 | 30.2 | 0.0881 | 1.038 | 22.9 | 29.8 | 0.0991 | 0.988 |
| 10.0 | 26.0 | 35.2 | 0.0658 | 1.152 | 25.4 | 34.2 | 0.0755 | 1.071 |
| 12.0 | 30.3 | 40.8 | 0.0548 | 1.260 | 28.8 | 39.0 | 0.0624 | 1.158 |
| 14.0 | 35.6 | 46.4 | 0.0508 | 1.329 | 33.1 | 44.0 | 0.0561 | 1.229 |
| 16.0 | 40.8 | 51.5 | 0.0503 | 1.347 | 37.7 | 48.7 | 0.0538 | 1.268 |
| 18.0 | 45.5 | 55.8 | 0.0511 | 1.342 | 42.1 | 52.9 | 0.0534 | 1.283 |
| 20.0 | 49.7 | 59.5 | 0.0525 | 1.329 | 46.2 | 56.5 | 0.0539 | 1.288 |

**TABLE 53.** Transport Properties of Freon-23 in the Single-Phase Region (*Continued*)

| | $T=393.15$ K | | | | $T=403.15$ K | | | |
|---|---|---|---|---|---|---|---|---|
| $p$ | $\eta \cdot 10^6$ | $\lambda \cdot 10^6$ | $a \cdot 10^6$ | Pr | $\eta \cdot 10^6$ | $\lambda \cdot 10^6$ | $a \cdot 10^6$ | Pr |
| 0.1 | 19,23 | 18,43 | 9,9000 | 0,903 | 19.69 | 18.97 | 10.3300 | 0,912 |
| 0.5 | 19.26 | 18.99 | 1,9920 | 0.890 | 19.71 | 19,52 | 2.0700 | 0.901 |
| 1.0 | 19.32 | 19,56 | 0,9940 | 0.883 | 19.77 | 20.1 | 1.0380 | 0.889 |
| 2.0 | 19.52 | 20.8 | 0.4940 | 0.872 | 19.97 | 21,2 | 0.5160 | 0.880 |
| 3.0 | 19.85 | 22.0 | 0.3240 | 0,874 | 20.3 | 22.4 | 0.3410 | 0.880 |
| 4.0 | 20.3 | 23.3 | 0.2390 | 0.881 | 20.7 | 23.7 | 0.2530 | 0.881 |
| 5.0 | 20.8 | 24.8 | 0.1883 | 0.888 | 21.2 | 25.0 | 0.1998 | 0.883 |
| 6.0 | 21.4 | 26.3 | 0.1532 | 0.906 | 21.8 | 26.5 | 0.1645 | 0.900 |
| 8.0 | 23.0 | 29.6 | 0.1095 | 0.957 | 23.2 | 29.6 | 0.1195 | 0.936 |
| 10.0 | 25,1 | 33,5 | 0.0846 | 1.018 | 25.0 | 33.0 | 0.0931 | 0,984 |
| 12.0 | 27,9 | 37.7 | 0.0699 | 1.087 | 27.3 | 36,8 | 0.0774 | 1,032 |
| 14.0 | 31,4 | 42.1 | 0.0619 | 1.149 | 30,3 | 40,7 | 0.0679 | 1.090 |
| 16.0 | 35.4 | 46,4 | 0,0580 | 1,197 | 33.7 | 44,7 | 0.0627 | 1.134 |
| 18,0 | 39.4 | 50,4 | 0.05ö4 | 1.224 | 37.2 | 48.4 | 0.0600 | 1,165 |
| 20,0 | 43.2 | 54.0 | 0.0554 | 1.252 | 40,8 | 51.9 | 0.0589 | 1,187 |

**TABLE 53.** Transport Properties of Freon-23 in the Single-Phase Region (*Continued*)

| | $T=413,15$ K | | | | $T=423,15$ K | | | |
|---|---|---|---|---|---|---|---|---|
| $p$ | $\eta \cdot 10^6$ | $\lambda \cdot 10^6$ | $a \cdot 10^6$ | Pr | $\eta \cdot 10^6$ | $\lambda \cdot 10^6$ | $a \cdot 10^6$ | Pr |
| 0.1 | 20,1 | 19,51 | 10,7300 | 0,918 | 20,6 | 20,0 | 11.1200 | 0,931 |
| 0.5 | 20.2 | 20.0 | 2.1500 | 0,911 | 20,6 | 20,6 | 2.2500 | 0,911 |
| 1.0 | 20.2 | 20.6 | 1.0800 | 0.897 | 20.7 | 21,1 | 1.1230 | 0.907 |
| 2,0 | 20,4 | 21.7 | 0.5390 | 0.887 | 20.8 | 22.2 | 0.5630 | 0,890 |
| 3,0 | 20,7 | 22,9 | 0.3590 | 0,880 | 21.1 | 23,3 | 0.3750 | 0,885 |
| 4.0 | 21,1 | 24.1 | 0.2670 | 0,881 | 21.5 | 24.5 | 0,2810 | 0,882 |
| 5,0 | 21.5 | 25.4 | 0.2120 | 0,884 | 21,9 | 25,7 | 0,2240 | 0,884 |
| 6,0 | 22.1 | 26.7 | 0.1749 | 0.896 | 22.4 | 27,0 | 0.1856 | 0,890 |
| 8,0 | 23.4 | 29.6 | 0.1287 | 0.922 | 23,6 | 29,7 | 0.1377 | 0.910 |
| 10,0 | 25.0 | 32.7 | 0.1012 | 0.956 | 25,1 | 32,6 | 0.1093 | 0,940 |
| 12,0 | 27.0 | 36.1 | 0.0844 | 0.999 | 26.8 | 35.7 | 0.0914 | 0.969 |
| 14.0 | 29,5 | 39.7 | 0.0739 | 1.042 | 29.0 | 38.9 | 0.0798 | 1,008 |
| 16.0 | 32,4 | 43.3 | 0.0676 | 1.081 | 31.5 | 42.2 | 0.0725 | 1.043 |
| 18,0 | 35.6 | 46.8 | 0.0639 | 1.117 | 34,3 | 45,5 | 0.0681 | 1.072 |
| 20,0 | 38,8 | 50.0 | 0.0619 | 1,142 | 37.2 | 48.6 | 0.0653 | 1.100 |

**TABLE 53.** Transport Properties of Freon-23 in the Single-Phase Region (*Continued*)

| | T=433.15 K | | | | T=443.15 K | | | |
|---|---|---|---|---|---|---|---|---|
| $p$ | $\eta \cdot 10^6$ | $\lambda \cdot 10^6$ | $a \cdot 10^6$ | Pr | $\eta \cdot 10^6$ | $\lambda \cdot 10^6$ | $a \cdot 10^6$ | Pr |
| 0.1 | 21.0 | 20.6 | 11.5300 | 0.934 | 21.5 | 21.1 | 11.9700 | 0.946 |
| 0.5 | 21.1 | 21.1 | 2.3300 | 0.923 | 21.5 | 21.6 | 2.4100 | 0.921 |
| 1.0 | 21.1 | 21.6 | 1.1660 | 0.913 | 21.5 | 22.1 | 1.2090 | 0.912 |
| 2.0 | 21.3 | 22.7 | 0.5860 | 0.901 | 21.7 | 23.2 | 0.6100 | 0.905 |
| 3.0 | 21.5 | 23.7 | 0.3910 | 0.890 | 22.0 | 24.2 | 0.4080 | 0.899 |
| 4.0 | 21.9 | 24.9 | 0.2940 | 0.888 | 22.3 | 25.3 | 0.3070 | 0.892 |
| 5.0 | 22.3 | 26.0 | 0.2340 | 0.893 | 22.7 | 26.4 | 0.2460 | 0.890 |
| 6.0 | 22.7 | 27.2 | 0.1951 | 0.898 | 23.1 | 27.6 | 0.2060 | 0.889 |
| 8.0 | 23.9 | 29.8 | 0.1461 | 0.906 | 24.1 | 30.0 | 0.1543 | 0.899 |
| 10.0 | 25.2 | 32.5 | 0.1167 | 0.926 | 25.4 | 32.5 | 0.1239 | 0.919 |
| 12.0 | 26.8 | 35.4 | 0.0981 | 0.952 | 26.8 | 35.2 | 0.1045 | 0.937 |
| 14.0 | 28.7 | 38.4 | 0.0858 | 0.980 | 28.5 | 38.0 | 0.0915 | 0.960 |
| 16.0 | 30.9 | 41.4 | 0.0776 | 1.010 | 30.5 | 40.8 | 0.0826 | 0.987 |
| 18.0 | 33.4 | 44.4 | 0.0722 | 1.042 | 32.7 | 43.6 | 0.0766 | 1.014 |
| 20.0 | 36.0 | 47.3 | 0.0688 | 1.065 | 35.0 | 46.3 | 0.0725 | 1.032 |

**TABLE 53.** Transport Properties of Freon-23 in the Single-Phase Region (*Continued*)

| | T=453.15 K | | | | T=463.15 K | | | |
|---|---|---|---|---|---|---|---|---|
| $p$ | $\eta \cdot 10^6$ | $\lambda \cdot 10^6$ | $a \cdot 10^6$ | Pr | $\eta \cdot 10^6$ | $\lambda \cdot 10^6$ | $a \cdot 10^6$ | Pr |
| 0.1 | 21.9 | 21.7 | 12.4100 | 0.948 | 22.4 | 22.2 | 12.8100 | 0.961 |
| 0.5 | 21.9 | 22.2 | 2.5000 | 0.936 | 22.4 | 22.7 | 2.5900 | 0.946 |
| 1.0 | 22.0 | 22.7 | 1.2580 | 0.927 | 22.4 | 23.2 | 1.3050 | 0.932 |
| 2.0 | 22.1 | 23.6 | 0.6310 | 0.913 | 22.6 | 24.1 | 0.6570 | 0.921 |
| 3.0 | 22.4 | 24.7 | 0.4250 | 0.904 | 22.8 | 25.1 | 0.4410 | 0.912 |
| 4.0 | 22.7 | 25.7 | 0.3200 | 0.897 | 23.1 | 26.1 | 0.3330 | 0.904 |
| 5.0 | 23.0 | 26.8 | 0.2570 | 0.891 | 23.4 | 27.2 | 0.2690 | 0.892 |
| 6.0 | 23.4 | 27.9 | 0.2150 | 0.892 | 23.8 | 28.2 | 0.2250 | 0.894 |
| 8.0 | 24.4 | 30.2 | 0.1624 | 0.898 | 24.7 | 30.4 | 0.1710 | 0.897 |
| 10.0 | 25.6 | 32.6 | 0.1310 | 0.912 | 25.8 | 32.7 | 0.1388 | 0.906 |
| 12.0 | 26.9 | 35.1 | 0.1106 | 0.928 | 27.0 | 35.1 | 0.1178 | 0.916 |
| 14.0 | 28.4 | 37.7 | 0.0970 | 0.946 | 28.4 | 37.5 | 0.1033 | 0.931 |
| 16.0 | 30.2 | 40.4 | 0.0876 | 0.966 | 29.9 | 39.9 | 0.0930 | 0.947 |
| 18.0 | 32.1 | 43.0 | 0.0809 | 0.986 | 31.6 | 42.4 | 0.0857 | 0.965 |
| 20.0 | 34.3 | 45.5 | 0.0762 | 1.012 | 33.6 | 44.8 | 0.0805 | 0.986 |

**TABLE 53.** Transport Properties of Freon-23 in the Single-Phase Region (*Continued*)

| | T=473.15 K | | | | | | | | |
|---|---|---|---|---|---|---|---|---|---|
| $p$ | $\eta \cdot 10^6$ | $\lambda \cdot 10^6$ | $a \cdot 10^6$ | Pr | $p$ | $\eta \cdot 10^6$ | $\lambda \cdot 10^6$ | $a \cdot 10^6$ | Pr |
| 0.1 | 22.8 | 22.8 | 13.2900 | 0.964 | 8.0 | 25.0 | 30.6 | 0.1805 | 0.895 |
| 0.5 | 22.8 | 23.3 | 2.7000 | 0.948 | 10.0 | 26.0 | 32.8 | 0.1470 | 0.901 |
| 1.0 | 22.9 | 23.7 | 1.3530 | 0.943 | 12.0 | 27.0 | 35.0 | 0.1253 | 0.904 |
| 2.0 | 23.0 | 24.6 | 0.6830 | 0.927 | 14.0 | 28.3 | 37.2 | 0.1101 | 0.916 |
| 3.0 | 23.2 | 25.6 | 0.4600 | 0.915 | 16.0 | 29.7 | 39.5 | 0.0993 | 0.929 |
| 4.0 | 23.5 | 26.6 | 0.3490 | 0.905 | 18.0 | 31.2 | 41.7 | 0.0909 | 0.947 |
| 5.0 | 23.8 | 27.6 | 0.2810 | 0.900 | 20.0 | 33.0 | 44.0 | 0.0851 | 0.966 |
| 6.0 | 24.2 | 28.6 | 0.2370 | 0.895 | | | | | |

# REFERENCES

## INTRODUCTION

0.1. Aleshin U. P., Arefieva L. H., Lugovtsev V. V. Speed of sound and isentropic index for Freon-12, -13, -12B1, -22, and ammonia.—In: Thermodynamic Properties of the Most Important Working Substances for Refrigeration Equipment.—Tr. VNI (Kh), Moscow, 1976, p. 7—12.

0.2. Altunin V. V., Spiridonov G. A. Method of analytical calculation of thermodynamic functions of compressed gases on arbitrary constant lines on the diagram of states.—Teploenergetika, 1969, No. 6.

0.3. Altunin V. V., Sakhabetdinov M. A. Comparative analysis of some equations of viscosity for gases and liquids.—In: Thermophysical Properties of Liquids. Nauka, Moscow, 1973, p. 123—130.

0.4. Altunin V. V. The use of digital computers for constructing accurate equations of state of pure substances from extensive measurements of thermodynamic properties.—In: Equation of State of Gases and Liquids. Nauka, Moscow, 1975, p. 117—134.

0.5. Altunin V. V. Thermophysical Properties of Carbon Dioxide. Izdatelstvo Standartov, Moscow, 1975.

0.6. Anisimov M. A. Investigation of critical phenomena in liquids.—Usp. Fiz. Nauk, 1974, v. 114, No. 2, p. 249—294.

0.7. Badielkes I. C. In: Refrigeration Engineering, Encyclopedia—Handbook. Gosstandart, Moscow, 1966, v. 1.

0.8. Bogdanov C. N., Ivanov O. P., Kupriyanova A. V. Properties of Working Substances, Thermocarriers and Materials Used in Refrigeration Engineering. LGU, Leningrad, 1972.

0.9. Bogdanov C. N., Ivanov O. P., Kupriyanova A. V. Refrigeration Engineering. Properties of Substances. Handbook. Mashinostroyeniye, Leningrad, 1976.

0.10. Vargaftik N. B. Handbook of Thermophysical Properties of Gases and Liquids. 2nd Edition, Nauka, Moscow, 1972; available in English translation as Handbook of Physical Properties of Liquids and Gases, Hemisphere, Washington, 1975.

0.11. Vaskov E. T., Panauty V. I. Thermodynamic Properties of Freon-21. Izd-vo LTI (Kh)P, Leningrad, 1968, 26 p.

0.12. Geller V. Z., Gorkin C. F., Kronberg A. V. Calculation of viscosity of some liquid freons of the methane and ethane series.—Kholod. Tekh. i Tekhnol., 1973, v. 18, p. 117—121.

0.13. Geller V. Z., Gorkin C. F., Zaporozhan G. B., Boitenko A. K., Peredri V. G. Thermal Conductivity of Freons 12, 13, 14, 22, 23, 12B1, 13B1 in a wide range of parameters of state.—Thermophysical properties of freons. Gosstandart SSSR, GSSSD, 1977, 1.

0.14. Zhokhovsky M. K. About equations of $p$, $T$ fusion curve.—Izmer. Tekh., 1976, No. 4, p. 49—52.

0.15. Ivanchenko C. I. A Study of Dynamic Viscosity of Freons of the Methane and Ethane Series. Author's Abstract of Candidate Thesis. OTIFI, Odessa, 1974.

0.16. Kesselman P. M., Kamenetsky V. R., Ykoub E. C. Transport Properties of Real Gases. Visha Shkola, Kiev, 1976.

0.17. Classifier of Freons and Their Properties. Izd-vo Standartov, Moscow, 1974.

0.18. Kletsky A. V. Thermophysical Properties of Freon-22. Izd-vo Standartov, Moscow, 1970; available in English translation from NTIS, Springfield, Va., as TT 70-50178 (A04), 72 p., 1971.

0.19. Kletsky A. V. The structure of corresponding equations of state of cryogenic agents.— In: Machines and Apparatus for Refrigeration, Cryogenic Engineering and Air-Conditioning. Leningrad, 1976, p. 169—174.

0.20. Kletsky A. V., Sagaedakova H. G. Review of Freon-22 viscosity data and approximation of their temperature and pressure functions.— In: Thermophysical Properties of Liquids. Nauka, Moscow, 1976, p. 117—120.

0.21. Klemenko A. P., Krasnoouky C. I., Kolesnik V. M. Thermodynamic properties of freons in the ideal gas state.—Kholod. Tekh. Tekhnol., 1973, v. 18, p. 110—114.

0.22. Klemenko A. P., Krasnoouky C. I., Kolesnik V. M. The Starling–Han generalized equation for the calculation of thermodynamic properties of freons and their mixtures using computers.—Kholod. Tekh., 1976, No. 8, p. 26—28.

0.23. Krasnoouky C. I., Kolesnik V. M. The BWR generalized equation for the calculation of $pvT$ variables of individual freons using computers.—Khim. Tekhnol., 1973, No. 5, p. 23—26.

0.24. Kronberg A. V. Viscosity equation for liquid freons F-21, 22, and 23.—Kholod. Tekh. Tekhnol., 1974, v. 20, p. 98—99.

0.25. Kurielov E. G., Onosovsky V. V. The use of equation of state suggested by Starling for the determination of parameters of the working substances for refrigerating machines.—Kholod. Tekh., No. 4, p. 31—33.

0.26. Moskviecheva V. H., Ogurechnikov L. A., Petin U. M. Mathematical modelling of thermophysical properties of freons.—Izd-vo SO AN SSSR, Ser. Tekh. Nauk, 1976, v. 3, p. 27—34.

0.27. Osipov O. A., Minkin V. I., Gardnovsky A. D. Handbook of Dipole Moments, 3rd Edition. Vieshaya Shkola, Moscow, 1971.

0.28. Perelshtein I. I. Tables and Diagrams of Thermodynamic Properties of Freons 12, 13, 22. VNI (Kh) I, Moscow, 1971.

0.29. Perelstein I. I., Parushin E. B. A method for the determination of thermodynamic properties of main cryogenic agents from experimental data.—Kholod. Tekh., 1976, No. 1, p. 27—30.

0.30. Rivkin C. L. A Study of Thermophysical Properties of Normal and Heavy Water. Author's Abstract of Doctorate Thesis. VTI, Moscow, 1965.

0.31. Rivkin C. L., Levin A. Y., Izraelevsky L. B., Kharitonov K. G. A study of the link between viscosity of liquids and their thermal properties.—Enzh. Fiz. Zh., 1971, v. 21, p. 405—410.

0.32. Sakhabetdinov M. A. Elaboration and Investigation of an Orthogonal Expansion Method for the Analysis of Experimental Data about Thermophysical Properties of Substances Using Computers. Author's Abstract of Candidate Thesis. MEI, Moscow, 1977.

0.33. Skripov V. P., Muratov G. N. Data about surface tension of liquids and their analysis using thermodynamic similarity.—Zh. Fiz. Khim., 1977, v. 51, No. 6, p. 1369—1372.

0.34. Tomanovskaya V. F., Kolotova B. E. Freons, Properties and Application. Khimiya, Moscow, 1970.

0.35. Tsvetkov O. B., Laptiev U. A., Polyakova N. A. Thermal conductivity of freons in a wide temperature and pressure interval.—Kholod. Tekh., 1974, No. 11, p. 39—43.

0.36. Tchaikovsky V. F., Geller V. Z., Goriekin C. F. Complex investigation of thermophysical properties of the most important and promising freons in the liquid and gaseous phases.— In: Thermophysical Properties of Liquids. Nauka, Moscow, 1976, p. 108—117.

0.37. Barho W. Die Molwärme der Fluor-Chlor-Derivate des Methans im Zustand idealer Gase.—Kältetechnik, 1965, Bd. 17, S. 219—222.

0.38. Basu R. S., Sengers J. V. Thermal conductivity of steam in the critical region. Proc. seventh sympos. on thermophys. prop., ASME, N.—Y., 1977.

0.39. Döring R., Löffler H. J. Thermodynamische Eigenschaften von Trifluorometan.—Kältetechnik, 1968, Bd. 20, S. 342—348.

0.40. Gordon D. T., Hamilton J. F., Fontaine W. E. An empirical equation for predicting the viscosity of liquid refrigerants.—ASHRAE Trans., 1969, v. 75, No. 1, p. 40—52.

0.41. Küper P., Löffler H. J. Eine neue Dampftafel für das Kältemittel 22.—Kältetechnik, 1971, Bd. 25, S. 47—51.

0.42. Morsy T. E. Eine neue zusfandgleichung für Trifluoromethan.—Kältetechnik, 1965, Bd. 17, S. 272—275.

0.43. Morsy T. E. Extended Benedict-Weber-Rubin Equation of state.—J. Chem. Eng. Data, 1970, v. 15, p. 256—265.

0.44. Nischiumi H., Saito S. An improved generalized BWR equation of state applicable to low reduced temperatures.—J. Chem. Eng. Japan., 1975, v. 8, p. 356—360.

0.45. Plank R. Handbuch der Kältetechnik. Bd. 4. Die Kältemittel. Brl., Springer Verlag, 1956. S. 364—382, 465—467.

0.46. Rombusch U. K., Giesen H. Neue Mollier H-lg P Diagramme für die Kälmittel R 11, 12, 13 und R 21.—Kältetechnik, 1966, Bd. 18, No. 2, S. 37—40.

0.47. Seshadri D. N. Modified equation of state for gases.—Indian J. Technology, 1975, v. 13, p. 19—24.

0.48. Starling K. E. Thermo data refined for LPG. Part I. Equation of state and computer prediction.—Hydrocarbon processing, 1971, v. 51, p. 101—104.

0.49. Starling K. E., Han M. S. Thermo data refined for LPG. Part 14. Mixtures.—Hydrocarbon processing, 1972, v. 5, p. 129—132.

0.50. Tauscher W. A. Correlation for gas thermal conductivity of halogenated methans.—AIChE Journal, 1971, v. 17, p. 1511—1512.

0.51. Thermophysical properties of refrigerants. Chlorodifluoromethane. Japanese Assoc. Refrig., 1975.

# CHAPTER 1

1.1. Altunin V. V. Method of calculating thermodynamic properties of real gas mixtures from a limited quantity of initial experimental data.—Teploenergetika, 1963, No. 4, p. 78—84.

1.2. Altunin V. V. Surface tension of freons-21, 22, and 23.—Thermophysical Properties of Freons. Gosstandart SSSR, GSSSD, 1977, 1.

1.3. Geller V. Z. Viscosity of Freon-21, 22, and 23.—Kholod. Tekh. Tekhnol., 1975, v. 22, p. 41—44.

1.4. Geller V. Z. Generalization of experimental data on thermal conductivity of Freons-21, 22, and 23.—Enzh. Fiz. Zh., 1977, v. 33, No. 1, p. 75—83.

1.5. Golubev I. F. Viscosity of Gases and Gaseous Mixtures. Phyzmatgiz, Moscow, 1959; available in English translation from NTIS, Springfield, Va., as TT 70-50022, 1959.

1.6. Golubev I. F., Gnezdilov N. E. Viscosity of Gaseous Mixtures. Izd-vo Standartov, Moscow, 1971.

1.7. Gurvich L. V., Khachuruzov G. A., Medvedev V. A., et al. Thermodynamic Properties of Individual Substances. Izd-vo AN SSSR, Moscow, 1962, v. 1—2.

1.8. Dorokhov A. P., Kiriyanenko A. A., Soloviov A. N. Surface tension of Freons.—PMTF, 1969, No. 1, p. 93—96.

1.9. Zhokhovsky M. K. About some laws governing the fusion of substances and their values for high pressure scale.—Izmer. Tekh., 1958, No. 2, p. 16—21.

1.10. Zhokhovsky M. K. Volume change during fusion under pressure.—Zh. Fiz. Khim., 1963, v. 37, p. 37, p. 2635—2639.

1.11. Zaalishvili Sh. D., Belousova Z. C., Kolesko L. E. Second virial coefficient of vapors and their mixtures. V. System Chloroform–Benzol.—Zh. Fiz. Khim., 1965, v. 39, p. 447—450.

1.12. Klimenko A. P., Krasnoouky C. I., Kolesnik V. M. The use of generalized equations for the calculation of thermodynamic properties of freons on computers.—Kholod. Tekh. Tekhnol., 1976, v. 24, p. 75—79.

1.13. Lapidus I. I., Niselson L. A., Seifer A. L. Fundamental thermodynamic properties and characteristics of halogenated monoselenium and methane.—Thermophysical Characteristics of Substances. Gosstandart SSSR, GSSSD, 1968, 1, p. 103—135.

1.14. Perelshtein I. I. Generalized equations of state and vapor pressure curve for freons.—Kholod. Tekh., 1967, No. 3, p. 27—33.

1.15. Perelshtein I. I., Parushin E. B. Generalized temperature dependence for saturated vapor pressure and density of boiling liquids.—In: Thermodynamic Properties of the Most Important Working Substances for Refrigerating Machines.—Tr. VNI (Kh) I, Moscow, 1976, p. 13—26.

1.16. Filipov L. P. Thermal conductivity of 50 organic liquids.—Vestn. MGU. Ser. Fiz. Mat. Yestestv. 1954, v. 8, No. 12, p. 45—48.

1.17. Filipov L. P. Method of calculation and prediction of properties of liquids and gases on the basis of theory of thermodynamic similarity.—In: Review of Thermophysical Properties of Substances, v. 2, Moscow, 1977.

1.18. Tsvetkov O. B., Laptiev U. A., Poliakova N. A. Thermal conductivity of gaseous freons at atmospheric pressure.—In: Machines and Apparatus for Refrigeration and Cryogenic Engineering and Air Conditioning. Izd-vo LTI (Kh) P, Leningrad, 1976, p. 179—182.

1.19. Aihara A.—J. Chem. Soc. Japan, 1949, v. 70, p. 384—386.

1.20. Babb S. E. Parameters in the Simon equation relating pressure and melting temperature.—Rev. Modern. Phys., 1963, v. 35, p. 400—413.

1.21. Bates O., Hazzard G., Palmer G. Thermal conductivity of liquids.—Ind. Eng. Chem., 1941, v. 33, p. 275—366.

1.22. Beckmann E., Liesche O. Ebulioskopisches verhaben von Lösungsmitteln bei verchiedenen. Drucken.—Z. Phys. Chem., 1914, Bd. 88, S. 23—34.

1.23. Braune H., Linke R. Über die innere Reibung einiger Gase und Dampfe III.—Z. Phys. Chem., 1930, Bd. A 148, S. 195—215.

1.24. Bridgman P. W. The phase diagramm of eleven substances with especial reference to the melting curve.—Phys. Rev., 1914, v. 3, p. 153—203.

1.25. Bridgman P. W.—Proc. Amer. Acad., 1941, v. 74, p. 12—28.

1.26. Coop I. E. The dielectric constants of ether-chloroform mixtures.—Trans. Farad. Soc., 1937, v. 33, p. 583—590.

1.27. Dolezalek F., Schulze A. Zur Theorie der binären Gemische und Konzentrierten Lösungen. IV.—Z. phys. chem., 1913, Bd. 83, S. 45—50.

1.28. Drucker C., Jumeno E., Kangro W. Dampfdrucke flüssiger Stoffe bei niedrigen Temperaturen.—Z. Phys. Chem., 1915, Bd. 90, S. 513—552.

1.29. Djalalian W. Measurement of the thermalconductivity of liquid refrigerants at low temperatures.—Bull. Inst. Intern. Froid. Annexe 2. Turin, 1964, p. 153—165.

1.30. Eucken A. Über des Wärmeleitvermögen, die spezifische Wärme und die innere Reibung der Gase.—Phys. Z., 1913, Bd. 14, S. 324—332.

1.31. Fort R. J., Moore W. R. Adiabatic compressibilities of binary liquid mixtures.—Trans. Farad. Soc., 1965, v. 61, p. 2102—2111.

1.32. Francis P. G., McGlashan M. L. Second virial coefficients of vapour mixtures.—Trans. Farad. Soc., 1955, v. 51, p. 593—599.

1.33. Gelles E., Pitzer K. S. Thermodynamic functions of the halogenated methanes.—J. Am. Chem. Soc., 1953, v. 75, p. 5259—5267.

1.34. Harrison D., Moelwyn-Hughes E. A. The heat capacities of certain liquids.—Proc. Roy. Soc., 1957, v. 239 A, p. 230—246.

1.35. Held E., Drunen F. The measurement of the thermal conductivity of liquids.—Proc. VII Intern. Congr. Appl. Mech., 1948, v. 3, p. 79—90.

1.36. Herz W., Rathman W. Physikalische Konstanten einiger als Lösungsmittel wichtiger chlorierter Kohlen-Wasserstoffe.—Chem. Ztg., 1912, Bd. 36, S. 1417—1418.

1.37. Herz W., Neukirch E. Zur Kenntnis Kritischer Grössen.—Z. Phys. Chem., 1923, Bd. 104, S. 433—450.

1.38. Hutchinson E. On the measurement of the thermal conductivity of liquids.—Trans. Farad. Soc., 1945, v. 41, p. 87—90.

1.39. JANAF thermochemical tables. 2nd ed./USA; Nat. Bur. Standards, NSRDS-NBS37, 1971.

1.40. Kudchadker A. P., Alani C. H., Zwolinski B. J. The critical constants of organic substances.—Chem. Rev., 1968, v. 68, p. 659—735.

1.41. Kuenan J. P., Robson W. G. Observations on mixtures with minimum or maximum vapor pressure.—Phil. Mag., 1902, v. 4, p. 116—132.

1.42. Lambert J. D. et al. Virial coefficients some of organic vapours.—Proc. Roy. Soc., 1949, v. A 196, p. 113—135.

1.43. Lambert J. D., Staines E. N., Woods S. D. Thermal conductivities of organic vapours.—Proc. Roy. Soc., 1950, v. A 200, p. 262—271.

1.44. Landolt—Börnstein. 6 Auflage. Bd. 2, Teil 5a. Springer Verlag, 1969, S. 195.

1.45. Mason H. Thermal conductivity of some industrial liquids from 0 to 100°C.—Trans. ASME, 1954, v. 5, p. 817—821.

1.46. Mathews J. H. The accurate measurement of heats of vaporization of liquids.—J. Amer. Chem. Soc., 1926, v. 48, p. 562—576.

1.47. Miller Ch. C. The Stokes—Einstein laws for diffusion in solution.—Proc. Roy. Soc., 1924, v. A 106, p. 724—749.

1.48. Pal A. K., Barua A. K. Intermolecular potentials and viscosities of some polar organic vapours.—Brit. J. Appl. Phys. (2), 1968, v. 1, p. 71—76.

1.49. Phillips T. W., Murphy K. P. Liquid viscosity of halogenated refrigerants.—ASHRAE Trans., 1971, v. 77, part. II, p. 146—156.

1.50. Phillips T. W., Murphy K. P. Liquid viscosity of halocarbons.—J. Chem. Eng. Data, 1970, v. 15, p. 304—307.

1.51. Rappanecker K. Über die Reibungskoeffizienten von Dämpfen und ihre Abhängigkeit von der Temperatur.—Z. phys. Chem., 1910, Bd. 72, S. 695—722.

1.52. Rex A. Über die Löslichkeit der Halogenderivate der Kohlenwasserstoffe in Wasser.—Z. phys. Chem., 1906, Bd. 55, S. 355—377.

1.53. Riedel L. Messung der Wärmeleitfähigkeit von organischen Flüssigkeiten, insbesondere von Kältemitteln.—Forsch. Geb. Ing.-Wes., 1940, Bd. 11, S. 340—347.

1.54. Rodgers A. S., Chao J., Wilhoit R. C., Zwolinski B. Ideal gas thermodynamic Properties of eight Chloro- and fluoromethanes.—J. Phys. Chem. Ref. Data, 1974, v. 3, p. 117—140.

1.55. Scatchard G., Raymond C. L. Vapor-liquid equilibrium. II Chloroform—Ethanol mixtures.—J. Amer. Chem. Soc., 1938, v. 60, p. 1278—1287.

1.56. Schmidt G. C. Binäre Gemische.—Z. Phys. Chem., 1926, Bd. 121, S. 221—253.

1.57. Schulze A. Über das Gleichgewicht in kondensierten systemen.—Z. phys. Chem., 1921, Bd. 97, S. 388—416.

1.58. Seshadri D. N., Viswanath D. S., Kuloor N. R. Thermodynamic properties of chloroform.—J. Indian Inst. Sci., 1968, v. 50, No. 3, p. 179—199.

1.59. Smyth C. P., Morgan S. O. The temperature dependence of the polarization in certain liquid mixtures.—J. Amer. Chem. Soc., 1928, v. 50, p. 1547—1560.

1.60. Staveley L. A., Tupman W. I., Hart K. R. Some thermodynamic properties of the systems acetone + chloroform.—Trans. Farad. Soc., 1955, v. 51, p. 323—343.

1.61. Stull D. R. Vapor pressure of pure substance organic compounds.—Ind. Eng. Chem., 1947, v. 39, p. 517—540.

1.62. Suhrmann R. Über die Druckabhängigkeit der Däpfing einer um ihre vertikale Achse schwingenden Scheibe.—Z. Phys., 1923, Bd. 14, S. 56—62.

1.63. Tauscher W. A. Messung der Wärmeleitfähigkeit flüssiger Kältemittel mit einem instationären Hitzdrathverfahren.—Kältetechnik, 1967, Bd. 19, S. 288—292.

1.64. Thorpe E., Rodger J. W. On the relations between the viscosity of liquids and their chemical nature.—Phil. Trans. Roy. Soc., 1897, v. A 189, p. 71—107.

1.65. Titani T. The viscosity of vapours of organic compounds.—Bull. Chem. Soc. Japans, 1933, v. 8, p. 255—267.

1.66. Tsakalotes D. E. Sur la hydrates des acides gras d'apres les mesures de viscosite de leurs solutions.—Compt. rend., 1908, v. 146, s. 1146—1149.

1.67. Vines R. G., Bennett L. A. The thermal conductivity of organic vapors.—J. Chem. Phys., 1954, v. 22, p. 360—366.

1.68. Vogel H. Über die Viscosität einiger Gase und ihre Temperaturabhängigkeit bei tiefen Temperaturen.—Ann. Physik, 1914, Bd. 43, S. 1235—1272.

1.69. Weber R. Untersuchungen über die Wärmeleitung in Flüssigkeiten.—Ann. Physik, 1895, Bd. 11, S. 1047—1060.

1.70. Williams J. W., Daniels F. The specific heats of binary mixtures.—J Amer. Chem. Soc., 1925, v. 47, p. 1490—1503.

1.71. Wright R. Densities of saturated vapours.—J. Phys. Chem., 1932, v. 36, p. 2793—2795.

# CHAPTER 2

2.1. Altunin V. V., Spiridonov G. A., Kaekin V. C. Equation of state and thermodynamic properties of Freon-21.—Teploenergetika, 1969, No. 1, p. 79—82.

2.2. Vargaftik N. B., Filipov L. P., Tarzimanov A. A., Urchak R. P. Thermal conductivity of Gases and Liquids. Izd-vo Standartov, Moscow, 1970.

2.3. Vukalovich M. P., Altunin V. V. Thermophysical Properties of Carbon Dioxide. Atomizdat, Moscow, 1965; available in English translation from Collet's Publishing Co., London.

2.4. Geller V. Z., Ivanchenko C. I., Kronberg A. V. Study of coefficients of dynamic viscosity of freons of the methane series.—Thermophysical Properties of Substances and Materials. Gosstandart SSSR, GSSSD, 1975, 8, p. 148—161.

2.5. Geller V. Z., Peredri V. G. Study of thermal conductivity of Freon-21 and Freon-14.—Kholod. Tekh. Tekhnol., 1974, v. 9, p. 102—106.

2.6. Geller Z. I., Nikulshin R. K., Piatnitskaya N. I. Viscosity of bromide-based freons.—Kholod. Tekh., 1969, No. 4, p. 60—65.

2.7. Gruzdev V. A., Shestova A. I. Thermal conductivity of liquid and gaseous Freon-21.—In: Study of Thermophysical Properties of Substances. Nauka, Novosibirsk, 1970, p. 144—149.

2.8. Gruzdev V. A., Shestova A. I., Selin V. V. Thermal conductivity of freons.—In: Thermophysical Properties of Freons. Nauka, Novosibirsk, 1969, p. 62—74.

2.9. Gruzdev V. A., Shumskaya A. I. Experimental Study of Heat Capacity of Vapors of Freons: F-11, F-12, F-13, F-21, F-22, and F-23.—Thermophysical Properties of Substances and Materials. Gosstandart SSSR, GSSSD, 1975, 8, p. 108—130.

2.10. Kalafati D. D., Raskazov D. C., Petrov E. K., Kaekin V. C. Experimental study of pvT variables for Freon-21.—Teploenergetika, 1968, No. 11, p. 80—83.

2.11. Kalafati D. D., Raskazov D. C., Petrov E. K., Kaekin V. C. Experimental Study of pvT Behavior of Freon-21.—In: Tr. Vses. Nauchno.—Tekhn. Konf. Termod., v. 2, Thermophysical Properties of Substances, 1969, p. 289—291.

2.12. Lavrov V. A., Sheludyakov E. P., Soloviov A. N. Study of heat capacity $c_p$ of liquid phase of Freon-21 and Freon-14B on the saturation line.—In: Thermophysical Properties of Freons. Nauka, Novosibirsk, 1969, p. 24—35.

2.13. Raskazov D. C., Babikov U. M., Filatov N. Y. Experimental investigation of viscosity for some freons of the methane series.—Tr. Mosk. Energ. Inst., 1975, No. 234, p. 90—96.

2.14. Tsvetkov O. B. Investigation of thermal conductivity of liquid freons.—Enzh. Fiz. Zh., 1965, v. 9, No. 6, p. 810—815.

2.15. Sheludyakov E. P., Komarov C. G., Kolotov N. L., Soloviov A. N. Experimental investigation of speed of sound in vapors and liquids.—In: Study of Thermophysical Properties of Substances. Nauka, Novosibirsk, 1967, p. 159—180.

2.16. Sheludyakov E. P., Kolotov N. L., Soloviov A. N. Study of speed of sound in freons at low frequencies using the stationary wave method in a resonator.—In: Thermophysical Properties of Freons. Nauka, Novosibirsk, 1969, p. 96—120.

2.17. Sheludyakov E. P., Buzhdan Y. M., Kolotov Y. L. Thermodynamic properties of Freon-21.—Thermophysical Properties of Substances and Materials. Gosstandart SSSR, GSSSD, 1971, v. 4, p. 100—124.

2.18. Sheludyakov E. P., Shilyakov A. A. Equation of state of Freon-21 and Freon-114B2.—Kholod. Tekh., 1971, No. 3, p. 35—38.

2.19. Shilyakov A. A., Sheludyakov E. P., Soloviov A. N. Experimental study of density of Freon-21 and Freon-114B2 using constant-volume piezometer.—In: Thermophysical Properties of Freons. Nauka, Novosibirsk, 1969, p. 6—24.

2.20. Shumskaya A. I., Gruzdev V. A. Constant-pressure heat capacity of vapors of freons.—In: Thermophysical Properties of Freons. Nauka, Novosibirsk, 1969, p. 35—43.

2.21. Shilyakov A. A. Experimental Study of Thermal Properties and Critical Parameters of Freons F-21 and F-114B2. Author's Abstract of Candidate Thesis. SO AN, SSSR, Novosibirsk, 1971.

2.22. ASHRAE handbook of fundamentals, 1968, v. 15, p. 213—224.

2.23. Benning A. F. McHarness R. C. Thermodynamic properties of fluorochloromethanes and —ethanes. Orthobaric densities and critical constants of three fluorochloromethanes and trifluorotrichloroethane.—Ind. Eng. Chem., 1940, v. 32, No. 6, p. 814—816.

2.24. Benning A. F., McHarness R. C. Vapor Pressure of three fluorochloromethanes and trifluorotrichloroethane.—Ind. Eng. Chem., 1940, v. 32, No. 4, p. 497—503.

2.25. Benning A. F., McHarness R. C. $p—v—T$ relations of three fluorochloromethanes and thifluorotrichloroethane.—Ind. Eng. Chem., 1940, v. 32, No. 5, 698—701.

2.26. Benning A. F., McHarness R. C. Orthobaric densities and critical constants of three fluorochloromethanes and trifluorotrichloroethane.—Ind. Eng. Chem., 1940, v. 32, No. 6, p. 814—816.

2.27. Benning A. F., McHarness R. C., Markwood W. H., Smith W. I. Heat capacity of the liquid and vapor of three fluorochloromethanes and trifluorotrichloroethane.—Ind. Eng. Chem., 1940, v. 32, No. 7, p. 976—980.

2.28. Challoner A., Powell R. The thermal conductivities of liquids.—Proc. Roy. Soc., 1956, Ser. A 238, p. 1212—1218.

2.29. Gallant R. W.—In: Physical properties of hydrocarbons.—Hydrocarbon processing, 1965, 2, p. 113—120.

2.30. Kamien C., Witzell O. Effect of pressure and temperature on the viscosity of refrigerants in the vapor phase.—ASHRAE Trans., 1959, v. 65, p. 663—674.

2.31. Liley P. The thermal conductivity of 46 gases at atmosphere pressure.—Proc. fourth symposium on thermophysical properties. April 1—4, ASME, N.Y., 1968, p. 323.

2.32. Makita T. The viscosity of freons under pressure.—Rev. Phys. Chem. (Japan), 1954, v. 24, No. 2, p. 74—80.

2.33. Markwood W., Benning A. Thermal conductances and heat transmission coefficients of freon refrigerants.—J. ASHRAE, 1943, No. 2, p. 95—99.

2.34. Masia A., Bracero V., Rienda B. Variacion de la conductividad calorifica con la presion en ocho derivados halogendos del metano.—Anal. real soc. esp fis. guim., 1964, v. A 60, No. 1—2, p. 89—108.

2.35. Powell R., Challoner A. New measurements of the thermal conductivities of several liquid refrigerants of the fluorochloro derivative types.—In: Proc. X Internat. Congr. Refrig., Kopenhagen, v. 1, 1959, p. 382—387.

2.36. Powell R., Jolliffee B., Tye R., Longton A. The thermal conductivities of some liquid refrigerants.—Bull. Inst. Internat. du Froid. Paris, 1966, Annexe 2, p. 79—88.

2.37. Rombusch U. K., Giesen H. Dampftafel des Kättemittels R 21 (Monofluordichlormethan CHFCl$_2$).—Kältetechnik, 1968, Bd. 20, S. 93—94.

2.38. Witzell O., Jonson J. The viscosity of liquid and vapor refrigerants.—ASHRAE Trans., 1965, v. 71, p. 30—36.

2.39. Woodline H. Chimie et industrie genie chemique.— 1969, v. 5, s. 101, s. 612.

2.40. Ziebland H. The thermal conductivity of toluene. New determinations and an appraisal of recent experimental work.—Internat. J. Heat Mass Transfer, 1961, v. 2, p. 273—279.

# CHAPTER 3

3.1. Altunin V. V., Gadetsky O. G. Equation of State and Thermodynamic properties of liquid and gaseous Freon 22.—Thermophysical Properties of Substances and Materials. Gosstandart SSSR, GSSSD, 1973, 7, p. 115—135.

3.2. Altunin V. V. Use of a new method for the analysis of measurements for generalizing experimental data on viscosity of Freon-22.—Thermophysical Properties of Substances and Materials. Gosstandart SSSR, GSSSD, 1975, 8, p. 130—141.

3.3. Altunin V. V. Thermophysical properties of Freon-22 on the liquid–vapor equilibrium line.—Thermophysical Properties of Freons. Gosstandart SSSR, GSSSD, 1977, 1.

3.4. Altunin V. V., Bondarenko V. F., Kuznetsov G. O. Present state of investigations of thermodynamic effects of mixing compressed gases. Gosstandart, SSSR, GSSSD, 1977 (Ser. Obzornaya Informatsiya).

3.5. Butierskaya C. T. Study of coefficient of dynamic viscosity of Freon-22.—In: Thermophysical Properties of Liquids. Nauka, Moscow, 1970, p. 73—76.

3.6. Butierskaya C. T. Experimental Investigation of Coefficient of Dynamic Viscosity of Freons 22, 114, 115, and C-318 in Liquid and Gaseous States. Author's Abstract of Candidate Thesis. LTI (Kh) P, Leningrad, 1971.

3.7. Geller Z. I., Nikulshin P. K., Pitnitskaya N. I. Viscosity of liquid freons on the saturation line.—Kholod. Tekh. Tekhnol., 1970, v. 10, p. 22—29.

3.8. Geller V. Z. Study of Thermophysical Properties of Mangishlak Oil and its Fractions. Author's Abstract of Candidate Thesis MEI, Moscow, 1968.

3.9. Geller V. Z., Ivanchenko C. I., Peredri V. G. Experimental study of coefficients of dynamic viscosity and thermal conductivity of difluorochloromethane.—Izv. Vuzov. Neft Gas, 1973, No. 8, p. 61—65.

3.10. Geller V. Z. Study of thermal conductivity of some freons of the methane series.—Thermophysical Properties of Substances and Materials. Gosstandart SSSR, GSSSD, 1975, 8, p. 162—176.

3.11. Geller V. Z., Karbanov E. M., Gunchuk B. V., Zakharzhevsky V. Y., Lapardin N. I. Investigation of viscosity coefficient of some liquefied gases near the saturation curve.—Gazov. Promst., 1976, No. 3, p. 32—33.

3.12. Gunchuk B. V., Karbanov E. M., Zakgarzhevsky V. Y. Investigation of viscosity coefficient of some freons on the saturation line.—Thermophysical Properties of Substances and Materials. Gosstandart SSSR, GSSSD, 1977, 11, p. 39—46.

3.13. Dorokhov A. R., Kiriyanenko A. A., Soloviov A. N. Surface tension of freons.—In: Thermophysical Properties of Freons. Nauka, Novosibirsk, 1969, p. 43—61.

3.14. Zhelezny V. P. Experimental study of surface tension of Freons-22 and 12B1.—Kholod. Tekh. Tekhnol., 1976, v. 24, p. 68—71.

3.15. Zenkevich V. B. Experimental Study of Thermophysical Properties of Liquid Fuels and Lubricants. Author's Abstract of Candidate Thesis, MEI, Moscow, 1961.

3.16. Kletsky A. V. Elasticity curve of Freon-22.—Enz. Phyz. Zh., 1964, v. 7, p. 41—43.

3.17. Kletsky A. V. Experimental study of vapor pressure curve and specific volumes of Freon-22.—Kholod. Tekh., 1964, No. 4, p. 37—40.

3.18. Kletsky A. V. Thermodynamic properties of Freon-22.—Kholod. Tekh., 1964, No. 6, p. 70—72.

3.19. Kletsky A. V. *h*-lg *P* Diagram for Freon-22.—Kholod. Tekh., 1965, No. 3, p. 71—72.

3.20. Kletsky A. V., Butierskaya C. T. Coefficient of dynamic viscosity of Freon-22.—Kholod. Tekh., 1973, No. 6, p. 31—33.

3.21. Kletsky A. V., Tsuranova T. N. Density and equation of state of liquid Freon-22.—Thermophysical Properties of Substances and Materials. Gosstandart SSSR, GSSSD, 1975, 8, p. 79—83.

3.22. Lagutina L. M. Experimental study of *pvT* behavior of Freon-22.—Kholod. Tekh., 1966, No. 12, p. 25—28.

3.23. Lagutina L. M. Thermodynamic Study of Difluorochloromethane Using Acoustic and Piezometric Methods. Author's Abstract of Candidate Thesis, MIFI, Moscow, 1967.

3.24. Latieshev V. P., Gittelson E. P. Volatility of vapors of Freons-12, 22, and 142.—Kholod. Tekh., 1968, No. 8, p. 32—34.

3.25. Novikov I. I., Lagutina L. M. Experimental study of speed of sound propagation in saturated and superheated vapors of difluorochloromethane.—PMTE, 1967, No. 2, p. 147—149.

3.26. Peredri V. G. Study of Thermal Conductivity of Freons of the Methane and Ethane Series. Author's Abstract of Candidate Thesis, OTIPP, Odessa, 1975.

3.27. Perelshtein I. I. Speed of sound and isentropic index in superheated vapors of Freon 12, 13, and 22.—Kholod. Tekh., 1973, No. 3, p. 21—28.

3.28. Sadiekov A. Kh., Gabdrakhmanov R. G., Briekov V. P., Mukhamedziyanov G. Kh. Experimental study of thermal conductivity coefficient for some freons of the methane series.—Tr. Kazan. Khim. Tekhnol. Inst., 1971, v. 47, p. 35—39.

3.29. Tkachev A. G., Butierskaya C. T., Agaev N. A. Study of viscosity of Freons 22, 114, 115, and C318.—Kholod. Tekh., 1971, No. 4, p. 39—42.

3.30. Tsvetkov O. B. Thermal conductivity of liquid freons of the methane and ethane series.—Kholod. Tekh., 1965, No. 4, p. 28—31.

3.31. Tsvetkov O. B., Polyakov N. A. Thermal conductivity of freon-22.—Thermophysical Properties of Substances and Materials. Gosstandart SSSR, GSSSD, 1975, 8, p. 177—182.

3.32. Tchaikovsky V. F., Geller V. Z., Bondar G. E. Experimental study of coefficient of dynamic viscosity for Freon-22 at low temperatures.—Thermophysical Properties of Freons. Gosstandart SSSR, GSSSD, 1977, 1.

3.33. Tchernieva L. I. Study of thermal conductivity of Freon-22.—Kholod. Tekh., 1953, No. 3, p. 60—63.

3.34. Baehr H. D., Duicu T. N., Pollak R. A canonical equation of state for gaseous R 22 with enthalpy, entropy and pressure as variables.—In: Some thermophysical properties of refrigerants and insulants. Paris, 1973, p. 15—20.

3.35. Benning A. F., Markwood W. H. The viscosities of freon refrigerants.—Refr. Engng., 1939, v. 37, p. 243—247.

3.36. Bier K., Ernst G., Maurer G. Flow apparatus for measuring the heat capacity and the Joule–Thomson coefficient of gases.—J. Chem. Thermodyn., 1974, v. 6, p. 1027—1037.

3.37. Booth H. S., Swinehart C. F. The critical constants and vapor pressures at high pressure of some gaseous fluorides of group IV.—J. Amer. Chem. Soc., 1935, v. 57, p. 1337—1342.

3.38. Donaldson A. B. On the estimation of thermal conductivity of organic vapors. Data for some freons.—Ind. Eng. Chem. Fundament., 1975, v. 14, p. 325—328.

3.39. Djalalian W. H. Messungen der Wärmeleitzahl von Flüssigkeiten mit einer stationären Hitzdrathmethode.—Kältetechnik, 1966, Bd. 18, S. 410—415.

3.40. Ernst G., Büsser J. Ideal and real gas state heat capacities $c_p$ of $C_3H_8$, i—$C_4H_{10}$, $C_2F_5Cl$, $CH_2ClCF_3$, $CF_2ClCFC_2$ and $CHF_2Cl$—J. Chem. Thermodyn., 1970, v. 2, p. 787—791.

3.41. Gelles E., Pitzer K. Thermodynamic functions of the halogenated methanes.—J. Amer. Chem. Soc., 1953, v. 75, p. 5259—5267.

3.42. Grassman P., Straumann W., Widmer F., Jobst W. Measurements of thermal conductivities of liquids by an unsteady state method.—In: Progress in Internat. res. on thermod. transp. prop. Princeton, 1962, p. 447—453.

3.43. Grassman P., Jobst W. A non-steady-state method for the thermal conductivity of liquid and gases.—In: Proc. XI Internat. congr. of refr. Münich, 1963; London, Pergamon Press, 1965, v. I, p. 301—305.

3.44. Hajjar R. F., MacWood G. E. Second virial coefficients and the force constants $\epsilon_0$ and $r_0$ of six halogensubstituted methans.—J. Chem. Phys., 1968, v. 49, p. 4567—4570.

3.45. Hajjar R. F., MacWood C. E. Determination of the second virial coefficients of six fluorochloromethans by a gas balance method in the range 40 to 130°C.—J. Chem. Eng. Data, 1970, v. 15, p. 3—6.

3.46. Haworth W. S., Sutton L. E. The second density virial coefficients of some polar gases.—Trans. Farad. Soc., 1971, v. 67, p. 2907—2914.

3.47. Kokernak R. P., Feldman C. L.—ASHRAE Journal, 1971, v. 13, No. 7, p. 59—61.

3.48. Kumagai A., Iwasaki H. Compressibility factor of chlorodifluoromethane.—Preprint of 13th High pressure conference of Japan, 1971, p. 146—148.

3.49. Latto B., Hesoun P., Ashrani S. C. Absolute viscosity and molecular parameters for R 13, R 500, R 12 and R 22.—In: Proc. fifth sympos. on thermophys. prop. ASME, N.Y., 1970, p. 177—185.

3.50. Martin J. J. Equations of state.—Ind. Eng. Chem., 1967, v. 59, p. 34—52.

3.51. Masia A. P., Bracero A. V., Rienda B. J. Variacion de la conductividad calorifica con la presion en ocho deviados halogenados del metano.—Anales real soc. esp. fis. quim., v. A 60, No. 1—2, 1964, p. 89—108.

3.52. Meyer K. J. Über den Zusammenhang zwischen der Schallgeschwindigkeit und der Wärmeleitfähigkeit bei flussigen fluor-chlor-derivaten des Methans und Athans.—Kältetechnik, 1969, Bd. 21, S. 270—275.

3.53. Miyahara Y., Richardson E. G.—J. Acoust. Soc. Amer., 1956, v. 28, No. 6, p. 1016—1018.

3.54. Morsy T. E. Die specifische Wärme von Tetrafluomethan, Difluormonochlormethan, Tetrafluodichloräthan und Pentafluormonochloräthan in der Gasphase.—Kältetechnik, 1966, Bd. 18, S. 373—374.

3.55. Neilson E. F., White D. The heat capacity, heat of fusion, heat of transition and heat of vaporization of Chlorodiffluoromethane between 16 K and the boiling point.—J. Amer. Chem. Soc., 1957, v. 79, p. 5618—5621.

3.56. Oguchi K., Matsushita Y., Sagara T., Watanabe K., Tanishita T. Pressure-volume-temperature properties of R 22 in liquid and gaseous states.—Preprint XIV Congress internat. du froid. Moscow, 1975, Sec. B I.

3.57. Puranasamriddhi D. Messung der Wärmeleitfähigkeit von Flüssigkeiten und Flüssigkeitsgemischen.—Kältetechnik, 1966, Bd. 18, S. 445—450.

3.58. Sale P. Mesure de la conductivite thermique de fluides frigorines par la methode.—Bull. Inst. Internat. Froid. Annexe 2. Paris, 1964, p. 145—151.

3.59. Srichand M., Tirunarayanan M. A., Ramachandran A. Studies of the viscosity of mixtures of R 12 and R 22 vapors.—ASHRAE Journal., 1970, v. 12, p. 61—66.

3.60. Steinle H. Über die Oberflächenspannung von Kältemitteln, Kältemaschinenölen und deren Gemischen.—Kältetechnik, 1960, Bd. 12, S. 334—339.

3.61. Suther H., Cole R. H. Dielectric and pressure virial coefficients of imperfect gases. III. $CClF_2H$.—J. Chem. Phys., 1971, v. 54, p. 4988—4989.

3.62. Tanishita I., Oguchi K., et al. Pressure-volume-Temperature properties of R-503.—In: Some thermophysical properties of refrigerant and insulants. Paris, 1973, p. 25—31.

3.63. Weissman H. B., Meister A. G., Cleveland F. E. Infrared spectral data and assignments for $CHCl_2F$ and $CHClF_2$ and potential energy constants and calculated thermodynamic properties.—J. Chem. Phys., 1958, v. 29, 72—77.

3.64. Widmer F. Messung der Wärmeleitfähigkeit von Flüssigkeiten, insbesordere Kältemitteln nach einem instationaren Verfahren.—Kältetechnik, 1962, Bd. 14, p. 38.

3.65. Yakobson V.—Kholod. Tekh., 1960, v. 37, p. 55—58.

3.66. Zander M. Pressure-volume-temperature behavior of chlorodifluoromethane in the gaseous and liquid states.—In: Proc. fourth sympos. thermophys. prop., ASME, N.Y., 1968, p. 114—123.

# CHAPTER 4

4.1. Geller V. Z., Peredri V. G. Study of thermal conductivity of Freons F-13 and F-23.—Izv. Vuzov. Energet., 1975, No. 2, p. 113—116.

4.2. Geller V. Z. Thermal conductivity of some liquid freons at low temperatures.—Thermophysical Properties of Substances and Materials. Gosstandart SSSR, GSSSD, 1979, 9, p. 147—161.

4.3. Raskazov D. C., Petrov E. K., Spiridov G. A., Ushmaekin E. R. Study of $pvT$ behavior of Freon-23. —Thermophysical Properties of Substances and Materials. Gosstandart SSSR, GSSSD, 1975, 8, p. 4—16.

4.4. Raskazov D. C., Krukov L. A. Experimental study of isothermal Joule–Thomson Effect of Freon-23.—Thermophysical Properties of Substances and Materials. Gosstandart SSSR, GSSSD, 1975, 8, p. 84—99.

4.5. Raskazov D. C., Spiridonov G. A., Krukov L. A. Experimental study of caloric properties of Freon-23.—Thermophysical Properties of Substances and Materials. Gosstandart SSSR, GSSSD, 1975, 8, p. 100—108.

4.6. Raskazov D. C., Petrov E. K., Ushmaekin E. R. Experimental study of density of Freon-23 in liquid phase.—Tr. Mosk. Energ. Inst., 1975, v. 234, p. 52—58.

4.7. Raskazov D. C., Babikov U. M., Filatov N. Y. Experimental study of viscosity of Freon-23.—Tr. Mosk. Energ. Inst., 1974, No. 129, p. 62—69.

4.8. Raskazov D. C., Babikov U. M., Filatov N. Y. Experimental study of viscosity of Freon-23.—Thermophysical Properties of Substances and Materials. Gosstandart SSSR, GSSSD, 1975, 8, p. 142—148.

4.9. Raskazov D. C., Babikov U. M., Filatov N. Y. Experimental viscosity study of some freons of the methane series.—Tr. Mosk. Energ. Inst., 1975, 234, p. 90—96.

4.10. Raskazov D. C., Babikov U. M., Filatov N. Y. Experimental study of viscosity of freons at atmospheric pressure.—Tr. Mosk. Energ. Inst., 1977, 336, p. 54—57.

4.11. Sagaedakova N. G. Experimental Study of Viscosity of Freons-12B1, 13B1 and 502 in a Wide Region of State Parameters. Author's Abstract of Candidate Thesis. LT (Kh) P, Leningrad, 1977.

4.12. Timoshenko N. I., Kholodov E. P., Yamnov A. L. Refractive index, polarization, and density of Freon-23.—Thermophysical Properties of Substances and Materials. Gosstandart SSSR, GSSSD, 1975, 8, p. 17—26.

4.13. Timoshenko N. I., Kholodov E. P., Tatarinova T. A. Virial coefficients and thermodynamic properties of Freon-23 from Refractometric Measurements.—Thermophysical Properties of Substances and Materials. Gosstandart SSSR, GSSSD, 1975, 8, p. 27—40.

4.14. Ushmaekin E. R. Experimental Investigation of Density of Freons-13 and 23. Author's Abstract of Candidate Thesis, MEI, Moscow, 1976.

4.15. Khodieva C. M., Gubochkina I. V. Boundary liquid–gas curves near the critical points for Freon-13 and 23.—Zh. Fiz. Khim., 1977, v. 51, p. 1708—1711.

4.16. Tsiklis D. C., Prokhodov V. M. Phase equilibrium in systems containing fluorine compounds.—Zh. Fiz. Khim., 1967, v. 41, p. 2195—2199.

4.17. Shavandrin L. M., Raskazova T. U., Chashkin U. R. Investigation of temperature–density parameters of boundary curve for Freon-23 using quasi-static thermographs.—Tr. Khim. Khim. Tekhnol., 1975, v. 4, p. 100—104.

4.18. Belzile J. L., Kaliguine S., Ramalho R. S. PVT study of trifluoromethane by the Burnett method.—Canad. J. Chem. Eng., 1976, v. 54, p. 446—450.

4.19. Buckingham A. D., Graham C. The density dependence of the refractivity of gases.—Proc. Roy. Soc., 1974, v. A 337, p. 1609—1615.

4.20. Copeland T. G., Cole R. H. Dielectric and pressure virial coefficients of imperfect gases.—J. Chem. Phys., 1976, v. 64, p. 1741—1746.

4.21. Dymond J. H., Smith E. B. Second virial coefficient of gases.—Trans. Farad. Soc., 1964, v. 60, p. 1368—1374.

4.22. Eiseman B. J. Multicomponent Refrigerants.—Bull. Inst. Internat. du Froid, 1965, v. 45, p. 181—193.

4.23. Elchardus E., Maestre M. Les fluides frigorigenes fluores.—La revue generate du froid, 1964, Annexe 55, No. 8, p. 949—970.

4.24. Henne A. I. Trifluoroform.—J. Chem. Soc., 1937, v. 59, p. 1200—1202.

4.25. Hou Y. C., Martin J. J. Physical and thermodynamic properties of trifluoromethane.—AJChE Journal, 1959, v. 5, p. 125—129.

4.26. Lange H. B., Stein F. P. Volumetric behavior of a polar—nonpolar gas mixture.—J. Chem. Eng. Data, 1970, v. 15, p. 56—61.

4.27. Michel Alais. L'aplication de l'effet Peltier aux refrigerateurs menagers et aux techniques du chauffage et du conditionnement.—La reviue generale du froid. 1965, Annexe 56, No. 8, p. 963—967.

4.28. Morsy T. E. Trifluormethan (R 23). Thermodynamische Eigenschaften und Dampftafel.—Kältetechnik, 1966, Bd. 18, S. 203—206, 347—349.

4.29. Riedel L. Die Berechnung unbekanter thermischer Dater mit Hilfe des erweiterten Korrespondenzprinzips.—Kältetechnik, 1957, Bd. 9, S. 127—134.

4.30. Ruff O., Bretschneider O., Luchsinger W., Mittschitsky G. Vom Jodoform bis fluoroform.—Ber. dtsch. Chem. Ges., 1936, v. 69, A. S. 299—309.

4.31. Seger G. Die thermischen Eigenschaften aller-Fluor-Chlor-Derivate des Methans.—Beihefte Z. ver. deutsch. Chem., 1942, Nr. 43.

4.32. Thomas W., Zander M. Die Füllmenge verflüssigter Gase in Behältern.—PTB—Mitt., 1966, v. 76, Nr 5, S. 425—428.

4.33. Valentine R. H., Brodale G. E., Giauque W. F. Trifluormethane.—Entropy, low temperature heat capacity, heats of fusion and vaporization and vapor pressure.—J. Phys. Chem., 1962, v. 66, p. 392—395.

4.34. Wagner W. Thermodynamische Eigenschaften von Trifluormethane (R 23).—Kältetechnik, 1968, Bd. 20, S. 238—240.

4.35. Wenzel L. A., Balaban S. M. Effect of pressure on heat capacity of the $N_2$—$CHF_3$ system.—Ind. Eng. Chem. Fundament., 1970, v. 9, p. 568—574.